What's Around Design?

Américo Mateus · Gabriel Patrocinio ·
Susana Leonor
Editors

What's Around Design?

Strategic and Speculative Biodesign for a Sustainable Future: Volume 1

Springer

Editors
Américo Mateus
CIAUD - Research Center in Architecture, Urbanism and Design
ISMAT - Instituto Superior Manuel Teixeira Gomes - Ensino Lusófona
Portimão, Portugal

Gabriel Patrocinio
CIAUD - Research Center in Architecture, Urbanism and Design
ISMAT - Instituto Superior Manuel Teixeira Gomes - Ensino Lusófona
Portimão, Portugal

Susana Leonor
ISMAT - Instituto Superior Manuel Teixeira Gomes - Ensino Lusófona
Portimão, Portugal

ISBN 978-3-032-08162-9 ISBN 978-3-032-08163-6 (eBook)
https://doi.org/10.1007/978-3-032-08163-6

This Springer imprint is published by the registered company Springer Nature Switzerland AG
The registered company address is: Gewerbestrasse 11, 6330 Cham, Switzerland

Foreword 1: Design's Responsibility and Accountability Towards our Future

By Mugendi M'Rithaa

"We do not inherit the Earth from our ancestors, we borrow it from our children"...

This proverb, attributed to various traditions, emphasizes responsibility, accountability, and sustainability, challenging notions of entitlement and reckless abandon in our generation. It reminds us that we must justify our consumption to future generations, underscoring the urgent need for mindful resource use. This reality is starkly illustrated by *Earth Overshoot Day*—the date by which the earth's biocapacity budget is already exhausted, and the resources the planet can no longer sustainably provide for the entire year. Consumption patterns are unsustainable, with a significant number of countries depleting available resources in their respective natural capacity at varying rates.

Similarly, as challenged by myriad conscientious and socially responsible thinkers like Victor Papanek, Ivan Illich and Ernst Friedrich Schumacher, our noble discipline must move beyond superficial aesthetics and material excess. Instead, it should embrace deeper concerns that enhance human well-being. This ethos aligns with Ubuntu philosophy, which highlights human interconnectedness and our shared humanity: "I am because we are." As designers shape the world, they must recognize the transformative power of their work, influencing daily life, infrastructure, and societal structures. With this power comes responsibility—design must be ethical, empathetic, participatory, sustainable, and inclusive, addressing climate change, social inequality, and technological challenges.

Ethical design acknowledges that every decision has consequences, from material choices to digital energy consumption and public space accessibility. The focus should shift from short-term profits to long-term benefits, ensuring sustainability and social good. For instance, product design should prioritize sustainable materials and ethical labour praxis. In technology, AI and digital interfaces must safeguard privacy, prevent misinformation, protect personal data and guarantee operational integrity and security. Ethical design ensures that innovation does not come at the cost of environmental harm or human rights violations.

Beyond environmental concerns, design must address social justice by fostering inclusivity. Whether through accessible urban planning or user-friendly technology, thoughtful design can enhance equality. Liveable and walkable cities, accessible and efficient public transportation, and habitable green spaces contribute to convivial environments, while digital platforms should incorporate accessibility features for persons with disabilities. By designing with empathy, we can bridge societal gaps and empower marginalized communities.

To enforce accountability, policies and regulations must guide ethical design practices. Governments and organizations should implement sustainability certifications and fair labour standards, ensuring responsible production. Education also plays a vital role—designers should be trained from the outset to consider ethics, sustainability, and inclusivity in their designerly endeavours.

As the African proverb avers, "Tomorrow belongs to those who plan for it today"—the future depends on present choices, and design has the power to shape that future meaningfully. Businesses, designers, and policymakers must commit to ethical and sustainable solutions that benefit both people and the planet. By prioritizing responsibility and accountability, design can drive positive socio economic and socio technical transformation, creating an arguably more desirable equitable, resilient, and thriving world not just for ourselves but more importantly, for future generations.

The Author

Mugendi Kanampiu M'Rithaa is a transdisciplinary industrial designer, educator, researcher, and consultant at Machakos University, Kenya. Educated across Kenya, the USA, India, and South Africa, he specialises in socially conscious design, Universal Design, sustainability, and co-design. He is a founding member of the Pan-Afrikan Design Institute and a former President of the World Design Organization.

Foreword 2: Educating Designers for a Challenging Future

By Lorenzo Imbesi

In an era marked by unprecedented global challenges - climate extremes, mass migrations, social inequalities, and geopolitical conflicts - design education stands at a critical juncture. The question posed by this volume, "What's around design?" is not merely academic curiosity, but an urgent call to examine the expanding boundaries of design practice and pedagogy in our interconnected yet fractured world.

The contributions gathered here from the international conference held in Portimão represent more than scholarly discourse; they embody the collective intelligence of a global design community grappling with fundamental questions about education, responsibility, and futures. As Hannah Arendt reminds us in The Human Condition, human existence is inherently plural - we are never alone but always in the company of others. This plurality becomes both our challenge and our greatest resource in preparing designers for an uncertain tomorrow.

Today's design students inherit a world where traditional boundaries - disciplinary, cultural, geographical - are increasingly fluid. The rhizomatic nature of contemporary knowledge, as conceptualized by Deleuze and Guattari, finds its expression in design education that must embrace multiple entry points, non-hierarchical structures, and horizontal connections. Knowledge lives, they argue, "only within the mobile borders, in the intersections, in the interferences, where everything is a question of contamination." This contamination, far from being a threat, becomes the source of innovation and cross-cultural understanding.

The internationalization of design education cannot be separated from the urgent need to address global crises. The European Higher Education Area's emphasis on academic freedom, institutional autonomy, and student mobility creates fertile ground for collaborative approaches to sustainability, social innovation, and peaceful coexistence. Through academic exchange programs like Erasmus - active since 1987 - we have witnessed how educational mobility fosters not only knowledge exchange but cultural understanding across the Bologna Process framework. Yet internationalization should go beyond mere student exchange - it requires fundamental shifts in pedagogical approaches that prepare designers to work across cultures, disciplines, and worldviews.

Cumulus Association's global network of nearly 400 institutions spanning continents, embodies this vision of shared responsibility. From product design to service design, from sustainability to social innovation, to name a few, the thousands of design courses represented across our network demonstrate the remarkable scope and diversity of contemporary design education. The international space of knowledge we inhabit transcends geographical boundaries, creating new possibilities for collaborative research and education. Yet numbers alone cannot capture the depth of transformation required.

As the Cumulus Kyoto Design Declaration of 2008 articulated, we need designers capable of "human-centered development" who can "deliver economic, ecological, social and cultural benefits to all people."

The concept of "sharing is caring" included in the Declaration becomes particularly relevant as we face global conflicts and environmental crises. Design education must cultivate not just technical competencies, but ethical sensibilities rooted in care and quality. As Robert Pirsig articulated in Zen and the Art of Motorcycle Maintenance, "Care and Quality are internal and external aspects of the same thing. A person who sees Quality and feels it as he works is a person who cares. A person who cares about what he sees and does is a person who's bound to have some characteristic of quality." This relationship between caring and quality becomes fundamental to preparing designers who can address complex global challenges with both technical excellence and ethical commitment. Pierre Lévy's notion of collective intelligence offers a framework for understanding how design communities can collaborate across borders to address challenges that no single institution or nation can solve alone.

In a world where borders are multiplying against the free movement of people and consequently the sharing of knowledge through students and researchers, rethinking borders as contact areas helps us reimagine our times and build spaces for knowledge sharing. Design education should create spaces where different design cultures can meet productively. This is not about blurring differences but about learning to work within and across them, fostering understanding precisely through creative tension and productive dialogue. To this end, Cumulus has launched the Design Culture(s) for Peace initiative, exemplifying how design education can become a vehicle for fostering dialogue and understanding across cultural divides.

The path forward demands courage to embrace uncertainty, wisdom to learn from diverse perspectives, and commitment to preparing designers who see themselves not as isolated practitioners but as contributors to collective intelligence.

The future of design education lies not in what we can predict, but in how well we prepare our students for the unpredictable - equipping them with both critical thinking and collaborative capabilities for a world that urgently needs both.

References

Arendt, H. (1958). The human condition. University of Chicago Press.

Lévy, P. (1997). Collective intelligence: Mankind's emerging world in cyberspace (R. Bononno, Trans.). Plenum Trade. (Original work published 1994)

Deleuze, G., & Guattari, F. (1987). A thousand plateaus: Capitalism and schizophrenia (B. Massumi, Trans.). University of Minnesota Press. (Original work published 1980)

Pirsig, R. M. (1974). Zen and the art of motorcycle maintenance: An inquiry into values. William Morrow and Company.

The Author

Lorenzo Imbesi is Full Professor of Design at Sapienza University of Rome, where he is the Director of Sapienza Design Research and the Chair of the PhD program in Service

Design for the Public Sector. Currently, he serves as the President of the Italian Society of Design SID (societaitalianadesign.it), the President of the Global Cumulus Association (cumulusassociation.org), and a member of the Executive Committee of the European Academy of Design (eadresearch.org).

Introduction: What's around (the expanded fields of) Design?

By Américo Mateus (ISMAT_CIAUD), Gabriel Patrocínio (ISMAT & IFHT/UERJ/CIAUD), and Susana Leonor Mateus (ISMAT/LIRA).

1. Introduction: A New Design for the 21st Century

Design has always been a reflection of its time, shaped by the challenges, needs, and values of society. In this initial quarter of the 21st century, as we confront unprecedented ecological, social, and economic crises, Design is evolving through paradigmatic changes, acquiring relevance to businesses and governments as a paramount instrument for sustainable growth (or degrowth) and innovation. Moving into transdisciplinarity, Design paired with such disciplines as anthropology, psychology, management, political sciences, economics, and lately, biology – from the principles of ecology to biomimetics. Design research, user experience design, services design, systems design, strategic design, biodesign. Artificial Intelligence turned into an inevitable tool, as much as a new challenge to be confronted and explored. Themes such as Biodesign, Circular Territories, Speculative Biofutures, Nature-based Solutions, and Strategic Design for Green Transition have emerged as transformative frameworks, becoming part of this expanded field to be explored.

2. The Expanding Scope of Design Bridging the Industrial Revolution to the Industry 4.0 Era

As a profession, Design is relatively new, but the human ability to transform the natural world and produce "artificial" objects, as stated by Herbert Simon in his book The Sciences of the Artificial arises from our capabilities of problem-solving and decision-making – or else the ability to synthesize things. He defines design as a discipline fundamentally concerned with how things ought to be to achieve human goals, rather than merely analyzing how things are. Constrained by the external environment, humans structure the world to serve functional needs (Simon 1969).

Design emerged as an Applied Art during the Industrial Revolution (18th–19th centuries) due to radical shifts in production, technology, and social organization. As industries grew, businesses recognized the need for artists with specialized roles in shaping products for usability and market appeal, reconciling functionality with aesthetics. The birth of Industrial Design, however, is deeply connected with the Arts and Crafts movement, by the end of the 19th Century, which reacted against the loss of craftsmanship, and furthermore to the rise of the most iconic movement of the beginning of the 20th Century: the Bauhaus School. A pivotal moment for modern design, Bauhaus merged art, craft, and industry, promoting functional, minimalist aesthetics suited for mass production.

This legacy has shaped Design in the 20th Century, and crossed Germany's borders to reach the rest of Europe, the Americas, the Middle East, and Asia, carried by its former students, and determining what became known as the International Style. The Bauhaus became a movement whose continued influence reaches into the 21st Century with its functionality and minimalism, and even its form follows function motto inherited from Sullivan (1896) – although this latter has been challenged by philosophers and academics (Petroski 1994; Fowler & Highsmith 2001; Reid 2007).

The Post-WW2 Industrial Design and consumer culture, fueled by mass production and consumer goods industries, introduced planned obsolescence and other unsustainable industrial practices posed ethical dilemmas for design, leading to an exhaustion of natural resources and thereupon to the terrible consequences we are facing today. It also has led to a transition from manufacturing-based economies to information and service-based economies, with the introduction of personal computing and the Internet revolution, with sustainability concerns gaining traction, and lately leading to eco-design principles and circular economy ideas.

Nowadays, living in what is considered the Fourth Industrial Revolution Era (Industry 4.0), we are facing new challenges and new paradigms. Industry 4.0 integrates AI and automation in design and manufacturing; 3D printing & digital fabrication, enabling mass customization; IoT (Internet of Things), leading to smart products and environments; biodesign and biomimicry, merging design with biology for sustainable solutions; and circular design strategies, addressing climate and waste crises (Culot et al. 2020). Under this new panorama of complexity, Bauhaus' functionalist ideals have moved to the Industry 4.0's systemic, AI-driven landscape, and design has transformed from aesthetics and utility to a powerful force shaping technology, sustainability, and human experience. The traditional role of Design shifted from its aesthetics and functional operational role towards a more tactical, systemic and strategic participation, positioned at the fuzzy front end of innovation (Patrocínio, Filgueiras & Nunes 2022).

3. Design in the Age of Crisis

We inherited a world crisis, but also the warnings from 1960s and 1970s thinkers towards the understanding of limited resources and the inevitable crisis that was approaching. The limits to growth was the name of the report from the discussions led by the Club of Rome in the 1960s, (Meadows, Randers & Meadows, 1972), that started to propose a new approach for the global economy. Boulding (1966) was another voice introducing the concept of a closed system - the spaceship Earth, also explored by Buckminster Fuller in his Operating manual for Spaceship Earth (1969). This closed system was further explored by the economist Georgescu-Roegen (1971)

and later by Stahel (1982). The same concerns were shared by Papanek in Design for the real world (1971) and further in his less well-known book, The green imperative (1995).

In this later book, Papanek advocates for a Green Design, stating that design should offer a positive and unifying response to the crisis, bridging "human needs, culture and ecology" (Papanek 1995, p. 29). He also identifies six concernings factors in the design process: [1] the choice of materials, [2] the manufacturing process, [3] packaging, [4] the finished product itself, [5] transportation, and [6] discard or waste (idem, p. 29–32). His concerns coincide with the idea of a closed loop economy, advocated by Stahel (1982), which became the ground for what we know today as Circular Economy.

The idea of a Circular Economy was explored, in the beginning of this century, by Braungart and McDonough in their groundbreaking work Cradle to Cradle (2002), where a conceptual model of a Circular Economy is explained by two cycles – one natural or biological and another artificial or technological. The first concerns biodegradable materials, and the latter to technological or non-degradable products. "Nature operates according to a system of nutrients and metabolisms in which there is no such thing as waste" (Braungart & McDonough 2002, p. 92). The same should apply to technological products, and so the authors speculate about a 'reincarnation of objects' as a metaphor to a 'cradle-to-cradle' cycle, where objects are upcycled or have their components reintroduced into manufacturing as raw materials (idem, p. 103).

Design has a paramount role to play in Circular Economy, as indicated by the Ellen MacArthur Foundation while proposing six levers of approach to the design process (EMF 2023):

1. Observe and interpret the system;
2. Envision circular futures;
3. Create the conditions for collaboration;
4. Build circular design capabilities;
5. Rewrite the rules;
6. Develop tools to design and evaluate.

To achieve those goals, consonant to the first factor – the choice of materials – appointed by Papanek (1995), designers are increasingly pairing with biologists and material engineers to develop new biodegradable materials, predominantly to replace energy-intensive and pollutant petroleum-based plastics. The field of biomaterials and biodesign is flourishing, as biomimetics also thrives delving into nature-based solutions (Patrocinio et al. 2023).

Communities and the urban environment are also targeted by designers with participatory strategies, as encouraged by Ezio Manzini (2020), and also being rethought as circular territories (Carvalho 2023). Facing the current scenario of extreme climate changes and social and political turmoil – a real wicked problem of gargantuan proportions –, design is transitioning to provide a different mindset and innovative strategies.

4. Emerging Themes in Design

This book brings together the research and thoughts of selected scholars about the expanded roles of Design in this first quarter of century. Five topics, covered through different approaches, have been pinpointed as transformative frameworks, as mentioned above, namely:

- Biodesign: Merging biology and design for sustainable innovation, the name 'biodesign' has been used by both fields of design and medicine. In a truly transdisciplinary approach, there are quite extraordinary examples where these two fields collaborate in the development of health and safety products. From the field of design, it frequently refers to the research with biomaterials or with biomimetics, the process of observing living systems and learning from nature.

- Circular Territories: Designing for closed-loop systems and regional sustainability became a paramount necessity currently. Design needs to face the challenges imposed by the socio-economic-environmental crisis, and develop innovative solutions for the territories with a community-based participatory approach in mind.
- Speculative Biofutures: Envisioning future ecosystems and bio-integrated societies is yet another challenge faced by our society, where design has a valuable role to play. Once again lining up or leading transdisciplinary teams, design is expected to bringing its particular mindset and tools to leverage the processes and outcomes.
- Nature-Based Solutions: One of the main approaches of biodesign, as mentioned above, is biomimetics, when designers observe nature and apply biological and ecological principles to leverage systemic design.
- Strategic Design for Green Transition: Design's mindset and ability to envisage systems and strategies has turned into a valuable asset, driving policy and business innovation. All the frameworks mentioned, which configure the current expanded fields of design, are indeed strategic for the sought goal of the Green Transition.

The discussions within these topics hope to bring fresh contributions for the advancement of the field of design in front of the complexity of this century's challenging demands.

5. Conclusion: the Future of Design as a Discipline

There is a significant role of education and research in shaping future designers, and a need for a shift towards transdisciplinary collaboration. Design is no longer a field where an individual author has predominance, but rather a joint team effort. In his book The Green Imperative, Victor Papanek gives advice for design education in the 21st Century, stating that "all design education must be based on ecological methods and ideas." He goes further saying that "this will include studies in the scientific method, as well as in biology, anthropology, cultural geography and related fields" and also "social and human ecology and philosophy and ethics." He calls it the "social-economic-political matrix within which design operates", and singles out design as a player of "the key role of synthesis" in this context (Papanek 1995, p. 48).

The ethics of design was a subject of deep concern for the Japanese designer Kenya Hara. Since the late nineteenth century, from William Morris to the Bauhaus, design assented its base over the concept of social democracy. Our societies are now moved, though, by economic principles, with great implications to design:

"Basically, the concept of design was conceived and developed in no small measure on the premise of idealistic social ethics. Now, within the intense magnetic field of economic principle, the purer the concept, the less able it is to live up to its ideal." (Hara 2007)

Nowadays, when we see truth being constantly perverted and defeated, ethics is without any doubt our greatest challenges, and must be taught and discussed with the new generations. Acknowledging this, the World Design Organization launched a campaign for the 2025 World Industrial Design Day (29 of June), with the motto: Let's Get Ethical

- One Building Block at a Time (WDO 2025). Design should focus all of its user-centric, community-centric and empathic tools, aligned with sustainable and circular thinking, to help our society confront its challenges.

The next decades will be marked by our ability to deal with the current social-economic-environmental challenges – first as human beings, then as design practitioners. Furthermore, we have the challenge to properly understand, and learn how to train and use A.I. to effectively help us obtain the best outcomes out of complex processes. All of it without losing sight of the ethical and humane (empathic) perspective and consequences of what we design. We design for people, and sometimes its easy to forget that while we may call them 'users', 'costumers', or even 'consumers' of products, we need to have, first and foremost, an empathic posture.

Biodesign, Circular Territories, Speculative Biofutures, Nature-based Solutions, and Strategic Design for Green Transition are among the transformative trends shaping Design. They are on the leading edge of the emerging transformative frameworks, but these expanded fields need to be explored making use of transdisciplinary, empathic and, primarily, ethical approaches. It is imperative that we embrace these design-led innovation frameworks to undertake the contemporary challenges.

It's our mission to put Design at the forefront to fulfill its ethical pursuit in this unsettling times.

References

Boulding, K. E. (1966) The Economics of the Coming Spaceship Earth. In: Jarrett, H., Ed., Environmental Quality in a Growing Economy, Resources for the Future/Johns Hopkins University Press, Baltimore, 3–14.

Braungart, M.; Mcdonough, W. (2002). Cradle to cradle: Remaking the Way We Make Things. Albany, California: North Point Press.

Buckminster Fuller, R. (1969). Operating Manual for the Spaceship Earth. Carbondale: Southern Illinois University Press.

Carvalho, I. M. I. (2023). As cidades circulares como estratégia territorial de sustentabilidade na região do Algarve. Unpublished Master Thesis. Portimão: ISMAT.

Culot, G., Nassimbeni, G., Orzes, G., & Sartor, M. (2020). Behind the definition of Industry 4.0: Analysis and open questions. International Journal of Production Economics, 226, 107617.

EMF (Ellen MacArthur Foundation) (2023). The six levers designers can use to unlock the potential of circular design. Available at: [https://ellenmacarthurfoundation.org/news/the-six-levers-designers-can-use-to-unlock-the-potential-of-circular-design]. Accessed: 04.02.2025.

Fowler, M., & Highsmith, J. (2001). The agile manifesto. Software development, 9(8), 28–35.

Georgescu-Roegen, N. (1971). The Entropy Law and the Economic Process. Cambridge: Harvard University Press.

Hara, K. (2007). Designing Design. Zürich: Lars Müller Publishers.

Manzini, E. (2020). Politics of the Everyday (Designing in Dark Times). London: Bloomsbury.

Meadows, D. H., Randers, J., & Meadows, D. L. (1972). The limits to growth. New York: Universe Books.
Papanek, V. (1971) Design for the Real World: Human Ecology and Social Change. New York: Pantheon.
Papanek, V. (1995) The green imperative: ecology and ethics in design and architecture. London: Thames & Hudson.
Patrocínio, G., Filgueiras, E., & Nunes, J. M. (2022). A New Role for Design in the 21st Century From Operational to Tactical and Strategic. In Virtual and Augmented Reality for Architecture and Design (pp. 88-119). CRC Press.
Patrocínio, G., Paoliello, C., Mateus, A. Leonor, S. (2023). Co-Creating Greener Futures: Developing and Transferring Innovative Bio-Design Modules for Education to Accelerate the Green Transition. COCOON Erasmus Plus Project Deliverable 2.1. Available at: [https://www.cocoon.bio/_files/ugd/f434f2_8d3cc2220d2044ddab07963d78a95909.pdf]. Accessed 20.02.2025.
Petroski, H. (1994). The Evolution of Useful Things: How Everyday Artifacts-From Forks and Pins to Paper Clips and Zippers-Came to be as They are. Vintage.
Reid, G. W. (2007). From concept to form in landscape design. John Wiley & Sons.
Simon, H. A. (1969). The sciences of the artificial. Cambridge, MA.
Stahel, W. R. (1982). The product-life factor. An inquiry into the nature of sustainable societies: The role of the private sector (Series: 1982 Mitchell Prize Papers), NARC. Available at: [https://www.quebeccirculaire.org/data/sources/users/4/32217.pdf]. Accessed 04.02.2025.
Sullivan, L. H. (1896). "The Tall Office Building Artistically Considered," Lippincott's Magazine, March 1896, 404-409.
WDO (World Design Organization). (2025). World Industrial Design Day. Available at: [https://wdo.org/programmes/widd/]. Accessed 25.02.2025.

Contents

Speculative Biofutures Inspiration

Nature-Based Solutions Inspirations

Strategic Design for Green Transition Inspirations

Biodesign Inspiration

Prologue 1: The Challenge of Biodesign

By Emma van der Leest, Blue CityLab, Netherlands

When we talk about technology, our minds often jump to artificial intelligence, smartphones, or space travel. But technology has always existed in more humble forms like agriculture, fermentation, and textiles. In fact, biology has always been the foundation of innovation. Today, we are reimagining this connection: learning from microbial intelligence to build computers, lighting our homes with microbial fuel cells, and dyeing textiles with the help of pigment producing fungi and bacteria.

The 20th century was shaped by oil and industrial mass production; the 21st century belongs to biology. Yet, this biological transition is still unfolding slowly. Bio-based materials face complex hurdles: economic competition with cheap petrochemicals, fragmented regulations, and industries impatient for scalable solutions. What we gained in speed and affordability through synthetic chemistry came at an environmental cost. It's time to reframe the false binary of 'nature versus technology'. Nature *is* our greatest technologist, efficient, regenerative, and circular.

As a designer, I've seen biodesign move from speculative experiments in studios to practical applications in fashion, construction, healthcare, and even cosmetics. From microbial coatings to fungal leather, we now have tools and partners, scientists, engineers, open facilities and farmers, to scale living materials. But progress requires collaboration. When I started as a student in 2012, lab doors were often closed to designers. I knocked anyway, driven by the belief that material innovation depends on transdisciplinary teamwork. Back then, design in the lab was novel. Now, art schools are building their own bio-labs, and awards like the Bio Art & Design Award, universities unite like in Portimão, and collaborations between creatives and scientists emerge.

Today, my projects still revolve around this interface. In some cases, like the Microbial Vending Machine, the speculative design becomes a near-future forecast because nature already has the answers. Whether we're growing building materials, cultivating microbial dyes, or restoring coral reefs, it's not just about *new* materials, it is about rethinking systems of production, value, and care.

Biodesign is no longer garage biology. But if we want nature-positive impact to go mainstream, we must design not just products, but pathways: pathways for policy, public acceptance, funding, and education. This is not a solo endeavour. It's a movement that thrives on co-creation, humility, and deep listening, especially to Mother Nature.

Introduction to Part 1: Biodesign Innovations: Transforming Textiles and Wearable Technologies

by Carla Paoliello, ISMAT, Portugal, and Derya Irkdaş Doğu, Izmir University of Economics, Turkey - Conference Track Chairs.

Biodesign is rapidly emerging as a transformative field, bridging biological systems to create sustainable materials and products. This section brings together six papers highlighting key advancements in this domain, spanning applications in skin therapeutics, biomaterials, textiles, and other wearable products. Despite their diverse approaches, these studies share a common focus on material innovation, interspecies collaboration and sustainability. They explore how biological processes can be harnessed to create functional, adaptive and aesthetically engaging materials while promoting ecological consciousness. Each paper highlights the interdisciplinary nature of Biodesign to push the boundaries of textile and wearable technologies.

The "Living Therapeutic Skin" (LTS) project presents a novel concept that integrates engineered microbes into wearable skin artefacts to detect and treat eczema flare-ups. This development signifies a shift in healthcare design, where biological materials interface directly with the human body, creating dynamic, responsive interfaces that redefine the boundaries between medicine and wearable technology.

"Biodesign in Textile Innovation" explores bacterial cellulose (BC) as a sustainable textile material. The study delves into BC's properties, aesthetic applications, and functional enhancements, demonstrating its potential for creating biodegradable and visually appealing textiles. The authors highlight how Biodesign can merge functionality with aesthetic expressions through experiments with natural dyes and laser-cutting techniques.

The "Lucid Life|Marama Ora" project investigates Biodesign's creative potential through interspecies collaboration, where bioluminescent algae are integrated into textiles and artefacts. This work follows the previously mentioned principles by emphasising the symbiotic relationship between living organisms and human-made objects. Beyond function, these luminous textile materials create interactive and awe-inspiring experiences that evoke emotional and cultural connections, reinforcing Biodesign as a sensory and experiential innovation medium.

"Designing Customizable Malleability and Padding in Fungal Textiles" extends the discussion into fashion and soft materials, presenting fungal textiles as an alternative to traditional applications. Unlike conventional textiles that maintain homogeneity, fungal textiles offer dynamic, localised properties, making them adaptable to various applications, from fashion to wearable healthcare.

Moving into architectural applications, "Building-By-Growing with Textile Logic" demonstrates the scalability of Biodesign principles. The project employs microorganisms—mycelium and microalgae—to develop new biofabricated materials that address environmental concerns in construction. By integrating biological processes into industrial systems, these materials contribute to a circular economy, promoting material-driven design that supports sustainable solutions.

Finally, "MYO: Mycelium Technology for Kids" introduces Biodesign into educational and interactive experiences. By integrating mycelium into children's toys and merging interactive digital experiences with biological materials, this initiative promotes sustainability and proposes biobased materials to foster cultural shifts in material appreciation, reinforcing that sustainable design must extend beyond industry to societal norms and values from an early age.

Together, these six papers illustrate the transformative potential of Biodesign in textiles and wearable applications. They challenge conventional material paradigms, demonstrate the adaptability and potential of living materials, and open new avenues for sustainability. As these fields intersect, Biodesign stands poised to redefine the relationship between humans, materials, and the environment, fostering a future where biological and technological systems coexist harmoniously.

Living Therapeutic Skin (LTS): Design Opportunities and Challenges of On-Skin Living Artefacts

Raphael Kim[1,2(✉)], Fanzhe Zeng[1], Tom Ellis[3], and Elvin Karana[1]

[1] Delft University of Technology, Delft, The Netherlands
raphael@biodesign.academy, {F.Z.Zeng,E.Karana}@tudelft.nl
[2] Biodesign Academy, London, England
[3] Imperial College London, London, England
t.ellis@imperial.ac.uk

Abstract. The Living Therapeutic Skin (LTS) is a novel living material currently in development as part of a European project. Integrating engineered microbes to detect and treat eczema flare-ups, LTS offers significant promise for managing this prevalent skin condition. However, as a microbial material in its early developmental stage, LTS faces challenges related to social acceptance when it is embedded in on-skin living artefact for daily human use. To address these challenges in the further development of these materials, we conducted a workshop employing boundary objects to illustrate four hypothetical LTS applications in everyday contexts. Our participatory approach engaged a multidisciplinary group, including a dermatologist, scientists, biodesigners, and eczema patients. Analysis of the workshop data revealed several important factors affecting the wearability and acceptance of on-skin living artefacts. This paper elaborates on these factors to explore the potential implications of LTS, examining its design prospects and hurdles while discussing possible avenues for a broader range of human-skin interfaces.

Keywords: On-skin living artefacts · Boundary objects · Workshop

1 Introduction

The Living Therapeutic Skin (LTS) is a microbial living material that is envisioned as a wearable artefact for the dynamic and responsive treatment of skin-based disorders (NextSkins 2023). It is composed of a bacterial cellulose hydrogel housing responsive yeast cells. Through synthetic biology and protein engineering methods (e.g., Ng et al. 2020; Gilbert et al. 2021), this therapeutic skin is intended to be designed into three layers: an interactive layer with sensory and response functions, a core layer housing living cells, and a barrier layer for containment and hydration. LTS is proposed to be engineered to detect clusters of Staphylococcus aureus bacteria during Atopic Dermatitis (eczema) flare-ups and releases compounds inhibiting bacterial growth. LTS design team follows a multidisciplinary and collaborative material development approach (Karana

A. Mateus et al. (Eds.): WAD 2024, What's Around Design?, pp. 5–20, 2026.
https://doi.org/10.1007/978-3-032-08163-6_1

et al. 2015; Barati and Karana 2019) to design the living material as a form of living artefact for human interaction in the everyday context (Karana et al. 2020). In its use, it is envisioned to provide cues to users concerning its active state or indicating the severity of the skin condition. Yet, as this innovative material integrates living components which are meant to be on human-skin, tensions arise concerning user experience and societal acceptance, as well as managing the end-of-life phase of these wearable designs (e.g., Lu & Lopes 2018; Jones & Brown 2019).

Understanding the user experience of LTS, a semi-developed material, presents significant challenges. Firstly, similar to other collaborative material development projects with semi-developed samples (Barati et al. 2019), LTS is not yet accessible for physical engagement due to its unfinished state. This limitation restricts eLective communication and collaboration between material and biological scientists, designers, and other crucial stakeholders such as medical experts and potential end-users. Secondly, even with more advanced samples, living materials require strict safety protocols for handling within collaborative design groups, making it diLicult to present them to potential end-users. To address these challenges and obtain feedback in the early development stages, we propose a workshop utilising diverse boundary objects (Star & Griesemer 1989), designed by our multidisciplinary living material development team.

To comprehensively delve into the multifaceted implications of LTS, we organised a workshop with stakeholders including a dermatologist, scientists, biodesigners, and eczema patients. Our LTS boundary objects portrayed four carefully designed hypothetical scenarios and their corresponding conceptual living artefacts showcasing potential LTS applications in everyday life. The participatory approach aimed to foster interactive discussions and vision expansion. Our analysis conducted on the rich dataset extracted from the workshop revealed several pivotal emergent themes, including aspects like material experience (Giaccardi & Karana 2015), ethical implications of integrating living components into wearables, technological feasibility, and overall social acceptance (Brown & Miller 2018). These themes serve as a foundational basis for exploring the wide-ranging implications of LTS within the design and Human-computer Interaction (HCI) community, particularly in the realm of on-skin interface designs.

2 Related Works

2.1 On-Skin Interfaces in Design

On-Skin Interfaces refer to technologies designed to be worn directly on the body, specially designed to adhere directly to the human skin, enabling seamless interaction between humans and computers (e.g., Ogata et al. 2013; Restrepo-Villamizar et al. 2021). These interfaces leverage advancements in materials, sensors, and miniaturized electronics to create unobtrusive, convenient, and often biometrically sensitive devices. Current on-skin interfaces designs exhibit a wide range of interaction patterns, for example, input modalities such as touching (e.g., Weigel et al. 2015; Kao et al. 2016), biometric data (e.g., Kao et al. 2016; Dong et al. 2022), body fluids (e.g., Reeder et al. 2019; Ikeda et al. 2022), and environmental factors like chemicals (e.g., Kao et al. 2017), and output modalities, for instance, morph and colour changes (e.g., Kao et al. 2017; Weigel at el. 2017), olfactory (e.g., Wang et al. 2020), and facial expressions (e.g., Wang et al. 2022).

However, keeping the balance between functionality and compactness of on-skin interfaces has always been a challenge; the pursuit of more complex functions often necessitates larger and heavier electronics, components, and power supplies—a juxtaposition that conflicts with the desirable minimal size and weight of these interfaces. Multiple efforts have been addressed to tackle this challenge, encompassing the invention of new and miniature components (e.g., Dementyev 2015; Wang et al. 2020), the development of novel materials serving as sensors and actuators (e.g., Kao et al. 2017; Kao et al. 2018), and more.

In acknowledging the multifaceted factors influencing the wearability of on-skin interfaces and recognizing the distinct design challenges and opportunities they present, Liu et al. (2016) have compiled a guideline. This guideline serves as a compass for designers, enabling them to efficiently navigate the myriad considerations when developing their own iterations of on-skin interfaces. Liu's exploration encompassed an in-depth analysis of wearability factors encompassing both the physiological aspects (such as location, movements, and individual body traits) and technical elements (such as covering weight, attachment methods, accessibility, interaction interfaces and aesthetics). This holistic approach provides a roadmap for designers seeking to strike the delicate balance between functionality and minimalism in the realm of on-skin interfaces.

2.2 On-Skin Biomaterials

The concept of biomaterials, usually defined as synthetic or natural substances capable of interacting with living systems (Achermann et al. 2014), has a long history dating back to traditional healing practices. Seaweed, for instance, has been used as a wound dressing for centuries by various indigenous cultures (Thurstan et al. 2018). Similarly, herbal remedies have been employed in both oral and topical forms for millennia in many countries across the world (Halberstein 2005). With the advancement of science and technology, microorganisms, once considered invisible and intractable, have become increasingly valuable for biomaterial development (Rodrigo-Navarro et al. 2021). The healthcare sector stands to benefit from the remarkable characteristics of lab-grown biomaterials, particularly their non-toxicity, biocompatibility, and biodegradability properties (Wang 2023). Specifically, bacterial cellulose, produced by bacteria and yeasts, emerge as a promising material for wound dressing, drug delivery, and tissue engineering due to its low cytotoxicity, excellent biocompatibility, and ability to foster cell growth and tissue formation (Cacicedo et al. 2020; Han et al. 2020).

In parallel, within the context of biological-HCI (Pataranutaporn et al. 2018), wearable biomaterials are beginning to be considered computational materials for human-microbe interactions. Among these advancements, an innovative and intriguing realm emerged recently; using computationally engineered microbes (also called living bits (Pataranutaporn et al. 2020) as interacting material or computers (Pataranutaporn et al. 2018) in on-skin interfaces. For example, the Living Tattoo Project involves printing engineered cells on a bilayer elastomeric sheet. These cells can sense and respond to chemicals on the user's skin, displaying their reactions through fluorescence and acting as a skin health monitor (Liu et al. 2018). These living interfaces, harnessing the intelligence and adaptability of the living microbes, are self-contained systems that offer dynamic and responsive user experiences without the need for electronics or external power. Moreover, livingness, as a surfaceable design quality (Karana et al. 2020; Kim et al. 2023), also fosters unconventional interactive and aesthetic engagement with wearers, potentially promoting an understanding and appreciation of microbes' diverse needs, scales, agencies, and temporalities, thereby encouraging practices of care and bridging the cognitive gap that has long separated humans from nature (Karana et al. 2023).

Moving forward, in this paper, we delve deeper into the realm of biological-HCI, where our focus shifts towards exploring living therapeutic skin as a nascent but promising wearable biomaterial. By harnessing existing wearability factor guidelines, we aim to delineate a distinct set of design opportunities and challenges inherent to Living Therapeutic Skin (LTS).

2.3 Exploration of LTS Through Boundary Objects

Boundary objects serve as common reference and knowledge points for various stakeholders in collaborative settings (Star & Griesemer 1989). These objects can take multiple forms, depending on objectives of the design process (e.g., Halpern 2012; Zimmermann & Pluchinotta 2020). They may be tangible artefacts, such as assembly drawings that facilitate knowledge exchange between manufacturing and design engineers (Carlile 2002), heritage products embodying perspectives of both craftspeople and designers (Suib et al. 2020), or machines that bridge understanding among managers, engineers, and assemblers (Bechky 2003). Boundary objects can also be more abstract, such as design methods, narratives, or metaphors (Koskinen 2005; Enninga & van der Lugt 2016). For instance, in designing a digital healthcare platform, patient stories can serve as open boundary objects and design goals as metaphorical semi-open boundary objects (Islind et al. 2019). More recently, boundary objects can also encompass artefacts that integrate living organisms and processes (Ramirez-Figueroa 2018).

The LTS project, as a biodesign endeavour (Myers 2012), involves stakeholders such as scientists, designers, medical experts, patients, and microorganisms. Current challenges stem from the semi-final development stage of LTS and safety considerations of genetically engineered organisms, restricting LTS to laboratory settings. This creates a knowledge and experiential gap between scientists developing the material and downstream stakeholders like medical experts and users. Furthermore, both the knowledge boundary and the physical boundary (the laboratory door) contribute to a separation between the novel living material and the everyday practicalities. Speculative scenarios have proven effective and provocative in design workshops for linking novel technology and design to reality (e.g., Huusko & Roto 2018; Nygrén 2019). Thus, we adopted speculative scenarios as boundary objects to bridge the gap between LTS and daily life, fostering critical discussions on LTS in various contexts. Through carefully designing and crafting LTS boundary objects, our objective is not only to align all involved actors, but also facilitating real-life engagement with LTS beyond laboratory settings, fostering a more intimate, experiential and inclusive participation for all.

3 Methodology

To explore the diverse perspectives on Living Therapeutic Skin (LTS), we designed and organised a workshop that brought together 15 stakeholders from various backgrounds, including a dermatologist, ten scientists, two biodesigners, and two Eczema patients, who were affiliated with the research project. The workshop spanned approximately 3 h. Four speculative scenarios and their corresponding artefacts have been designed, visualised and prototyped to explore various potentials of LTS and transform them into comprehensible narratives.

3.1 Context of Use: Eczema

Eczema, or atopic dermatitis, is a chronic inflammatory skin disease marked by persistent itching and eczematous lesions (McNally & Phillips 2000). It affects around 20% of children and 5% of adults globally (Asher et al. 2006). Patients experience intense itching, sleep disturbances, reduced productivity, and social challenges, along with anxiety, depression, and frustration (Simpson et al. 2016). The disease fluctuates between flare-ups and remissions, with periods of worsening followed by relative control (Girolomoni & Busà 2022). There is a noted correlation between the severity of flares and increased colonization of Staphylococcus aureus (S. aureus) on the skin (Girolomoni & Busà 2022). Therapeutic LTS materials are used during flares to detect S. aureus and release molecules that inhibit its growth. Eczema impacts both physical and psychological well-being. Treatments include topical medications, oral drugs like Methotrexate, and phototherapy (Ross 2023). Psychological therapies, such as muscle relaxation, social competence training, and hypnotherapy, have also improved patient quality of life (Ersser et al. 2014).

3.2 Designing the LTS Boundary Objects

When designing the Living Therapeutic Skin (LTS), we integrated the initial material development directions provided by Tom Ellis Group at Imperial College London—such as potential thickness, transparency, and colour indications—with the psychological implications of eczema. For the latter, we placed special emphasis on crafting the material's texture to enhance the experiential qualities of LTS, focusing on three key aspects: (1) the material should exhibit vitality and naturalness, highlighting the living quality of LTS; (2) it should convey a sense of healing, reflecting the therapeutic nature of LTS; and (3) it should foster meditative interactions between patients and the textural interface, aiming to alleviate negative emotions associated with eczema flares. To guide our design process, we drew inspiration from textures in nature, clustering them into three pairs: Water Ripple and Wave Textures: Inspired by the dynamic aesthetics of water, these textures aim to make users feel calm and relieved; Shell and Scale Textures: Designed to resemble the protective layers of critters, these textures convey meanings of protection, security, and intimacy; Tentacle and Fin Textures: These exhibit more interactive qualities, prompting actions such as caressing, sliding, and squeezing to evoke calmness and relaxation during use (Fig. 1).

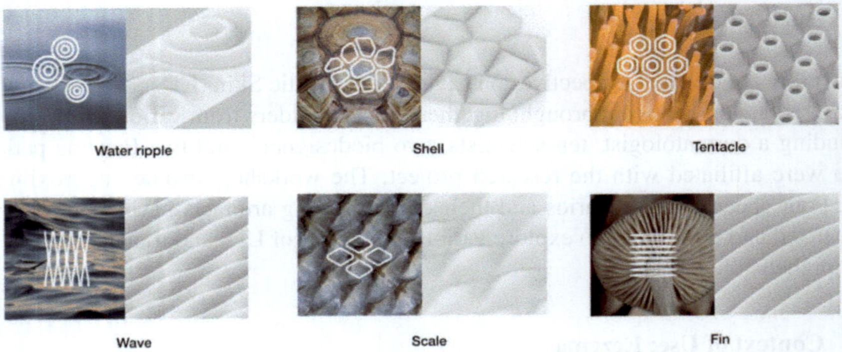

Fig. 1. Directions for LTS texture design

3.3 Design Scenarios and the Corresponding Artefacts

Based on the analysis of the eczema context and the initial development of LTS within the NextSkins project consortium, we created four scenarios that integrate the technology, design potentials, and daily narratives related to eczema. In the first scenario, named "daily patches", we delve into the ease of use and convenience aspects of LTS. LTS patches serve as ready-to-use products in pharmacies. The second scenario, situated in skin "therapy centres", addresses the possibilities of highly customised and responsive skin therapy through LTS. In "domestic integration", we shift the focus to the care practices of LTS as a daily routine. Finally, the "implantation" scenario speculates on the future extension of the technology- implanting LTS underneath the skin as a permanent treatment. Table 1 shows a summary of each scenario, and the accompanying illustrations.

Table 1. LTS scenarios developed for the workshop with diverse stakeholders.

Scenarios	Scenario A Daily patches	Scenario B Therapy Centers	Scenario C Domestic Integration	Scenario C Implantation
Descriptions	LTS patches could serve as prescription remedy for AD patients, usable for immediate relief or preventive purposes. Following General Practitioners (GP)'s prescription, patients are able to collect them in pharmacies. LTS patches will come with various shapes tailored to fit different affected areas of the patients' body. Patients will be informed how to store the patches and the viability of the LTS through instructions on the package.	LTS patches could be used in skin therapy centres for eczema patients. These centres store and maintain the patches in highly controlled settings, maintaining their viability. They also track patients' progress by recording the microbial interface during visits, aiding in future treatments. Additionally, these patches can be tailored to individual needs by analyzing customers' skin microbiome composition. This customization enhances their effectiveness, providing targeted care based on unique skin requirements.	Design Vision with the scenario is to integrate LTS material seamlessly into eczema patients' domestic life, offering not only effective treatment for patients' symptoms, but also for the LTS's care during its dormant states, at home. The design concept shapes the LTS material like a roll, making it familiar and user-friendly, allowing customization for optimal coverage. It includes a domestic incubator to maintain LTS material viability during dormant periods, ensuring it's ready for use.	Eczema is a long-lasting skin condition, especially severe cases require ongoing treatment. There may be a development of an option for patients to use implants, that would embed LTS-like patches under the skin, for continuous therapy. LTS patches use living microbes and thrive by using the body's temperature and moisture, functioning like "secondary organ" under the skin, to sense and treat eczema without electronic and/or manual intervention.
Illustration	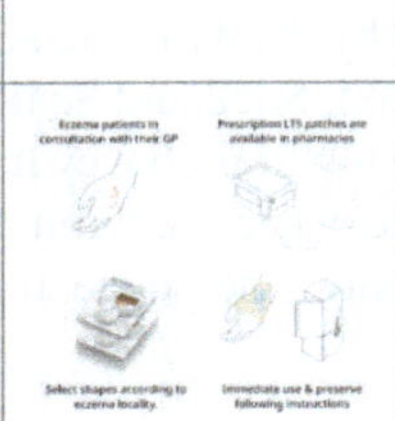	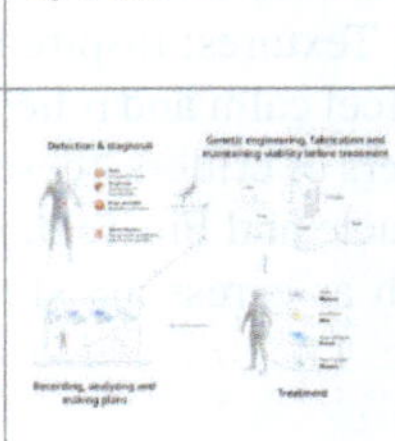	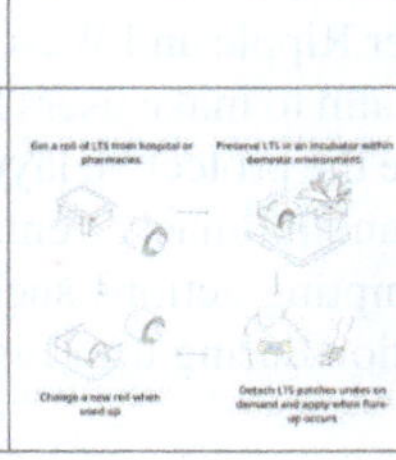	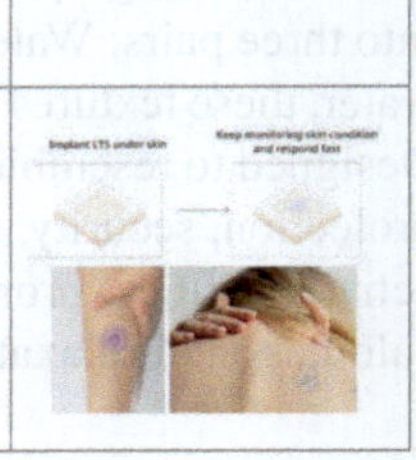

These scenarios and their visuals are implemented into the workshop as boundary objects. Although we use the term "speculative", these scenarios are situated in familiar contexts including pharmacy, therapy centres, home and hospital, which creates common vessels, within which participants from various communities will have different experiences as internal contents (Star & Griesemer 1989). For instance, medical experts who prescribe LTS and patients accepting LTS might have different experiences, perspectives and concerns in a therapy centre or hospital contexts, and scientists and designers might have different visions for the future development of LTS.

Apart from the graphic representations of four scenarios, we also designed and prototyped tangible artefacts following the central concepts embodied in the four scenarios (Fig. 2), aiming to establish an experiential, sensory relationship with the speculative scenarios for all workshop participants (Coughlan et al. 2007). We used bio-silicone in the design of our tangible boundary objects due to its close resemblance to the LTS (revealed in informal discussions with the scientists within NextSkins) in terms of sensorial and performative qualities (Karana et al. 2015). It is a safe material for physical engagement. Moulds, crafted through 3D printing and Lego bricks, were employed to shape the LTS objects and engrave nature- inspired textures as mentioned in the previous part.

For the Daily Patches scenario, alternative LTS samples with textures were crafted into ergonomic shapes tailored to different body sites, enabling participants to wear them and intimately experience the material (2, A). In the Domestic Integration scenario, we shape the material into long-stripe form and wound it around the knob of an incubator, inviting participants to test the convenience of the using and caring interaction (2, B). For the Therapy Centers scenario, a digital-physical simulation was implemented. The temporal changes in the responsive interface facilitated by microbes, which is designed to reflect users' real-time skin condition, were simulated through animation. An iPad was used to play the animation with a piece of material sample on top of the screen, creating the illusion of microbial responses being directly displayed on the material's surface (2, C). In the Implantation scenario, digital simulations were predominantly employed to render the temporal changes of the interface beneath the skin (2, D).

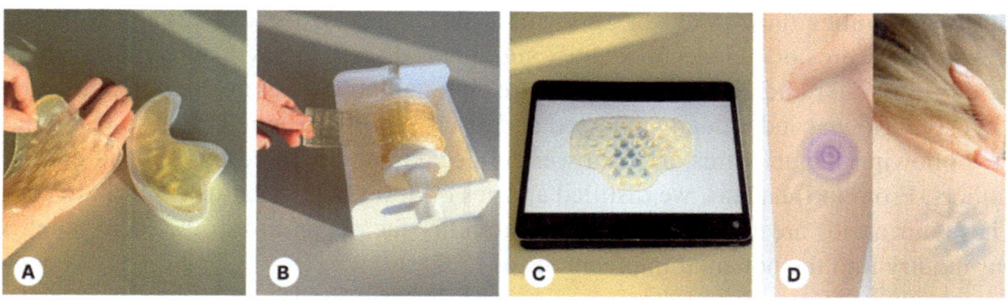

Fig. 2. LTS tangible boundary objects.

4 Workshop

The workshop consists of three main parts. Firstly, the workshop kicked off with a short joint presentation by the first and second authors, outlining the main aims behind the session and introducing the LTS technology. The workshop was then divided into two parts, namely ice-breaker activity and scenarios exploration.

4.1 Ice-Breaker Activity

The main aim of the activity was to assess how participants perceived LTS technology in treating eczema. This involved considering key factors such as its efficacy, comfort, ease of use, safety, and environmental impact. Participants collaborated in groups of 3 or 4 and were tasked with rating LTS based on these five aspects, using a 7-point Likert scale provided in a worksheet. For instance, under the efficacy section, the worksheet prompted them to answer the question, "How do you perceive LTS in terms of its efficacy?" ranging from -3 (not very effective) to $+3$ (very effective), and similar questions were posed for the other factors. Each question on the worksheet included a box below the scale, allowing participants to briefly explain their rating decisions. Overall, this exercise aimed to engage participants and set the stage for subsequent activities, fostering a more receptive atmosphere among the participants.

4.2 Scenario Explorations

Here, we have delved further into the potential applications of LTS through four scenarios and related tangible artefacts. Following the presentations of the scenarios, working in the same assigned group, the participants were asked to assess each scenario, using a 7-point Likert scale provided on separate worksheets. This time, there were four criteria with which their responses were recorded: Plausibility of the scenario to occur within 5 years; Plausibility of the scenario to occur within 10 years; Social acceptance; and Ease of application. In addition to the ratings, a separate box was provided in the worksheet for the participants to briefly elaborate on the scenarios, highlighting the potential challenges and opportunities they would bring to the stakeholders of the technology. Note that these scales were not intended to identify significant discrepancies or agreements among participants, but rather to support and inspire our qualitative assessment. Following that, each scenario was individually assessed and discussed in a group discussion (Fig. 3).

Following the activities, all data were collected and analysed by the lead and co-authors. From the analysis, we distilled a series of design opportunities and challenges. These intricacies are rooted in various wearability factors and form the focal point of our inquiry into the potential implications of LTS.

Fig. 3. Workshop activities: participants exploring boundary objects and completing worksheets

5 Findings

5.1 General Assessment of LTS Technology

Here, we offer a summarised representation of the Likert scale findings regarding the five factors (Fig. 4). These visuals aim to encapsulate participants' views on each factor. Additionally, we outline concise explanations of the findings, capturing the general sentiment associated with each factor. To enhance comprehension, we occasionally include direct quotes from participants to illustrate our points.

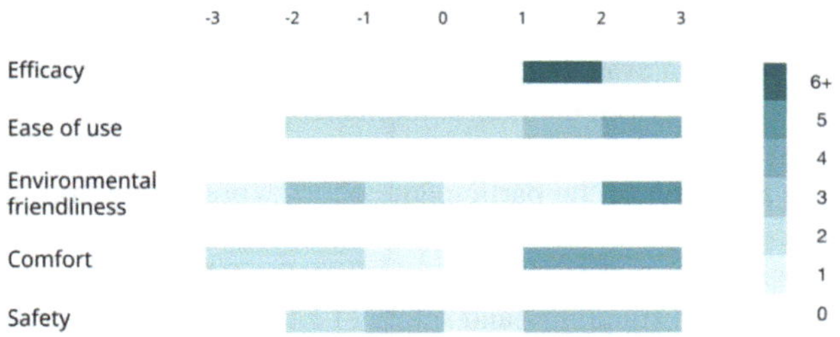

Fig. 4. Perceptions of LTS

Efficacy

Most participants generally regard the LTS artefact as relatively effective, with some expressing it as better than existing treatments. However, concerns were raised regarding its *temporal behaviour*, particularly about the speed at which the LTS artefact would function or its promptness in treating the affected skin areas—given its microbial interactions, many participants speculated that it might take a while to produce noticeable effects. Moreover, the complexity of different elements might influence sensing or the therapeutic response, posing another challenge. The efficacy was also questioned in terms of scale, as the human skin covering the entire body presents a significantly larger surface area than the LTS patch, potentially limiting direct contact and interaction. Nonetheless, there were positive perspectives expressed as well, some participants believed the patch might offer a cooling and calming effect.

Ease of Use

Regarding the ease of use of LTS, the majority of participants perceived it to be relatively straightforward. They mentioned that the artefact seemed intuitive to use, describing it as a "stick and go" concept or using phrases like "easy as 1,2,3." However, a smaller subset of participants didn't find it easy to use. They expressed concerns that using LTS would necessitate *proactive care practices* to maintain the artefact.

Environmental friendliness

The majority of participants viewed the LTS material as relatively environmentally friendly. Their reasoning stemmed from being informed that the material was crafted from renewable compounds and organic substances. Additionally, some participants were aware that it was designed to be biodegradable and manufactured with reduced dependence on fossil energy. However, a minor concern surfaced when one participant found it challenging to envision the optimal *end-of life* scenarios for the LTS material. Despite this, the overall consensus leaned towards perceiving LTS as a product aligned with "green chemistry," fostering sustainability.

Comfort

Participants had diverse opinions regarding the comfort of the LTS. Some found it to be relatively comfortable and expressed positive remarks, highlighting the pleasant *tactile experience* it provided. They mentioned feeling a nice sensation akin to a "cold patch and cloth," preferring it over the sensation of a band-aid. They also appreciated its potential cooling effect, attributing it to its non-dry nature and the presence of hyaluronic acid. Conversely, those who didn't find LTS comfortable cited several reasons. They believed that the design of LTS might compromise comfort, potentially limiting movement and causing it to fall oL when sweating. Additionally, there were concerns about potential unpleasant odours arising from the microbial makeup of the material.

Safety

As for safety, a portion of the participant cohort expressed minor doubts about the safety of *using genetically modified organisms* (GMOs). They felt uncertain about the idea and were hesitant as it didn't seem entirely safe. Additionally, some participants raised concerns about the suitability and safety of LTS for use on open skin or wounds. They highlighted the potential risk of microbes entering the skin, suggesting it might pose a safety issue, despite this being the intended working principle of the LTS material.

5.2 LTS Application Scenarios

We used the same method to visualise data from the workshop for the four scenarios (Fig. 5). Participants praised the *Daily Patch* for its convenience, home storage, and travel-friendliness. It was likened to a reassuring, easy-to-use bandage. A dermatologist noted its *effectiveness* for mild to moderate eczema in skin folds during flares. However, participants raised concerns about potential allergic reactions from repeated use, the environmental impact of single-use packaging, and the delay between prescription and relief. They also mentioned the inconvenience of frequent pharmacy visits and the wastefulness of non-reusable patches. While not widely expected in the next five years, most participants felt daily patches could gain traction within a decade due to their high social acceptance and ease of use.

Noteworthy opportunities identified include the prospect of receiving treatment from *specialised therapy centres*, allowing for fine-tuned and personalised approaches tailored to individual patients. A dermatologist participant emphasised the suitability of such treatments for individuals with severe eczema. Furthermore, one participant highlighted the social aspect, emphasising the increased potential for meaningful human interactions, providing a more personal *engagement experience.* However, drawbacks include the time-consuming nature of booking and visiting centers, along with significant costs that may hinder widespread adoption. Most participants felt therapy centres were unlikely to become widespread within five years but slightly more likely within a decade. They received high ratings for social acceptance and ease of application due to controlled, expert-assisted indoor treatments.

In the *Domestic Integration* scenario, the opportunities are quite practical, offering an in-house solution. This setup presents the chance for patients to establish a stable ritual or routine, adding a potentially enjoyable aspect to their treatment experience. Moreover, it opens up the opportunity to consider the *care of other- than-human entities* - the microbes integral to the therapy material. However, potential drawbacks exist including the risk of overuse, incorrect usage, or improper storage by patients, necessitating detailed guidelines or instructions. Managing these aspects could be perceived as an additional hassle on top of the primary goal of treating the disease. Regarding the domestic integration of LTS and its perception, the majority of participants expressed a belief that its plausibility in everyday life was fairly high, both in the next five years and the subsequent decade.

No substantial opportunities or advantages were identified regarding the *Implantation* scenario, as participants expressed strong criticism of this scenario. They deemed it highly unfeasible in the near future and considered it more as a long-term solution for eczema. While the idea had a futuristic appeal, envisioning hybrid cyborgs and aliens, it was perceived as invasive and raised concerns about potential resistance from the patients, anaphylactic shock, negative immune reactions, and new infections. The safety and practicality of using color displays within implanted skin layers were also questioned. Overall, participants felt this scenario presented more challenges than benefits, with low plausibility in the next five to ten years and low social acceptability due to its invasive nature in treatment situations.

Fig. 5. Data visualisation for the scenarios

6 Discussion and Conclusions

This paper presents the findings of a workshop, focusing on Living Therapeutic Skin (LTS), exploring the potential barriers to the societal acceptance of this emerging technology, supporting its further development toward socially relevant directions. Herein, we succinctly summarise our findings, highlighting both design opportunities and challenges for LTS.

6.1 Design Opportunities

One of the most compelling opportunities for designers working with Living Therapeutic Skin (LTS) material and technology lies in starting with a *moderate, small-scale* approach, such as employing a patch format. This strategy follows the principle of introducing a novel living material through a familiar application (Karana et al. 2018). This format offers significant advantages in terms of wearability, portability, comfort, and ease of use. Notably, discussions with participating dermatologists underscore the empowering potential of daily patches for patients in managing therapies, as offering flexibility and choice is crucial in enhancing their experience. Furthermore, when examining these findings in light of the wearability factors outlined by Liu et al. (2016), it becomes evident why the daily patch stands out as the most viable and accessible application of LTS material. The biological characteristics of the material, coupled with the absence of conventional bulky electronics, make it inherently lightweight, portable, and flexible. Additionally, the daily patch format aligns with a regenerative ecology mindset (Karana et al. 2023) by utilising organic matter for an on-skin interface, which can potentially return to nature to complete its life cycle. This sustainable approach adds a crucial dimension to the design considerations of living materials.

Moreover, integrating LTS into domestic settings, including care and maintenance routines, opens up possibilities for *therapeutic care* designs that extend beyond physical treatment to encompass mental well-being. While the feasibility of such concepts in everyday situations remains untested, they present intriguing avenues for exploration. Looking ahead, the prospect of therapy centres transcends the artefact or object level, hinting at new opportunities for interaction design. This suggests a shift beyond human-to-object interaction towards more social levels, potentially paving the way for innovative designs extending beyond conventional realms of therapeutic objects.

6.2 Design Challenges

Mutualistic care has been identified as a crucial principle in designing living artefacts (Karana et al. 2020), and affection is an essential dimension of care (de La Bellacasa 2017). These affections, however, are not always positive; they can also manifest as stress, guilt, and regret upon perceiving the decline of the well- being of the living artefacts (Lu & Lopes 2018). Thus, while the emergence of an intimate bond with living artefacts holds the potential for cultivating sensibilities toward other living entities, it can also introduce complexities and challenges for users who engage with these artefacts on a daily basis (Zhou et al. 2024).

Some workshop participants articulated that constant notice and care for LTS might be engaging and intriguing, but they were also concerned this extra layer of responsibility might provoke negative feelings like worry and guilt and potentially adding to the burdens experienced by AD patients, who are already wrestling with anxiety and stress related to their sensitive skin. This juxtaposition raises critical questions: Does revealing the livingness of LTS material, aiming to form an intimate connection which is potentially able to foster a more sustainable behaviour and mend the longstanding cognitive separation of humans from nature (Karana et al. 2023), outweigh the potential burden it may impose on users, especially in a medical context? Conversely, could concealing this livingness alleviate the already heavy load on eczema patients, providing them with relief from an additional layer of responsibility? And if designers deem it necessary to surface the livingness, what approach can be adopted to establish a system that nurtures the microorganisms while simultaneously guiding users towards a more eLortless and seamless incorporation of caring practices into their daily lives? Further research is needed in the biodesign field to incorporate a broader range of perspectives from various stakeholders, making the concept of living artefacts acceptable and accessible to a wider audience.

Lastly, careful consideration is crucial when designing beyond the biological material. The LTS material is not intended to be sold or made available on its own; rather, it will be accompanied by appropriate packaging or other non-living accessories associated with the therapy. Addressing concerns raised by participants regarding the environmental impact of daily patch packaging, there is a need for a holistic view on sustainability when designing such artefacts.

Acknowledgments. This research was conducted as part of the NextSkins project, funded by the European Union's Horizon Europe research and innovation programme (Grant Agreement No. 101071159). Special thanks to the members of the Materials Experience Lab and the Centre of Design Research for Regenerative Material Ecologies (DREAM) at the Department of Sustainable Design Engineering, Delft University of Technology, for their valuable feedback and support. We are also deeply grateful to all the participants of the workshop, both within and beyond the NextSkins project, for generously sharing their time, insights, and expertise.

References

Achermann, Y., Kerns, P., Shirtliff, M.e., Barnes, l.: In vivo infection studies (2014)

Asher, M.I., et al.: Lancet **368**(9537), 733–743 (2006)

Barati, B., Karana, E.: Int. J. Des. **13**(3), 105–123 (2019)

Barati, B., Karana, E., Hekkert, P.: Int. J. Des. **13**(2), 21–38 (2019)

Bechky, B.A.: Sharing meaning across occupational communities: the transformation of understanding on a production floor. Organ. Sci. **14**(3), 312–330 (2003)

Brown, L., Miller, J.: Int. J. Hum.-Comput. Interact. **32**(3), 321–335 (2018)

Cacicedo, M.L., Pacheco, G., Islan, G.A., Alvarez, V.A., Barud, H.S., Castro, G.R.: Int. J. Biol. Macromol. **147**, 1136–1145 (2020)

Carlile, P.R.: A pragmatic view of knowledge and boundaries: boundary objects in new product development. Organ. Sci. **13**(4), 442–455 (2002)

Coughlan, P., Suri, J.F., Canales, K.: J. Appl. Behav. Sci. **43**(1), 122–134 (2007)

De la Bellacasa, M.P.: Matters of care: speculative ethics in more than human worlds, vol. 41. University of Minnesota press (2017)
Dementyev, A., Kao, H.L., Paradiso, J.A.: Sensortape: modular and programmable 3D-aware dense sensor network on a tape. In: Proceedings of the 28th Annual ACM Symposium on User Interface Software & Technolog, pp. 649–658 (2015)
Dong, M., et al.: J. Colloid Interface Sci. **610**, 913–922 (2022)
Enninga, T., van der Lugt, R.: Proj. Manag. J. **47**(2), 103–114 (2016)
Ersser, S. J., Cowdell, F., Latter, S., Gardiner, E., Flohr, C., Thompson, A. R., Drury, A.: Psychological and educational interventions for atopic eczema in children. Cochrane Database Syst. Rev. (1) (2014)
giaccardi, E., Karana, E.: Foundations of materials experience: an approach for HCI. In: Proceedings of the 33rd Annual ACM Conference on Human Factors in Computing Systems, pp. 2447–2456 (2015)
Gilbert, C., et al.: Nat. Mater. **20**(5), 691–700 (2021)
Girolomoni, G., Busà, V.M.: Therapeutic Adv. Chronic Disease **13**, 20406223211066730 (2022)
Halberstein, R.A.: Medicinal plants: historical and cross-cultural usage patterns. Ann. Epidemiol. **15**(9), 686–699 (2005)
Halpern, M.K.: Across the great divide: boundaries and boundary objects in art and science. Public Understand. Sci. **21**(8), 922–937 (2012)
Han, Y., et al.: Biomed. Mater. **15**(3), 035022 (2020)
Huusko, M., Wu, Y., Roto, V.: Structuring and engaging: the roles of design fictions in a co-design workshop. In: Proceedings of the 30th Australian Conference on Computer-Human Interaction, pp. 234–241 (2018)
Ikeda, S., et al.: Results Chem. **4**, 100502 (2022)
Islind, A.S., Lindroth, T., Lundin, J., Steineck, G.: Health Technol. **9**, 425–438 (2019)
Jones, R., Brown, S.: Ethical considerations in wearable design incorporating living organisms. In: Proceedings of HCI International Conference, pp. 275–289 (2019)
Kao, C.H.L., Bamforth, M., KIM, D., Schmandt, C.: Skinmorph: texture-tunable on-skin interface through thin, programmable gel. In: Proceedings of the 2018 ACM International Symposium On Wearable Computers, pp. 196–203 (2018)
Kao, C.H.L., Nguyen, B., Roseway, A., Dickey, M.: earthtones: chemical sensing powders to detect and display environmental hazards through color variation. In: Proceedings of the 2017 Chi Conference Extended Abstracts on Human Factors in Computing Systems, pp. 872–883 (2017)
Kao, H.L., Holz, C., Roseway, A., Calvo, A., Schmandt, C.: Duoskin: rapidly prototyping on-skin user interfaces using skin-friendly materials. In: Proceedings of the 2016 ACM International Symposium on Wearable Computers, pp. 16–23 (2016)
Karana, E., Barati, B., Rognoli, V., Zeeuw van der Laan, A.: Material driven design (MDD): a method to design for material experiences. Int. J. Des. **9**(2), 35–54 (2015)
Karana, E., Blauwhoff, D., Hultink, E. -J., Camere, S.: When the material grows: a case study on designing (with) mycelium-based materials. Int. J. Des. **12**(2), 119–136 (2018)
Karana, E., Barati, B., Giaccardi, E.: Living artefacts: conceptualizing livingness as a material quality in everyday artefacts. Int. J. Des. **14**(3) (2020)
Karana et al., 2023.Karana, E., Mcquillan, H., Rognoli, V., Giaccardi, E.: Living artefacts for regenerative ecologies. Res. Direct.: Biotechnol. Des. **1**, E16 (2023)
Kim, R., Risseeuw, C., Groutars, E.G., Karana, E.: Surfacing livingness in microbial displays: a design taxonomy for HCI. In: Proceedings of the 2023 CHI Conference on Human Factors in Computing Systems, pp. 1–21 (2023)
Koskinen, K.U.: Role of metaphoric boundary objects in the development of a company's strategic vision. Int. J. Manage. Concepts Philosophy **1**(2), 156–167 (2005)

Liu, X., Vega, K., Maes, P., Paradiso, J.A.: Wearability factors for skin interfaces. In: Proceedings of the 7th Augmented Human International Conference 2016, pp. 1–8 (2016)
Liu, X., et al.: 3D printing of living responsive materials and devices. Adv. Mater. **30**(4), 1704821 (2018)
Lu, J., Lopes, P.: Integrating living organisms in devices to implement care-based interactions
Mcnally, N., Phillips, D.: Geographical studies of atopic dermatitis. atopic dermatitis: the epidemiology, causes and prevention of atopic eczema, pp. 71–84 (2000)
Myers, W.: Bio design. museum of modern art (2012)
Nextskins Project: Retrieved september 19, 2024 https://www.nextskins.eu/home#ltm
Ng, I.S., Keskin, B.B., Tan, S.I.: Biotechnol. J. **15**(8), 1900228 (2020)
Nygrén, N.A.: Scenario workshops as a tool for participatory planning in a case of lake management. Futures **107**, 29–44 (2019)
Ogata, M., Sugiura, Y., Makino, Y., Inami, M., Imai, M.: Senskin: adapting skin as a soft interface. In: Proceedings of the 26th Annual ACM Symposium on User Interface Software and Technology, pp. 539–544 (2013)
Pataranutaporn, P., Ingalls, T., Finn, E.: Biological HCI: towards integrative interfaces between people, computer, and biological materials. In: Extended Abstracts of the 2018 CHI Conference on Human Factors In Computing Systems, pp. 1–6 (2018)
Pataranutaporn, P., Vujic, A., Kong, D.S., Maes, P., Sra, M.: Living bits: opportunities and challenges for integrating living microorganisms in human-computer interaction. In: Proceedings of the Augmented Humans International Conference, pp. 1–12 (2020)
Ramirez Figueroa, P.C.: Bio-material probes: design engagements with living systems (Doctoral Dissertation, Newcastle University) (2018)
Reeder, J.T., et al.: Waterproof, electronics-enabled, epidermal microfluidic devices for sweat collection, biomarker analysis, and thermography in aquatic settings. Sci. Adv. **5**(1), EAAU6356 (2019)
Restrepo-Villamizar, J., Vos, S., Verhagen, E., Lallemand, C.: Crafting on-skin interfaces: an embodied prototyping journey. In: Proceedings of the 2021 ACM Designing Interactive Systems Conference, pp. 1129–1142 (2021)
Rodrigo-Navarro, A., Sankaran, S., Dalby, M., Del Campo, A., Manuel, S.-S.: Engineered Living Biomaterials. Nat. Rev. Mater. 6 (2021). https://doi.org/10.1038/S41578-021-00350-8
Ross, G.: Treatments for atopic dermatitis. Australian Prescriber, **46**(1), 9 (2023)
Simpson, E.L., et al.: J. Am. Acad. Dermatol. **74**(3), 491–498 (2016)
Star, S.L., Griesemer, J.R.: Soc. Stud. Sci. **19**(3), 387–420 (1989)
Suib, S.S.S.B., van Engelen, J.M., Crul, M.R.: Int. J. Des. **14**(1), 113–133 (2020)
Thurstan, R.H., Brittain, Z., Jones, D.S., Cameron, E., Dearnaley, J., Bellgrove, A.: J. Appl. Phycol. **30**, 1821–1832 (2018)
Wang, H.: Biomaterials in medical applications. Polymers **15**(4), 847 (2023)
Wang, S., Fang, C.M., Yang, Y., Lu, K., Vlachostergiou, M., Yao, L.: Morphace: an integrated approach for designing customizable and transformative facial prosthetic makeup. In: Proceedings of the Augmented Humans International Conference 2022, pp. 58–67 (2022)
Wang, Y. & Amores fernandez, J., Maes, P.: On-face olfactory interfaces, pp. 1–9 (2020). https://doi.org/10.1145/3313831.3376737
Weigel, M., Lu, T., Bailly, G., Oulasvirta, A., Majidi, C., Steimle, J.: Iskin: flexible, stretchable and visually customizable on-body touch sensors for mobile computing. In Proceedings of the 33rd Annual ACM Conference on Human Factors in Computing Systems, pp. 2991–3000 (2015)
Weigel, M., Nittala, A.S., Olwal, A., Steimle, J.: Skinmarks: enabling interactions on body landmarks using conformal skin electronics. In: Proceedings of the 2017 CHI Conference on Human Factors in Computing Systems, pp. 3095–3105 (2017)

Zhou, J., Doubrovski, Z., Giaccardi, E., Karana, E.: Living with cyanobacteria: exploring materiality in caring for microbes in everyday life. In: Proceedings of the CHI Conference on Human Factors in Computing Systems, pp. 1–20 (2024)
Zimmermann, N., Pluchinotta, I.: Supporting interdisciplinary research projects via system dynamics boundary objects: an application to integrated urban water management. In: 38th International Conference of the System Dynamics Society (2020)

Biodesign in Textile Innovation: Emerging Applications of Bacterial Cellulose to Fashion

Ziqian Yu(✉), Valentina Rognoli, and Paola Bertola

Politecnico di Milano, Milan, Italy
ziqian.yu@polimi.it

Abstract. The growing emphasis on sustainability in the design industry has brought renewed attention to the potential of biodesign and biotextiles, particularly through the innovative use of bacterial cellulose (BC). BC is a biodegradable, versatile material produced by certain bacterial species. This study explores the potential and limitations of using BC in textile production, focusing particularly on its aesthetic qualities and how it performs in terms of comfort.

The paper provides an overview of bacterial cellulose, emphasising its unique properties, biodesign strategies, fabrication methods and emerging applications. Using a Research through Design (RtD) methodology, the study includes experimental work on producing BC textiles. The study also investigates the application of natural dyes and laser-cutting techniques to enhance the material's visual and tactile appeal.

The results suggest that BC holds significant potential for developing biotextiles, offering a balance of durability and aesthetic value. However, ensuring comfort remains challenging, highlighting the need for further interdisciplinary research and development.

Ultimately, this study encourages designers to reconsider textile production through the lens of biodesign. Incorporating sustainable resources and living systems into the design process establishes the foundation for future advancements in regenerative and responsible textile practices.

Keywords: Material design · Biodesign · Biotextile · Bacterial cellulose · Biofabrication

1 Introduction

In light of growing concerns over global environmental degradation and the rapid development of new manufacturing technologies for materials, the importance of circular textiles is becoming increasingly evident. In response, the European Union (EU) introduced its *Strategy for Sustainable and Circular Textiles* in 2023. The strategy aims to accelerate the development of bio-based textiles by establishing innovative legal frameworks that support and encourage the transition towards more sustainable and regenerative textile systems.

Bacterial cellulose (BC) is a regenerative material that is gaining popularity in fashion design. It is a natural, biodegradable substance that fully aligns with regenerative design

A. Mateus et al. (Eds.): WAD 2024, What's Around Design?, pp. 21–32, 2026.
https://doi.org/10.1007/978-3-032-08163-6_2

principles. It plays a particularly significant role in the evolving fields of biodesign and textile innovation. BC's inherent antimicrobial properties and compatibility with advanced textile manufacturing techniques further enhance its appeal. Consequently, the development of bacterial cellulose-based textiles, which combine aesthetic appeal, comfort and functional performance, is attracting increasing interest from disciplines such as material design, biodesign, materials science and bioengineering.

This paper presents an interdisciplinary study of the intersection between material and textile design, biodesign, materials science and bioengineering. Adopting a Research through Design (RtD) methodology, it explores the development of biotextiles by applying bacterial cellulose and biofabrication techniques.

The study also critically examines the challenges and limitations of early-stage bacterial cellulose during development and implementation, drawing on insights from experimental sample testing. The aim is to deepen understanding of the potential and constraints of bacterial cellulose within the textile design, particularly concerning aesthetic qualities and user comfort, and to lay the groundwork for future research and innovation in this emerging area of sustainable materials.

2 BC – Bacterial Cellulose

2.1 Bacterial Cellulose Origin and Definition

Bacterial cellulose (BC) was first synthesised in 1886 by Adrian John Brown, who produced it through a biosynthetic process using glucose as a carbon source in an aerobic environment. The bacteria involved in this process belonged to the *Acetobacter genus*, which is notable for its bluegrass-negative properties.

BC is a naturally occurring polymer produced by fermentation by specific bacterial species, such as *Acetobacter* (Fatima et al. 2021a, b). Although it has a similar chemical composition to plant-derived cellulose, bacterial cellulose is distinguished by its method of production and structural properties. Unlike plant cellulose, BC forms a highly pure nanoscale fibre network with a unique three-dimensional architecture that is free from the impurities typically found in plant-based cellulose (Ul-Islam et al. 2020). These structural differences result in outstanding mechanical strength, a high degree of crystallinity and excellent water retention capabilities. This makes bacterial cellulose a promising material for various applications (Fatima et al. 2021a, b).

2.2 Bacterial Cellulose Characteristics Within Textile

Bacterial cellulose (BC) is a highly distinctive material thanks to its diverse functionality. Its most significant applications include wound healing and skin tissue repair, and it is used in the food and packaging industries. BC has also gained recognition as a regenerative material, particularly in the field of circular textiles, and is being increasingly incorporated into design practices thanks to its environmentally friendly properties. This section explicitly explores BC's key attributes as a material for textile applications.

Fermentation culture involves cultivating microorganisms, such as bacteria or yeast, to facilitate fermentation. Unlike conventional fabrics, BC is produced through fermentation in a liquid medium. However, only specific bacteria, such as *Acetobacter*,

Achromobacter and *Rhizobium*, can synthesise bacterial cellulose (Fatima et al. 2021a, b).

BC is highly malleable, making it easy to modify. According to some scholars (Kaminski et al. 2020), BC forms on the culture medium's surface and takes the container's shape, as defined by its dimensions and the resulting gas-liquid contact area. This property enables BC to be cultivated directly into three-dimensional clothing patterns, facilitating a 'zero waste' design process (Kawasaki & Mizuno 2017). Furthermore, garments produced this way can fit the body seamlessly, almost like a second skin (Eryilmaz et al. 2013).

Although BC is chemically similar to plant-derived cellulose, it is unique because it contains no lignin, hemicellulose, or pectin. Its distinctive three-dimensional fibre network provides high water retention, excellent crystallinity, and outstanding tensile strength. Its tensile strength has been measured at 241.42 MPa, demonstrating strong resistance to stretching. Meanwhile, a maximum elongation at a break of 8.21% indicates its capacity to deform before breaking (Fatima et al. 2021a, b).

The hydrophilic nature of BC can be viewed as both an advantage and a challenge. Its ability to absorb moisture, which is a result of its hydrophilicity, also contributes to its potential antibacterial properties. It can efficiently draw moisture from the skin, thereby inhibiting bacterial growth in humid environments. This reduces the risk of infection and bacterial proliferation in areas prone to perspiration. However, its porous and highly absorbent nature poses a challenge for textile applications (Provin et al. 2021).

BC also exhibits antimicrobial properties, rendering it effective at killing or inhibiting the growth of harmful microorganisms. Most BC used in textiles today is derived from kombucha fermentation. The resulting bacterial broth contains a rich diversity of beneficial fungi and produces a material high in bioactive compounds with notable antibacterial and antioxidant effects. The dominant organism is *Acetobacter acetobacter*, often called 'blue bacteria', and it is one of the most abundant sources of microbial cellulose in nature (Antolak et al. 2021).

During kombucha fermentation, a Symbiotic Culture Of Bacteria and Yeast (SCOBY) forms a cellulose biofilm on the surface (Dufresne & Farnworth 2000). Kombucha-derived polymers have been shown to exhibit antibacterial activity against both Gram-positive and Gram-negative bacteria, including methicillin-resistant Staphylococcus aureus (Matei et al. 2020). Some scholars (Sreeramulu et al. 2000) confirmed this through an agar diffusion assay. Combining BC with antimicrobial agents such as nanosilver can significantly enhance wound healing by amplifying its antibacterial effects (Maneerung et al. 2008). *Acetobacter bacteria* metabolise glucose during fermentation into cellulose and gluconic acid, while yeast converts fructose into ethanol and carbon dioxide. The ethanol produced helps to protect the culture from contamination by foreign microbes (Sreeramulu et al. 2000).

Bacterial cellulose demonstrates excellent dyeability and readily absorbs various dyes over different intervals. It exhibits a high affinity for both plant-based natural dyes and reactive dyes (Galdino et al. 2021). Colour alteration of BC can occur in two main ways: by introducing natural pigments into the culture medium at the beginning of fermentation or by dyeing the material after drying. These methods greatly expand BC's aesthetic potential (Costa et al. 2022). Additionally, due to its high purity, BC requires less pre-treatment than plant cellulose before dyeing or further processing (Wood 2019).

BC's regenerative potential is evident in its properties and production process. As an exceptionally pure form of natural cellulose, it is non-toxic, requires minimal energy to produce and breaks down easily at temperatures above 280 °C (Fatima et al. 2021a, b; Provin et al. 2021). As it comes from a renewable source and can be composted after use, it provides a practical way to reduce textile waste (Wood 2019).

Another key advantage is that BC production does not rely on traditional cellulose sources like wood. It can be cultivated on various nutrient-rich substrates, further enhancing its appeal as an environmentally responsible material.

Despite its many benefits and its wide range of applications in medicine, design and textiles, bacterial cellulose has yet to achieve mainstream status within the textile industry. This is primarily due to current limitations, such as sourcing difficulties, inconsistent properties and appearance, and limited consumer acceptance.

2.3 Bacterial Cellulose Textile Fabrication: From Bacterial to Biotextile

The production of bacterial cellulose textiles involves two key stages: the cultivation of the bacteria that produce the cellulose and the subsequent processing of the textile, which includes washing, dyeing, softening and drying.

Kombucha, *saccharide bacteria* and sugar are commonly used as primary substrates in BC cultivation (Matei et al. 2020). Saccharide *bacteria* are currently considered the most effective strain for bacterial cellulose biosynthesis due to their accelerated growth rate, making them ideal for large-scale production and particularly valuable in design-focused research (Fatima et al. 2021a, b). Although glucose is the main component involved in BC synthesis, other sugars such as fructose, galactose, and sucrose have also been used in various experiments. Researchers have also explored using alternative, low-cost resources such as fruit juice, sugarcane, brewery residues and agricultural waste to reduce production costs (Ul-Islam et al. 2020). Water quality is critical in preparing the tea infusion for fermentation. Mountain spring and filtered water contain more biocompounds than mineral water, enhancing the fermentation process (Antolak et al. 2021).

Bacterial cellulose is cultivated in a specially prepared medium for a set period of time. Current cultivation methods include static, stirred and bioreactor-based cultures. The strain, nutrient composition, pH level and oxygen delivery all play crucial roles in determining the final properties of the material (Wang et al. 2019). In the context of material design, most contemporary production experiments favour the static culture method, in which bacterial cellulose (BC) forms at the air-liquid interface and takes the shape of the container holding the culture (Fatima et al. 2021a, b). While this approach is straightforward and capable of producing consistently structured cellulose with an adjustable thickness on a large scale, it does present limitations in the form of longer cultivation times and lower yields compared to bioreactor systems. These limitations mean that BC materials remain mainly in the development phase and have yet to be introduced to the market on a large scale.

Optimal fermentation conditions require the container to be refilled with a fresh nutrient solution and the environment to be kept aerobic, with temperatures ideally ranging from 25 to 30 °C and pH levels between 4 and 7 (Wang et al. 2019). Within 7–15 days, a biological film forms with a jellyfish-like appearance, known in Western cultures as SCOBY (Symbiotic Culture of Bacteria and Yeast). This culture thrives in a tea-based medium, gradually thickening as new layers of BC accumulate on the surface to form a suspended, structured film (Matei et al. 2020). Films that are grown without agitation in a medium containing 15 g/L of tea over six days demonstrate optimal characteristics in terms of comfort and appearance for fashion applications (Ng & Wang 2016). Material science's current focus is enhancing BC's properties through ex-situ modification.

One approach involves adding substances to the nanofibre network during production to create BC composites with specific properties. For instance, chitosan is added to the culture medium to enhance the material's antibacterial properties (Liang et al. 2019). As fermentation progresses, the culture medium produces various organic acids (Sreeramulu et al. 2000), which can enhance antibacterial activity. However, increased acidity may disrupt the balance of microbes on the skin. Therefore, once fermentation is complete, it is essential to thoroughly wash and soak the BC film to eliminate any microbial residues or undesirable odours from the growth medium, ensuring the material is safe for skin contact (Eryilmaz et al. 2013; Harmon 2017; Kaunas 2021).

Once the desired thickness has been achieved, bacterial cellulose can easily be dyed thanks to its high affinity for colourants. Natural dyeing techniques that use food waste can be employed during growth, eliminating the need for additional chemicals or water. This pre-colouring method is a potentially transformative step in creating sustainable biotextiles (Harmon 2017). Experiments using eucalyptus (*Eucalyptus globulus L.)* and onion (*Allium cepa L.*) dyes have demonstrated that BC can be dyed just as effectively as traditional natural fibres, resulting in a uniform beige colour and smooth texture in the wet BC films (Galdino et al. 2021). This highlights the potential for environmentally friendly, non-toxic and hypoallergenic colouration techniques that align well with modern biotechnological innovation.

As the functionalisation of BC continues to evolve, chemical modification methods can further enhance its antibacterial properties. For instance, adding antimicrobial agents such as silver nanoparticles during fermentation increases BC's effectiveness by introducing additional mechanisms that hinder bacterial growth (Fatima et al. 2021a, b). Submerging BC in a polymer solution can also result in a chitosan-cellulose composite suitable for applications such as wound care (Maneerung et al. 2008).

Another key objective in textile development is to improve BC's flexibility. Softening agents, including oils, are used to modify the material's mechanical behaviour. Directly applying oil to the BC surface or immersing the film in oil during drying can significantly enhance softness and consistency (Araújo et al. 2015). Adding stearic acid creates a smoother textile surface (Eryilmaz et al. 2013) and controlling oil absorption during drying further enhances the pliability required for textile applications.

Unlike moist BC dressings, which are commonly used in medical contexts, bacterial cellulose textiles are usually handled as dry sheets during the design process. Once the growth phase is complete, the materials are washed, dyed and dried to produce finished textiles (Kamiński et al. 2020). Some materials resemble durable, non-woven, leather-like fabrics (Mihaleva 2020) and display superior mechanical strength compared to conventional plant-based fabrics such as cotton (Fatima et al. 2021a, b).

2.4 Bacterial Cellulose Textile and Application

Suzanne Lee[1], founder of *BioCouture*, began exploring bio-alternatives at the intersection of design and bioscience. She used the Kombucha microbe to produce a bacterial cellulose material that is aesthetically pleasing and functional. Often referred to as 'plant leather' due to its resemblance to cowhide, this material combines Kombucha-derived cellulose with laser cutting techniques to improve design accuracy (Mihaleva 2020). The potential of BC as a textile has been clearly demonstrated, and its integration with laser technology creates new design possibilities in fashion. BC textiles can also be combined with novel technologies or materials, expanding their application in the textile industry.

Dry BC films were combined with flat textiles made from a 70:30 linen and polyester blend using stitching techniques in one trial. This resulted in garments such as Japanese-style jackets and polka-dot undershirts. These experiments confirmed that BC can bond with blended fabrics and improve their mechanical properties. Furthermore, the findings supported the feasibility of using BC as a replacement material in garment production (Costa et al. 2022).

Another study introduced a method of producing cellulose by synthesising stabilised hydrogel bacterial cellulose (HGBC) from kombucha. The resulting fabrics were used to create wristbands and T-shirts tested for skin compatibility over two weeks. Even after prolonged exposure to water and sweat, the material's microstructure remained intact. Participants rated the fabrics positively or neutrally in terms of key factors such as comfort, visual appearance, softness, skin irritation, odour, elasticity and moisture absorption. Importantly, no skin discolouration or discomfort cases were reported, which supports the material's suitability for use in clothing. Furthermore, adding glycerol during production converted the bacterial cellulose (BC) into a hydrogel, enhancing its softness and mechanical properties (Kamiński et al. 2020).

Josephine de Fijter[2], a researcher at the Eindhoven School of Design, developed bacterial cellulose textiles infused with probiotics for delivering skincare products. She created non-toxic, non-irritating, biodegradable, edible, skin-compatible biofilms using waste from local tea producers. The probiotics generated during fermentation were examined for their potential skincare benefits, including acne treatment. Her work also considered users' aesthetic expectations concerning colour, scent and texture, with the aim of eventual commercialisation.

This project demonstrates the functional value of probiotics in bacterial cellulose and the importance of considering user needs at an early stage in the design process. This proactive approach could lead to greater user satisfaction and product adoption. The striking blue bacterial cellulose (BC) material developed in this research is notable

[1] Fashion designer and biodesigner, founder of Biofabricated: https://www.launch.org/innovators/suzanne-lee/; https://www.biofabricate.co

[2] Prominent social designer and the founder of Studio Uncoated, based in the Netherlands: https://www.studiouncoated.com

for its uniqueness. It offers fresh visual possibilities in BC textiles and sparks further exploration into new dyeing techniques.

Bacterial cellulose films are suitable for garment prototyping as an alternative textile surface. Combining conventional fabrics with BC materials is a practical way to incorporate BC into clothing. However, verifying that these materials provide adequate comfort for real-world usage is still important. Improving the mechanical performance of BC remains challenging, given the demanding conditions garments must endure. Although research on bacterial cellulose textiles is still in its infancy, the material shows extraordinary potential for the future of the textile industry.

3 Methods

3.1 Research Through Design Approach

This study uses a Research through Design (RtD) approach to examine the production of bacterial cellulose (BC) textiles, paying particular attention to BC biomanufacturing and texturisation treatments. The experiment was structured around developing a static culture method for BC and was divided into three sequential phases: preparation of the culture solution, cultivation, and cleaning. The texturisation phase involved dyeing, drying, softening and laser cutting to produce fully developed BC textile samples.

3.2 Experimentation: Bacterial Cellulose Biomanufacturing

All containers and tools used for fermentation that directly interacted with the culture medium were sterilised. To prepare the medium, one litre of water was brought to a boil, and 150 g of organic white sugar was added, stirring until fully dissolved. Next, 20 g of black tea was added, and the mixture was simmered for 15 min. The resulting liquid was transferred to a plastic container.

Once cooled, 5 g of apple cider vinegar and 20 g of freshly brewed kombucha tea were added to the solution. A sterile gauze was placed over the mixture and secured in place. The container was stored in a shaded, well-ventilated area at a stable temperature of 20–30 °C with approximately 65% humidity and a pH of 4.5–5.0 to ensure optimal fermentation conditions.

After 14 days, a 6 mm-thick layer of BC mycorrhizal membrane was extracted. The resulting material was translucent and pale white, with a mild, acetic acid-like scent and a jelly-like texture. The BC film was cleaned by gently scraping off any surface residues and rinsing it under a continuous water flow. The water was changed every 4–8 h to reduce acidity and odour further.

3.3 Experimentation: Texturisation Treatment

To dye the BC naturally using purple kale, 300 g of purple kale was washed and boiled in 1.5 L of water with the addition of 10 g of salt. The mixture was simmered for 30 min, stirred every 10 min, and could be extended to increase dye intensity, resulting in approximately 1 L of pigment solution.

Once filtered, the dye was transferred to a container, and the BC film was immersed in it for 48 h before removal. The drying process was carried out in a cool, well-ventilated space, with the BC film rotated regularly and lightly treated with coconut oil to maintain flexibility. After drying, any excess oil was gently removed with a cotton cloth, yielding the final BC textile.

The textile design motifs were created using Adobe Illustrator and then laser cut in the School of Design prototype lab, *Politecnico di Milano*, enabling a wider variety of aesthetic possibilities.

4 Results

The biofabrication experiment yielded four samples, each measuring 12 cm by 12 cm and 2 mm thick.

Sample 1 (Fig. 1) displays an undyed, untextured BC sample, giving it a clean, minimalist appearance that showcases its natural qualities. This simplicity provides a valuable baseline for further exploration and experimentation.

Sample 2 (Fig. 2) has the exact dimensions and thickness but has a laser-cut textured pattern that enhances visual and tactile appeal while maintaining an undyed surface.

Sample 3 (Fig. 3) does not include a textured design like Sample 2, but it is dyed with purple kale to demonstrate the effect of natural dyeing on the material.

Sample 4 (Fig. 4) combines both techniques, it is dyed with purple kale and laser-cut, creating a surface with added depth and complexity that enriches the overall visual appearance.

Fig. 1: Sample 1, undyed BC sample

Fig. 2: Sample 2, undyed BC sample with laser-cut texture

Fig. 3: Sample 3, naturally dyed BC sample

Fig. 4: Sample 4, naturally dyed with laser-cut BC sample

5 Discussion

5.1 BC Textile Texture Discussion

Thickness and Translucency

Translucency is a material's property that allows light to pass through while diffusing it, meaning that objects on the other side are not clearly visible. In textiles, this property is often affected by the thickness of the material. Under indoor lighting at noon on a sunny day, the bacterial cellulose (BC) samples exhibited a degree of translucency. Future experimentation could involve cultivating BC textiles of different thicknesses and testing them in a laboratory to more precisely evaluate their light transmission properties.

Tactility and Smoothness

The smoothness of the BC textiles' surface lies somewhere between that of PVC and leather. In terms of softness, they are comparable to light calfskin, although they lack the suppleness of sheepskin. Suppleness is closely linked to softening treatments, and improper handling of BC can result in rigidity, rendering the material uncomfortable, particularly in garments that require a soft, flexible fit. Further research is necessary to evaluate the effects of softening agents such as glycerine, coconut oil and essential oils. This includes identifying the optimal stage for application, dosage, method and potential combinations. Samples 2 and 4 demonstrate that BC textiles can be intricately patterned and textured. However, additional testing is essential to assess the material's long-term durability and resistance to wear.

Elasticity and Extensibility

BC textiles exhibit minimal to no drape or stretch, limiting their suitability for specific applications. For instance, garments like sportswear, which require flexibility and responsiveness to body movement, may not benefit from BC in its current form. Further investigation is needed to explore ways to improve the elasticity and drapability of Scoby-based textiles during production.

Graphics and Texture

Visually, BC textiles have a texture with characteristics such as 'creases', 'grain' and 'asymmetry'. These features resemble those of leather, contributing to a perception of durability and tear resistance, as observed during initial testing. However, BC textiles are prone to creasing, affecting their appearance and producing a wrinkled or unkempt look.

While the leather-like quality of BC suggests robust durability, it may degrade over time. Prolonged exposure to sunlight or ultraviolet radiation can lead to cracking or tearing of the surface. Further studies are needed to evaluate the material's longevity and explore ways to extend its usable lifespan.

From a design perspective, BC offers opportunities for visual experimentation. Designers can influence the appearance of surfaces by drying cultured Scoby on materials with different textures, creating textiles with unique visual and tactile qualities that enhance their artistic appeal.

5.2 Limitation

The experimental phase of this study was conducted in a domestic setting, which introduced certain constraints. These included a limited number of culture samples and poor lighting, making it challenging to compare thoroughly with material composition, dyeing results and post-treatment outcomes.

Much of the material evaluation during the review phase relied on subjective perception due to the lack of rigorous scientific testing. Key performance indicators, such as material strength, light transmittance and washability, were not measured objectively. User research could be implemented through focus groups or structured surveys to strengthen future studies.

While BC textiles exhibit promising durability and tactile appeal qualities, consumer perception of their colour and texture may be mixed. Nevertheless, the material's inherent properties make it a promising candidate for use in garment production, and further development could increase its appeal to target users.

5.3 Suggestion for Future Work

Initially, future research could involve a comparative analysis of experimentally produced BC textiles and natural fabrics, such as cotton, silk and wool. This would include examining and contrasting the characteristics of these different materials, identifying the current limitations of BC textiles and determining which aspects, particularly in terms of comfort, require improvement to enhance BC performance in textile applications.

When evaluating the properties of BC textiles, it is essential to acknowledge that, while they have demonstrated antimicrobial capabilities, achieving a balance of functionality, comfort, and aesthetic appeal requires extensive interdisciplinary collaboration between biology and material science experts.

Cost, sustainability, and market acceptance will all play a part in shaping the future of bacterial cellulose in the textile industry, particularly advances in BC manufacturing technologies. Future studies could focus on developing and implementing a comprehensive evaluation framework based on existing BC textile prototypes to support the broader adoption of BC in textile design. This would involve defining clear, consistent criteria and metrics for assessing textile quality and performance. These criteria should encompass many factors, including aesthetic value, durability, comfort, usability, safety, and environmental impact.

6 Conclusions

Bacterial cellulose is an innovative and promising material that is attracting increasing interest in the design industry. Staying informed about scientific research advancements can significantly enhance the potential of bacterial cellulose textiles, particularly in terms of aesthetics, comfort, functionality and sustainability. Further investigation supported by medical and material sciences expertise is essential in areas such as texture treatment and material performance.

In conclusion, the potential applications of BC textiles are extensive. To realise this potential, collaboration extending beyond the field of design is crucial. Such cooperation will be key to continuously improving the functionality and performance of bacterial cellulose textiles and expanding their relevance across diverse domains.

Acknowledgements. This paper is based on my Master's thesis, which I carried out within the Design Department at the Politecnico di Milano with the support of the university's on-campus prototyping laboratory. I conducted all experiments and processed all data. I would like to express my sincere gratitude to Professors Paola Bertola and Giulia Tomasello for their invaluable guidance throughout the theoretical and practical phases of the thesis. Thanks to Dr Angelica Vandi for her support in the writing process and to Marisa Satsia for her insightful assistance during the experimental work.

References

Antolak, H., Piechota, D., Kucharska, A.Z.: Kombucha tea—a double power of bioactive compounds from tea and symbiotic culture of bacteria and yeasts (SCOBY). Antioxidants **10**(10), 1541 (2021). https://doi.org/10.3390/antiox10101541

Araújo, S., Silva, F.M.D., Gouveia, I.C.: The role of technology towards a new bacterial-cellulose-based material for fashion design. J. Indust. Intell. Inform. **3**(2), 168–172 (2014). https://doi.org/10.12720/jiii.3.2.168-172

Costa, A.F.D.S., et al.: Bacterial cellulose: characterisation of a biomaterial for apparel products application. Res. J. Text. Appar. **26**(4), 532–545 (2022). https://doi.org/10.1108/RJTA-04-2021-0048

Dufresne, C., Farnworth, E.: Tea, kombucha, and health: a review. Food Res. Int. **33**(6), 409–421 (2000). https://doi.org/10.1016/S0963-9969(00)00067-3

Fatima, A., et al.: Bacterial cellulose: history, synthesis, and structural modifications for advanced applications. In: Khan, S.B., Kamal, T. (eds.). Bacterial Cellulose: Synthesis, Production, and Applications, p. 26, 1st ed.. CRC Press (2021). https://doi.org/10.1201/9781003118756

Fatima, A., Yasir, S., Khan, M.S., Manan, S., Ullah, M.W., Ul-Islam, M.: Plant extract-loaded bacterial cellulose composite membrane for potential biomedical applications. J. Bioresources Bioprod. **6**(1), 26–32 (2021). https://doi.org/10.1016/j.jobab.2020.11.002

Eryilmaz, J., Cass, G., Akdemir, Ö., Şener, D., Hamitbeyli, A., Çobanoğlu, Ö.: Microbial cellulose production by *Acetobacter xylinum* and its application for the fashion and textile industry. In: Proceedings of the International Istanbul Textile Congress 2013, pp. 1–5. Istanbul, Turkey (2013)

Galdino, C.J., Jr., et al.: The future of sustainable fashion: bacterial cellulose biotextile naturally dyed. Chem. Eng. Trans. **86**, 1333–1338 (2021). https://doi.org/10.3303/CET2186223

Harmon, J.: Homegrown: Investigating design potential of bacterial cellulose. In: International Textile and Apparel Association Annual Conference Proceedings, vol. 74, no. 1. Iowa State University Digital Press (2017). https://doi.org/10.31274/itaa_proceedings-180814-216

Kamiński, K., et al.: Hydrogel bacterial cellulose: a path to improved materials for new eco-friendly textiles. Cellulose **27**(9), 5353–5365 (2020). https://doi.org/10.1007/s10570-020-03128-3

Kawasaki, K., Mizuno, D.: In-between of two and three dimensions: Development of fashion pattern cutting for bio-fabric. In: Karana, E., Giaccardi, E., Nimkulrat, N., Niedderer, K., Camere, S. (eds.), Proceedings of the International Conference of the Design Research Society Special Interest Group on Experiential Knowledge: Alive. Active. Adaptive (EKSIG 2017), pp. 79–89. TU Delft Open(2017)

Kaunas, R.: Sustainable strategies on cultivation, modification and rehydration of bacterial cellulose to overcome hornification for industrial applications: a review. J. Bioresource Bioprod. **6**(1), 26–32 (2021). https://doi.org/10.1016/j.jobab.2020.11.002

Liang, J., Wang, R., Chen, R.: The impact of cross-linking mode on the physical and antimicrobial properties of a chitosan/bacterial cellulose composite. Polymers **11**(3), 491 (2019). https://doi.org/10.3390/polym11030491

Maneerung, T., Tokura, S., Rujiravanit, R.: Impregnation of silver nanoparticles into bacterial cellulose for antimicrobial wound dressing. Carbohyd. Polym. **72**(1), 43–51 (2008). https://doi.org/10.1016/j.carbpol.2007.07.025

Matei, B., Diguţă, C.F., Matei, F., Popa, O.: Antimicrobial potential of kombucha bacterial biopolymer. Sci. Bull. Seri. F. Biotechnol. **24**, 133–137 (2020)

Mihaleva, G.: Bio matter in creative practices for fashion and design. AI Soc. **36**(4), 1361–1365 (2021). https://doi.org/10.1007/s00146-020-00957-5

Ng, M.C.F., Wang, P.W.: Natural self-grown fashion from bacterial cellulose: a paradigm shift design approach in fashion creation. Des. J. **19**(6), 837–855 (2016). https://doi.org/10.1080/14606925.2016.1208388

Provin, A.P., Dos Reis, V.O., Hilesheim, S.E., Bianchet, R.T., De Aguiar Dutra, A.R., Cubas, A.L.V.: Use of bacterial cellulose in the textile industry and the wettability challenge—a review. Cellulose **28**(13), 8255–8274 (2021). https://doi.org/10.1007/s10570-021-04059-3

Rocha, A.V., Sarubbo, L.A.: Review - Bacterial cellulose: an ecofriendly biotextile (n.d.)

Sreeramulu, G., Zhu, Y., Knol, W.: Kombucha fermentation and its antimicrobial activity. J. Agric. Food Chem. **48**(6), 2589–2594 (2000). https://doi.org/10.1021/jf991333m

Ul-Islam, M., Ullah, M.W., Khan, S., Park, J.K.: Production of bacterial cellulose from alternative cheap and waste resources: a step for cost reduction with positive environmental aspects. Korean J. Chem. Eng. **37**(6), 925–937 (2020). https://doi.org/10.1007/s11814-020-0524-3

Wang, J., Tavakoli, J., Tang, Y.: Bacterial cellulose production, properties and applications with different culture methods – a review. Carbohyd. Polym. **219**, 63–76 (2019). https://doi.org/10.1016/j.carbpol.2019.05.008

Wood, J.: Bioinspiration in fashion—a review. Biomimetics **4**(1), 16 (2019). https://doi.org/10.3390/biomimetics4010016

Lucid Life | Marama Ora

Christopher Bellamy(✉)

Central Saint Martins - University of the Arts, London, England
chris@biocrafted.com

Abstract. Lucid Life (Marama Ora in Tahitian) is a practice-based research project, which brings together scientific and traditional knowledge, to explore interspecies and intercultural collaboration on the island of Mo'orea, French Polynesia. Living materials' ability to grow, adapt, and heal has been used symbiotically by humans for thousands of years, however, they are now replaced by the more reliable and repeatable inert materials which dominate our day to day lives. This project sought to explore how biodesign could be used to create a bridge between indigenous and scientific communities, and combine their deep understandings of nature. Using bioluminescent algae, similar to those in symbioisis with corals, a contemporary living material was developed. This ethereal material emits light in response to touch. The micro-organisms are encapsulated in a way which allows them to live, sequester carbon, and emit light for at least six months; needing only sunlight in return. In collaboration with three different Polynesian artisans, who have a deep understanding of the symbiotic relationships in the reefs around their islands, a series of artifacts were co-created using the bioluminescent living material. A drum, a swimsuit, and a necklace were made; combining local materials and practices, with the bioluminescent material. Each of these objects was found to create awe-inspiring and transcendent experiences, inspiring those who experience them to live more symbiotically with other cultures and species. With living materials, similar to Polynesian traditional knowledge which exists only in living memory, there is a risk that it may die. However it is this fragility that allows it to adapt, react, and interact – and is what makes this temporal state of livingness so beautiful.

Keywords: Bioregional design · Engineered living materials · Traditional knowledge · Decolonisation · Interspecies design

1 Introduction

Lucid Life | Marama Ora is a practice based research project applying bio-regional design principles on the Island of Mo'orea, French Polynesia. The island is home to an indigenous population of around 10,000 people, but also visiting marine biologists at two international field research stations located on the island.

The island has a rich indigenous polynesian culture, but also a strong presence of western scientific research. These two cultures work hard to co-exist harmoniously, however this project explored how these two worlds of science and traditional knowledge could be more deeply integrated through biodesign [1].

A. Mateus et al. (Eds.): WAD 2024, What's Around Design?, pp. 33–43, 2026.
https://doi.org/10.1007/978-3-032-08163-6_3

2 Research Process

Bioprospecting, the practice of seeking novel and valuable organisms within a bioregion, is a field of research with high potential in French Polynesia [2]. This discipline was identified as an area interest to local indigenous, scientific, and global communities, and as such was chosen as starting point for this research project.

A bioprospecting permit was obtained from the government of French Polynesia to allow the collection, analysis, and transportation of bioloigcal matter [3, 4]. The permit granted rights to research the microorganisms found in French Polynesia, however, no guidance was provided on how indigenous traditional knowledge should be treated as part of this project. With the guidance of experts at the platform "Local Contexts", permission was sought from local indigenous leaders and community organisations, as well as individual collaborators, to gain social approval.

64 interviews were conducted with marine scientists, anthropologists, government officials, marine businesses, cultural ambassadors, material researchers, biologists, conservationists, ecologists, teachers, and local community members. The learnings from these were summarised into a 5 min documentary [5] (Fig. 1).

Fig. 1. CHRISTOPHER BELLAMY Meeting with local artists and educators Naumi 'Mamie' Tapi and Tekoui 'Jérémie' Tamari. PHOTO CREDIT: CHRISTOPHER BELLAMY

3 Project Objectives

The goal of this project was to use biodesign as a bridge between the indigenous and scientific communities on the island of Mo'orea. The project worked with local artisans to find applications for the scientific research taking place on the island. This process reversed the typical relationship between visiting scientists and locals, and had the potential to generate value and engage both communities.

Based on the ethnographic research conducted, the following criteria were set for the project, to ensure it appropriately addressed the values and interests of all parties:

3.1 Living Materials

Within Polynesian culture, knowledge is transmitted orally, and is rarely documented. It only exists in living memory, which allows it to constantly adapt and evolve. The local community also had strong understanding of the symbiotic processes that exist with their island eco-system, and their role within them. In this spirit, it was chosen to work with living materials for this project.

Engineered living materials is a field of research of high potential and growing interest [6–8]. When combined with bioprospecting, there is the potential to create novel materials, with unique behaviours, by encapsulating micro-organisms into useable materialities. For this project, a living material would be developed, using micro-organisms from the island of Mo'orea, which could then be used and applied by local artisans.

3.2 Bioregional Design

The Polynesian community on Mo'orea have a strong culture of bioregional design. Their comprehensive understanding of the materials which grow around them is evident in all aspects of life. From food, to medicine, to building materials. The choice was made to work with local materials and micro-organisms wherever possible during this project.

Low-tech processes would also be used, so that materials can be made onsite, and be accessible to local communities. Biotechnology is often treated in the same industrial ways as other manufacturing processes, and done at scale. However, in the context of island communities, it is also possible for biotechnologies can be completed at small scales with minimal equipment.

3.3 The Sublime

The local community often described their symbiosis with nature, but also how this is disappearing with the arrival of modern technologies. As our society become more disconnected from nature, this project sought to reverse that trend by creating a transcendant and awe inspiring experience. This is possible through the beauty, surprise and responsiveness of living materials. The choice was made to work with bioluminescent micro-algae, as their production of light was a familiar, but always inspiring, experience for the local community [9, 10].

4 Material Development

Two bioluminescent algae species found in Mo'orea were identified as the model organism to use during the development of living materials in this project. While their production of light may not have a practical function this context, it served two purposes.

Firstly, the use of a light emitting organism allows the rapid development of a living material platform, as it is simple to tell it is alive. This platform can then be adapted to host other micro-organisms for functional applications in the future.

Secondly, the emission of light creates a beautiful and intriguing experience, creating interest in the project among the local community. Getting local engagement in the project is essential for its success, and so using this strong aesthetic supports that [11, 12] (Fig. 2).

Fig. 2. CHRISTOPHER BELLAMY Growing bioluminescent algae in the laboratory. PHOTO CREDIT: CHRISTOPHER BELLAMY

4.1 Living Organisms

The two algae species selected, pyrocystis fusiformis and pyrocystis lunula are commonly used, and well understood, allowing the project to develop within its tight timescales.

Pyrocystics fusiformis was found to be more sensitive to mechanical stimulation, however due to its larger cells, created a less dense light source. Both were found to be adequate in creating living materials, with pyrocystis lunula appearing to be more robust.

To culture these living organisms, experiments were conducted in light intensity, growth media, temperature, sterility, agitation, light timings, growth vessels and blackout, based on existing literature [13–15]

– Light: An inverted cycle of 18 h daylight and 6 h of darkness
– Environment: 100% blackout environment
– Light Intensity: 2450 lx cool white LEDs
– Media: Natural seawater + 50% guillards media - 22 micron filtered
– Vessels: Cell Culture flasks, conical flasks, or hanging grow bags
– Temperature: 20 celcius (16 minium, 26 maximum)
– Movement: Shaking a minimum of once every 3 days
– Ventiliation: No air transfer was found to be nessecary

4.2 Encapsulation

The material form used in this project was inspired by coral's symbiotic relationship with micro-algae. While the exact interaction between the host coral, and visiting dinoflagellate micro-algae is not yet fully understood, this project looked to achieve a similar spirit through encapsulation of bio-luminescent dinoflagellates in a skin like material which can be applied to the surface of objects.

To encapsulate the bioluminescent algae, they are mixed in a hydrogel, which is then coated in a hydrophobic gas permeable membrane. The final optimised recipes for each part of this are documented below. When encapsualted in this way, they are found to have a life span in excess of 6 months [13, 16, 17]. The hydrogel recipe is adjusted based on the manufacturing process, and varied between each artefact created in this project (Fig. 3).

Fig. 3. A photograph of the living material next to the corals which inspired it. PHOTO CREDIT: CHRISTOPHER BELLAMY

The hydrophobic gas permeable coating allows the material to breath, while also protecting it from the environment and contamination. Once encasulated the material can be put in harsh environments, such as Sodium Hydroxide or Hydrochloric Acid without it affecting the material (Fig. 4).

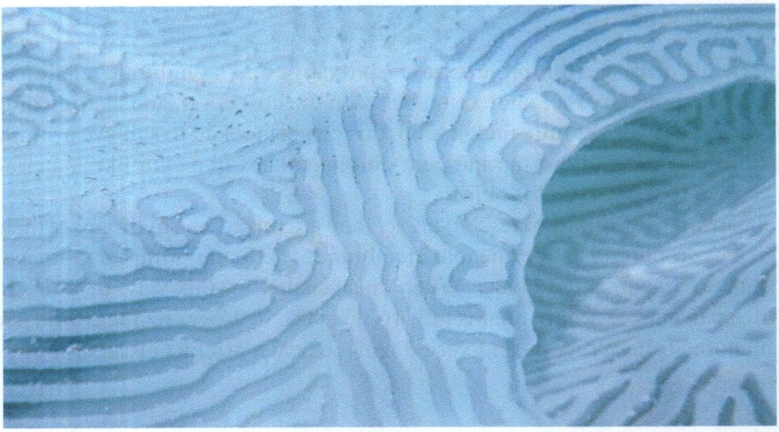

Fig. 4. A photograph of the living material showing it "breathing" underwater. Photo Credit: Christopher Bellamy

Transparent silicone was used in this project to create a robust, high quality coating in short timescales. There are sustainability concerns around the use of a crosslinking polymer in the material. As such, research was conducted into biodegradable alternatives, which hold potential for future development.

Various techniques were tested for encapsulation including bioprinting, microfluidics, casting, micro-pipetting and electronic discharge printing. Each of these processes has potential, but given the timescale of the project, the simplest process, casting, was deemed to be the most suitable (Fig. 5).

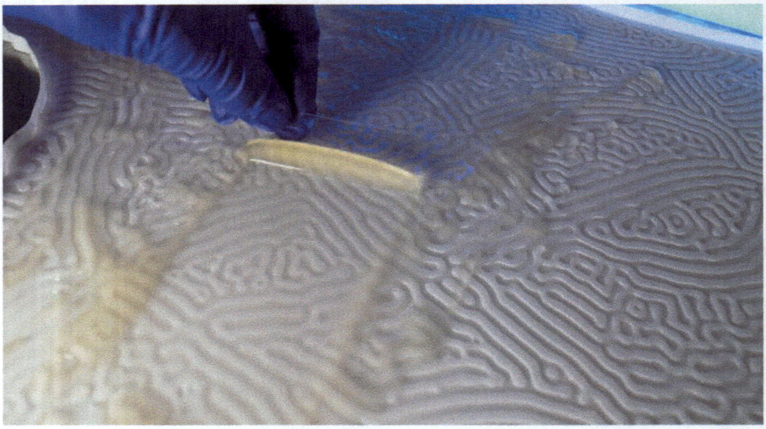

Fig. 5. A photograph showing the material being made in the laboratory. Photo Credit: Christopher Bellamy

4.3 Material Development in the Field

In the spirit of bioregional design, all materials and processes were developed in ways that were suitable to be performed by remote island communities [18].

Processes were developed so that these materials could be created outside of a lab environment. Using a combination of portable equipment, such as the Oxford Nanopore DNA sequencer, and improvisation (filters made from coffee filters and 2 litre water bottles), it was possible to complete all of the required processes on the island (Fig. 6).

Fig. 6. A photograph of algae being cultivated on the island by Christopher bellamy. Photo Credit: Christopher Bellamy

5 Design

The light transmission within alginates means that light cannot penetrate deeper than 0.5mm. In addition, to maintain structural integrity of the alginate, it was not possible to have any dimensions greater than 2mm in width.

As such, it is nessecary to have a pattern of equidistant lines, which infact emulates the reaction-diffusion patterns often found in nature [20]. Discovered by Alan Turing, these common patterns can be mathematically generated.

A script was developed to replicate this biological process in a graphical form. The script took graphical inputs and transformed them into feasible designs for the living material. This tool was able to translate indigenous designs in an impartial way, transforming traditional symbology and graphics without any interference.

5.1 The Swim Suit

The swimsuit was developed based on an idea proposed by collaborator Tekoui 'Jeremie' Tamari. Jeremie comes from the Marquesas islands, where night spearfishing is common profession and past time. Unfortunately many of his friends and relatives have died during the pursuit of this high risk activity.

Jeremie proposed that is the bioluminescent material could be made into a swimsuit, allowing the divers to be visible at night, and permitting rescue in case of any issues. The aesthetic and form of the design was intended to reflect Jeremie's deep connection to nature (Fig. 7).

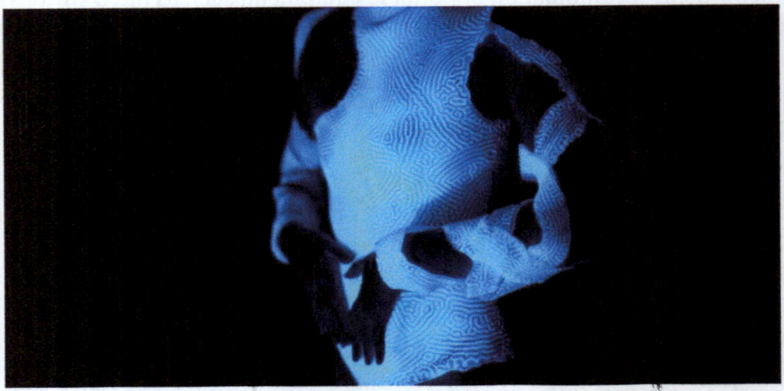

Fig. 7. A photograph of the swimsuit inspired by conversations with Tekoui 'Jérémie' Tamari. Photo credit: Christopher Bellamy

5.2 The Drum

The drum was developed in collaboration with Hinatea and Moe Colombani, who are local cultural ambassadors, dancers, and makers.

Dance is incredibly important to Tahitian society, and the drum is the driving beat behind this. Using a marine organism to make a drum which illuminated when played was a compelling idea. The design was based on a traditional style of drum using half coconut shells, inline with the spirit of bioregional design.

The patterns on the drums were designed by Hinatea, and represent Marees, the symbolic Tahitian gathering points.

A fully working drum was able to be developed in the timescale of the project (Fig. 8).

Fig. 8. A photograph of the bioluminescent living material illuminating when touched and the drum made in collaboration with Hinatea & Moé Colombani. Photo credit: Christopher bellamy

5.3 The Collier

The collier was developed with Tokainiua Jean-Daniel Devatine, an artist and head of design at the local art school. He was inspired by the marine origins of the bioluminescence, and wanted to see how it would interact with the irridesence of mother of pearl, another marine material often used by Tahitian craftspeople.

The fractured design is intended to be a distruptive response to the traditional polished curves of mother of pearl, while talking to vornoi patterns often found in nature as inspiration. Tokai hoped that the collier would illuminate in response to someone's breath, or heart beat (Fig. 9).

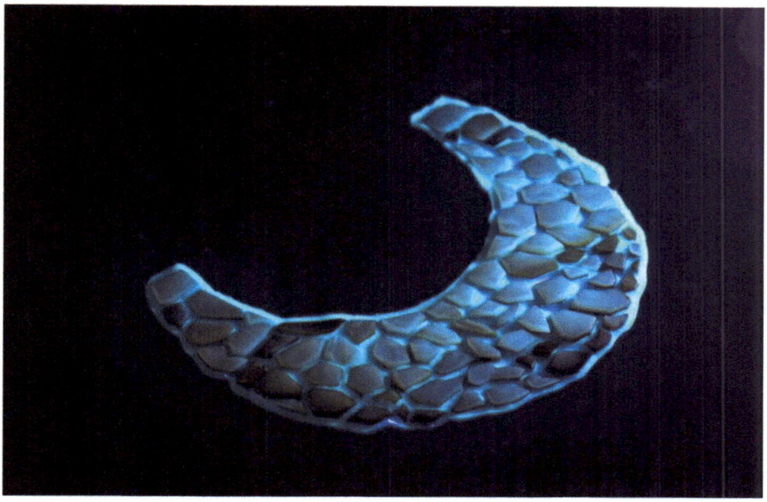

Fig. 9. A photograph of the collier made in collaboration with Tokainiua Jean-Daniel Devatine. Photo credit: christopher bellamy

6 Conclusions

This practice-based research project sought to use design as a bridge between indigenous and scientific communities on the Island of Mo'orea, French Polynesia. The outcome was the creation of three artefacts which demonstrated beautiful, practical, and inspiring objects which could be created when these two communities came together.

It is hoped that this project can inspire these communities, and similar communities, of the potential that can come from intercultural collaboration.

It has also showcased the potential of biodesign as a tool for applying biotechnologies, and scientific research in ways which are relevant to local communities.

There are three future workstreams which this project has inspired:

6.1 Decolonising Biodesign

During this project it became clear that there is a need to decolonise the practice of biodesign. The aesthetic of expensive equipment and laboratories makes it be perceived as inaccessible to many communities.

To build trust and give back to the community a week long biomaterial workshop was organised to educate young local artists and designers in biodesign, while also engaging them all in the living materials research.

6.2 Indigenous Intellectual Property

This project has sought to use best practice regarding traditional knowledge wherever possible. It generated valuable intellectual property, with significant commercial potential. Inline with the newly announced WIPO treaty in genetic resources and associated traditional knowledge, benefit sharing for the local community will be sought through the protection of this intellectual property [21].

Permission to use biocultural information and traditional knowledge has been managed through the Local Contexts platform, a hub supporting the responsible use of biological information and traditional knowledge [3].

6.3 Living Materials

The adaptability, responsiveness, and repairability of the state of livingness was highlighted during this project. The living material platform developed has huge potential for high impact applications, when applied with other micro-organisms.

Aknowledgements. Collaborating Artists:

Tekoui 'Jérémie' Tamari, Tokainiua Jean-Daniel Devatine, Hinatea & Moé Colombani, Naumi 'Mamie' Tapi.

Supporting Institutions: The Francis Crick Institute Making Lab, The Gump Station University of California, Tahitian History Society, Centre de Métiers d'Art de la Polynésie française, The Coral Gardeners, The Atitia Center, The Royal Commission for the Exhibition of 1851.

Tutors: Nancy Diniz - Course Leader MA Biodesign / Alice Taylor - Lecturer Biology and Living Systems / Jon Flint - Lecturer in Design and Fabrication / Thora Arnardottir – Associate Lecturer in Biodesign / Lorraine Archer – Environmental Scientist, Associate Lecturer / Paula Corsini – Specialist Grow Lab Technician / Shem Johnson – Specialist Grow Lab Technician / Barbara Paes – Specialist Grow Lab Technician.

Land, Material and Knowledge Acknowledgement: This project was completed in the UK and on the islands of Tahiti and Mo'orea. This project contains traditional and biocultural knowledge, and has been completed with the permission of cultural leaders, and each individual collaborator.

The project is registered with local contexts, project ID: 83cf1cc3-bdab-4125-acd8-5c5c5a05dc3f.

Biological samples were collected under Arrêté Number 7581 granted the 22nd August 2023.

References

1. Kong, B.: Community driven design of living technologies. Participatory Design Conference (2018)
2. Guézennec, J., Moppart, X.: Microbial mats in French Polynesia and their biotechnological applications (2021)
3. Local Contexts. https://localcontextshub.org/projects/83cf1cc3-bdab-4125-acd8-5c5c5a05dc3f
4. Project ID: Reservation ID: 85535. University of California Nature Reserve System – RAMS. https://rams.ucnrs.org/
5. Lucid Life | Marama Ora Project Video – biocrafted.com/lucid-life
6. Thomsen, M.R., Tamke, M., Mosse, A., Tyse, G.: Designed substrates for living architecture performance. Construction, Energy, Environment & Sustainability Conference (2021)
7. Karana, Barati, Giaccardi: Living artefacts. Conceptualizing livingness as a material quality in everyday artefacts. J. Des. (2020)
8. Rodrigo-Navarro. Engineered living biomaterials. Nat. Rev. (2023)
9. Oktay. Designing with the more-than-human: temporalities of thinking with care (2023)
10. Karana, Barati, Giaccardi. Living artefacts. Conceptualizing livingness as a material quality in everyday artefacts. J. Des. (2020)
11. Ofer. Designing Direct Interactions with Bioluminescent Algae (2021)
12. Jalaal. Stress-Induced Dinoflagellate Bioluminescence at the Single Cell Level (2020)
13. Li, C., et al.: Ultrasensitive and robust mechanoluminescent living composites. Sci. Adv. J. (2023)
14. Fajardo. An "omic" approach to Pyrocystis lunula: New insights related with this T bioluminescent dinoflagellate (2019)
15. Barr. (2023) Sustainable Bioluminescent Applications for Textiles and Environments
16. Caldwell, G.: Immobilising microalgae and cyanobacteria as biocomposites: new opportunities to intensify algae biotechnol. Bioprocess. (2021)
17. Tong. Bio-coatings as immobilized microalgae cultivation enhancement: a review (2023)
18. Weber, R.: A sympoietic ocean: Design research within the marine holobiont (2023)
19. Rodenburg. Bioluminesence in product design (2023)
20. Wentworth, D.: On Growth and Form (1942)
21. World Intellectual Property Organisation. WIPO Member States Adopt Historic New Treaty on Intellectual Property, Genetic Resources and Associated Traditional Knowledge (2024)

Designing Customizable Malleability and Padding in Fungal Textiles

Jierui Fang(✉), Stefano Parisi, and Elvin Karana

Industrial Design Engineering, Delft University of Technology, Delft, The Netherlands
jierui@alumni.stanford.edu, s.parisi@tudelft.nl

Abstract. This paper presents our investigation into methods to create fungal textiles with variable cushioning and flexibility properties, for differing technical and experiential characteristics. Fungal materials have generated rising interest in design communities over the past several years, resulting in applications in packaging, building, insulation, fashion and more due to their circular lifecycle and tunable customization on a myriad of properties and scales. In the fashion industry, these novel materials have been used as leather alternatives with a broad range of applications to the human form. Imitating conventional textile behavior, these applications, however, usually retain the same characteristics throughout the breadth of the material, a departure from the dynamic living organism it was created from. We aim to bridge this gap by endowing a singular fungal textile with multiple properties, serving the dual function of highlighting its organic origins and reducing the material complexity often required to achieve such versatility in a single material. Specifically, we present our design exploration to create localized support and flexibility in fungal textiles. We explored two techniques of creating fungal textiles, investigating diverse fungal species, growth and manufacturing methods, and post-processing techniques. The new techniques and unique properties of these materials suggest promising applications in areas such as assistive wearables, enhancing both user and planetary wellbeing.

Keywords: Fungal Textiles · Sustainability · Biodesign · Material Driven Design · Customization

1 Introduction

Materials grown from living sources like fungi offer sustainable solutions as well as unique benefits for product applications in their customization, permeability, and sensory experiences (see, for an overview, Camere & Karana 2018; Karana et al. 2018, Manan et al. 2021). Mycelium, the assimilative body of fungi that grows below ground in large networks, provides nutrients, communication and support to the fruiting bodies above ground. It can be grown into a flexible material (see, for example, Raman et al. 2022; Danninger et al. 2022) or solid structures (see Elsacker et al. 2020; Jones et al. 2020 for examples) depending on the structure of pure mycelium or its colonization of different substrates such as agricultural waste, sawdust, and cloth. Its unique heat resistant and

A. Mateus et al. (Eds.): WAD 2024, What's Around Design?, pp. 44–57, 2026.
https://doi.org/10.1007/978-3-032-08163-6_4

hydrophobic properties have led to stiffer foam and composite applications in packaging (e.g., Ecovative), interiors (e.g., GROWN bio), sound insulation (e.g., MOGU), fitness accessories (Fraxinea), and more pliable ones in fashion (e.g., Mylium, Mylo, SQIM).

1.1 Fungal Textiles

Designers such as Aniela Hoitink have used fungi to grow mycelial discs and mold into custom clothing forms, leading to the development of MycoTex (NEFFA; Hoitink 2022). Helena Elston has combined mycelium with discarded textiles to create clothing that is compostable. Eden Power Corp, Hermes, Gucci and Stella McCartney have all used fungi based materials for their fashion collections. Many current consumer facing applications of fungal materials today are focused on fashion textiles at the luxury level, using industrial scale processes to produce leather alternatives (see Amobonye et al. 2023 for a table of companies). These fungal textiles hold the potential to be formed into custom shapes with differential patterning and properties for specific areas of flexibility or cushioning, facilitating novel material experiences and sustainable solutions to people's everyday products. However, current fungal textiles that have been created as leather alternatives maintain similar properties throughout without enabling more dynamic variation (Fig. 1). To achieve varying three-dimensional patterns within the same mono-material fungal textile, layering techniques can be used after material making, as demonstrated in a conceptual, never-commercialized woven yoga mat with three-dimensional patterns for user alignment and support (Lululemon Athletica 2021). However, this approach requires additional processing, rather than conformally fabricating the material in its final form and properties.

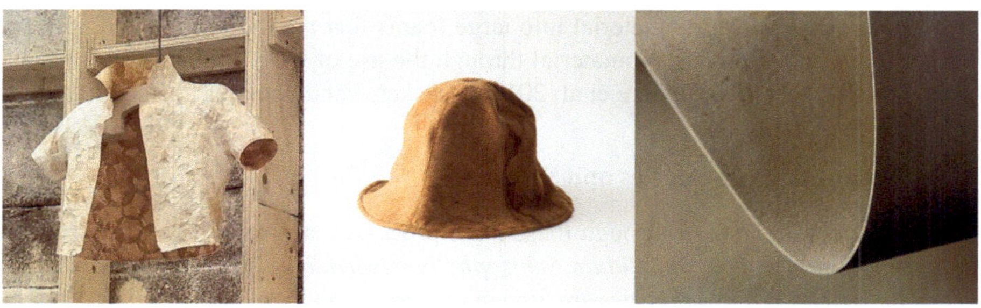

Fig. 1. Left: MycoTEX jacket by NEFFA, Middle: Amadou hat from Eden Power Corp, Right: Ephea® Radicle (©SQIM)

To bridge this gap, we present our explorations with various forms of fungi, both the hidden mycelia and the visible fruiting body towards the goal of differential patterning for varied stiffness, flexibility, and cushioning within the same material.

2 Literature Review

2.1 The Making of Fungal Materials: A Brief Overview

The body of research on fungal material making spans three categories (Gandia et al. 2021): (1) fungal material harvesting from the trama (an inner layer of the stiff fruiting body) of wild basidiocarps, (2) liquid state fermentation of pure mycelium, and (3) solid state fermentation through the use of substrates that could generate composite or mostly pure forms of mycelium depending on harvested area. The former two can lead to the production of more flexible, textile-like material, and the latter which would suit stiffer applications (Vandelook et al. 2021).

Within category 1, select fungal fruiting bodies, for example that of hoof fungi, have been explored as material, more commonly in traditional craft practices (Gandia et al. 2021). Craftspeople harvest this hoof fungi from beech forests yearly and carve off the top layer to expose the trama underneath which is then pounded and stretched. The resulting suede-like material, known as *amadou*, has shown promising absorptive and anti-inflammatory properties and has been fashioned into hats, wound patches, fly-fishing bait, and tinder by craftspeople (Peger 2001). Recently for category 2, submerged fermentation or liquid state fermentation using pure liquid cultures of mycelia, sometimes in bioreactors, has been emerging as a way to produce large quantities of flexible mycelial hides (Elsacker, Vandelook, & Peeters 2023) by first inoculation in static liquid culture and then growth in shaking incubators to allow for continued nutritional uptake (Adamatzky, et al. 2021). These require very controlled laboratory conditions but present a lot of potential in terms of scale and form. In category 3, mycelium material companies (like Ecovative) have used a form of solid-state fermentation to create aerial mycelium for leather alternatives. Aerial mycelium is where the mycelium is encouraged to grow further out of the substrate material into large foams that are then rolled and flattened alternately to create thicker biomaterial through the use of CO2 gradients, airflow, and other factors (Appels & Camere et al. 2019; Elsacker, Vandelook, & Peeters 2023).

2.2 Effect of Fungals Species and Substrates

Common species explored through these three processes are wood decaying white rot fungi including *Ganoderma lucidum*, *Schizophyllum commune*, *Trametes versicolor*, and *Pleurotus ostreatus*. These species are used in growing mycelium largely because of their ability to decay lignin found in woody materials to get to the cellulose and hemicellulose (Elsacker et al. 2020). Some species, like *Ganoderma lucidum* and *Fomes fomentarius* (used in the production of amadou as mentioned earlier), have health related properties for anti-inflammatory or anti-microbial effects used in Eastern medicine or in traditional Translyvanian craft practices (Ekiz et al. 2023; Pegler 2001). Common ways to create composite materials with mycelium include using bran, beechwood sawdust, rapeseed straw, hemp hurd, and even organic waste materials as substrates. Increasing sugar levels in the growth environment promotes faster growth, but the tradeoff in increased sugar concentration is the likelihood of faster growing contaminants also propagating, necessitating a rigorous sterile environment. In growing fungi with substrate material, more difficulty for the mycelium to break down the substrate could lead to stiffer material with increased chitin concentration (Haneef et al. 2017). In general, the presence, size and orientation of natural fibers influence the stiffness of the resulting fungal material (Manan et al. 2021; Elsacker et al. 2019).

2.3 Post-processing of Fungal Materials

Fungal materials exhibit different properties depending on its state, plasticization (using glycerol) (Appels & Wösten 2021), deacetylation, cross linking and tanning (Jones et al. 2021), and heat treatment (Appels & Camere et al. 2019) methods applied. These properties can be fluidly changed to confer various stiffness/resistance or flexibility qualities depending on the required material outcomes. Varying the percentage of glycerol, for example, can change the material from brittle, to natural, to elastomeric (Appels et al. 2020). Different coatings with oils and films can also be added afterwards to preserve and confer hydrophobic and antimicrobial properties (see Elsacker et al. 2020 for examples). Materials able to be kept alive (sans the baking or heat pressing treatments) (Chang et al. 2019) show responsiveness to mechanical and optical stimuli (Adamatzky et al. 2021) and self-healing capabilities when rehydrated or put in nutrient media (Elsacker, Zhang, & Dade-Robertson 2023) presenting a promising avenue for *living regenerative artifacts* (Karana et al. 2023; Giaccardi and Karana 2015).

3 Methodology

The Material Driven Design (MDD) method (Karana et al. 2015) helped guide our material explorations. MDD suggests material tinkering (Parisi et al. 2017) as an essential activity throughout the design process. Explorative material tinkering gave way to more systematic tinkering that then fed into how the resulting properties would best meet human needs through later formgiving and concept explorations. Over the course of 9 months, the first author undertook an extensive exploration of two most promising fungal textile processes using different fungal sources: fungi from slant cultures (Exploration 1) and fungi from fruiting bodies (Exploration 2). Work requiring inoculation (Exploration 1) was performed in a biosafety level 1 biolab, while work with fruiting bodies (Exploration 2) was mostly performed in a materials lab, with only the end processes for both being done in a biolab. Even though the two processes were conducted separately, they occurred in parallel and cross-informed each other's development pathways and our understanding of textile material possibilities from fungi. Below we share our protocols and tacit knowledge obtained through these processes.

3.1 Exploration 1: Material Experimentation with Grown Mycelia from Cultures

Petri dishes were inoculated with samples taken from slant cultures that had been kept in a fridge. After at least 5 days, incisions were taken from the petri dishes and inoculated into large Erlenmeyer flasks containing 200 mL of liquid media according to the protocol in (Elsacker, Zhang, & Dade-Robertson 2023). Based on previous literature two species were used initially, *G. lucidum* (Bae et al. 2021) and *S. commune* (Appels et al. 2020). However, the *G. lucidum* strain, while growing quite well on the petri dish, did not translate over to liquid cultures and was discontinued. Subsequent growth rounds were done with *S. commune*, testing out various liquid media compositions. The highest sugar content with 10% malt extract broth (MEB) yielded the fastest and most robust formation of the mycelial skin layer (with similar optimal growth on hydrogels seen in Gantenbein et al. 2023), but 8% MEB was chosen as a more moderate compromise because of concerns around contamination, as motivated in Table 1. See Fig. 2 for the overall process diagram.

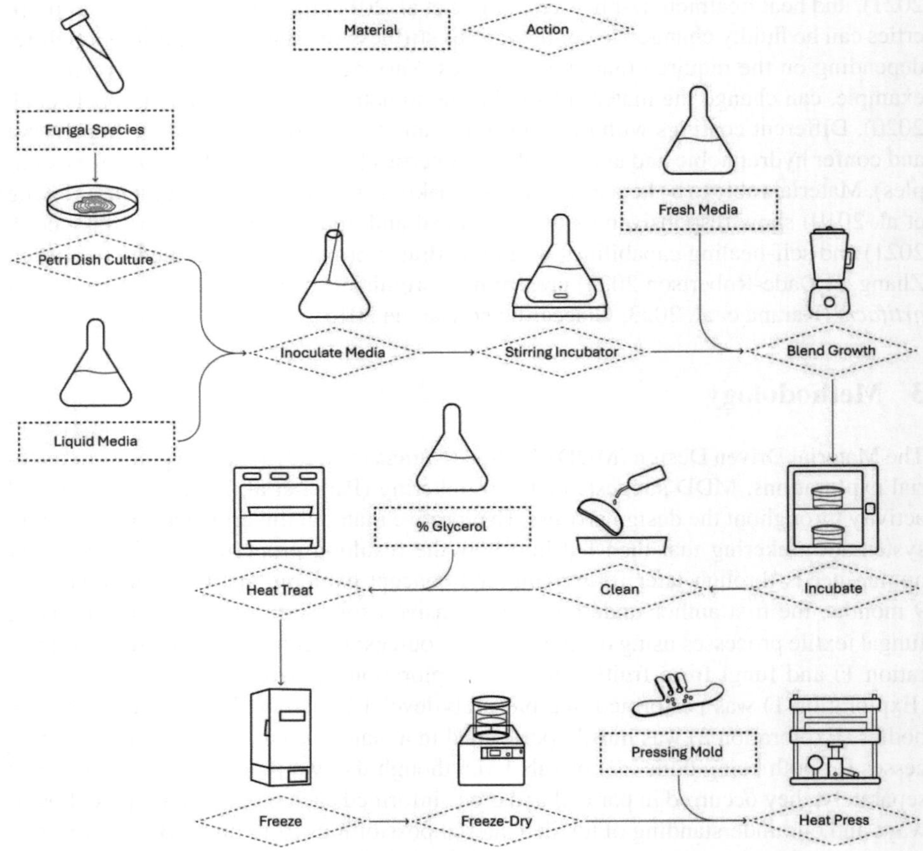

Fig. 2. Making process diagram of fungal textiles with liquid cultures

The flasks of liquid culture were incubated in static, stirred, and shaken conditions due to workarounds with equipment availability in the lab. It was found that stirring in the flask with a magnetic stirrer at 350 rpm produced the most consistent and even growth. After incubation in the flasks for at least 9 days, the cultures were blended with 100–400 mL of fresh liquid media in a Waring blender. It was found that 10 0mL of fresh media was sufficient while 400 mL led to slightly more growth, but had more danger of contamination and spillage in the blending process.

We observed that cultures containing the foam produced after blending yielded thicker and more cushion-like end mycelium material. Subsequent rounds also tried adding anti-foam which produced spottier and less consistent growth, confirming the helpful presence of foam. The new blended media was divided evenly over several new containers which included flat glass petri dishes, the same Erlenmeyer flasks, and tall crystallization glasses. The crystallization glasses were found to yield the best surface area to volume ratio for growth and provide sufficient height for the aerial mycelium to be contained in. Containers were incubated in both light and dark environments, with the dark environment yielding slightly less fruit body formation, although still some if the cultures were left too long.

After at least 5 days of incubation in the new containers, the resulting mycelium skins were removed from the culture, washed with water and cleaned with a cell scraper to remove the bottom gelatin-like layer that had formed. The mycelium was then deactivated at 60 degrees Celsius, either pressed and dried between parchment paper and a DIY press with plasticization afterwards, or directly in a percentage glycerol solution for 48 h. Mycelium in the varying percentages of glycerol followed material behavior similar to the ones obtained by Appels et al. (2020). 9% glycerol was chosen to be most applicable for a wearable context, in line with the conclusions in Table 1. Moisture trapped in the press made it more prone to contamination. Directly adding glycerol kept the material from sticking to surfaces and at approximately similar thicknesses, so conducting deactivation and plasticization at the same time without pressing was chosen as ideal. Samples prepared from that method were then freeze-dried and heat-pressed in a similar manner to those which will be outlined in Exploration 2. The resulting material can be seen in Fig. 3. The below table (Table 1) indicates various conclusions drawn from the variables iterated upon during this process.

3.2 Exploration 2: Material Experimentation with Blended Fruiting Bodies

In this exploration, we followed a similar method to the one presented by Nakauchi et al. (2023) of preparing mycelium pulp with a variety of different fruiting bodies. A Waring blender was used in place of an ultrasonicator due to lack of equipment availability. 200g of white button mushrooms were initially blended with varying glycerol percentages in 100mL of water and subsequently oven dried at 40 degrees Celsius. The resulting material was leather-like with significant shrinkage and curling done in molds. Bleaching was tried with store bought 3% hydrogen peroxide and UV light and also with 2% NaOH. 2% NaOH yielded faster results and a tan-yellow color, but the bleaching route was not pursued as the initial dark material color was deemed satisfactory and due to concerns about leftover chemical traces in the potential material applications. Because of the lack of predictability with oven drying, it was also discontinued and other methods for creating different textures and concept forms were pursued. See Fig. 4 for the overall process diagram.

Subsequent trials to replicate the foam variable shown in the grown process were made using freeze-drying with the blended pulp (Nakauchi et al. 2023). Mixtures were frozen after blending for at least 24 h and then freeze-dried at 0.303 mbar and −55 deg Celsius. Freezing at colder temperature -80 degrees Celsius versus 0 degrees Celsius in the 24 h before freeze-drying yielded a less moist material, most likely due to increased

Table 1. Takeaways from fungal textile explorations with grown mycelia from cultures.

Growth Media GPYM, 2% MEB, 8% MEB, 10% MEB	GPYM (Glucose Peptone Yeast Medium) shows slow and steady growth in separate chunks but starts to slow down at a smaller thickness. MEB shows faster growth from one central chunk but fruits earlier. Higher % MEB leads to thicker growth with less fruiting bodies, but no noticeable difference between 8% & 10%. 10% has more issues with contamination
Incubation Environment Static, Shaken, Stirred	Stirred incubation produced more consistent growth
Growth Container Erlenmeyer flasks, Glass petri dishes, Crystallization glasses	Tall petri dishes lead to more consistent and less deformed growth
Growth Composition Higher biomass, Coffee grounds, Activated charcoal	Higher biomass leads to faster and thicker initial growth. Substrate addition (coffee grounds) separates from material sections of pure mycelium. Activated charcoal promotes earlier fruiting and adds slight colored texture
Light No light, light	Presence or absence of light doesn't seem to make too much of a difference in 8% or 10% MEB in initial incubation. After blending, low light inhibits some fruiting
Foam No foam, Foam layer, Anti-foam	Layering the foam created after blending on top can lead to added thickness, more even, cushiony material. Material density still dependent on growth time. Need to have sufficient starting biomass for the foam to stay. Anti-foam causes uneven growth
Deactivation In paper press, In glycerol	Pressed drying helps prevent deformation but also leads to trapped moisture and potential contamination. All sugars should be washed off. Thin layers need a sheet between to prevent sticking to plate. Deactivation in glycerol maintains more thickness and prevents deformation
Glycerol Concentrations 3%, 6%, 9%, 16%	3% and 6% feels like the material is in danger of cracking, 16% feels too moist to the skin, 9% is ideal

sublimation of various oils present in the mixture. The resulting material had a memory foam-like quality which could be heat pressed to contain both cushion-like and leather-like qualities within the same piece (Fig. 5; Fig. 6). Varying the thicknesses, coatings, mold infills and patterns, temperatures, and durations for heat pressing was done to optimize for the two separate material qualities of padding and flexibility. In addition,

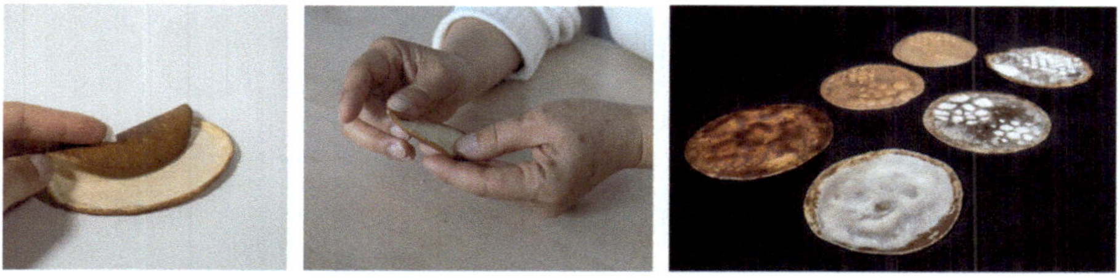

Fig. 3. Left: Flat sample of grown mycelia textile, Middle: Interaction with flat sample, Right: Pressed textures on grown samples

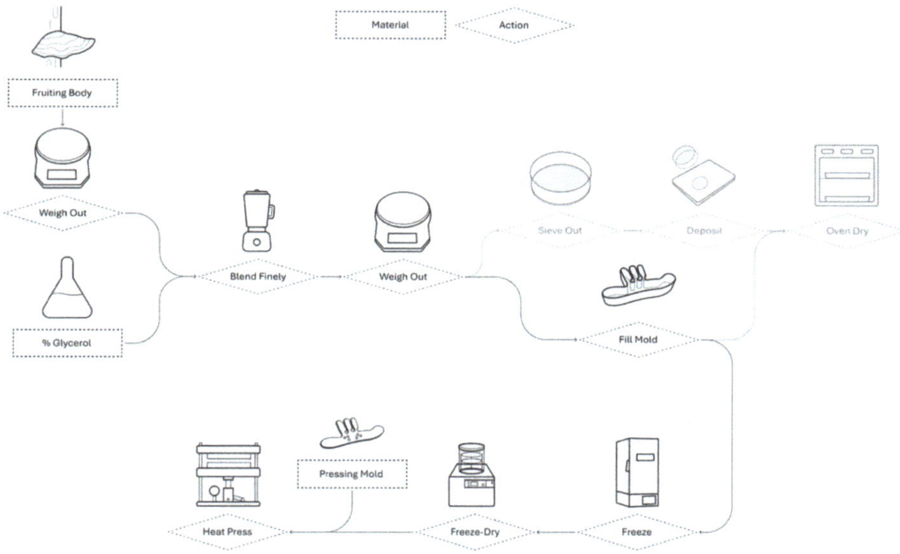

Fig. 4. Making process diagram of fungal textiles with blended fruiting bodies

all materials made using this fruitbody method displayed some property changes with surrounding environmental changes in humidity (e.g., if it was raining that day) with more "stickiness".

A variety of customizable patterns were tried out to test which ones separated from the pressing molds cleanly after heat pressing. We also evaluated how varying the thickness and weight of the pulp mixture in the freeze-drying molds, corresponding to benchmarking by either volume or density, might affect the durability and surface quality of the final pressed material. Varying the pressing molds and pressing factors (time, temp and coating) was looked into for the finer details of surface finish. Figure 5 above demonstrates the conclusions drawn during the freeze-drying and heat-pressing processes summarized at the end of Table 2.

The unpressed areas remained like a padded foam while the pressed areas were more leather-like for the white button mushrooms. Several bracket fungi were also blended

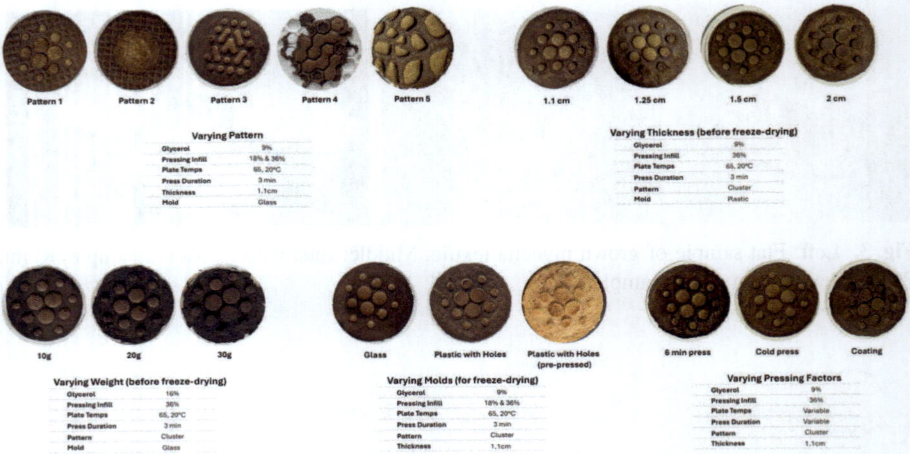

Fig. 5. Heat press experiment tracking for blended fruiting body (white button)

using this process which yielded much more padding and cotton-like properties both before and after heat pressing (Fig. 6).

4 Discussion

In this paper, we showed diverse techniques for differing qualities of padded support and flexibility within the same fungal material, with the parallel processes informing each other during development. During the material tinkering process, procedural knowledge and insights gained through the making of a foam-like quality in liquid culture led to similar aspirations in the fruiting body process which was achieved through freeze-drying. From there, different localized material properties could be obtained through heat pressing on both of the end results. Overall, the fruiting body process (Exploration 2) yielded a material that was easier to prototype with due to its quick turnaround time and simpler method to simulate a target thickness. Conversely, the inoculum-grown mycelium process (Exploration 1) produced more robust properties in elasticity and bending but was more difficult to simulate target thickness.

Reflecting first-hand on the experiential qualities of these fungal materials, several distinctive qualities stand out characterizing their materials experience (Karana et al. 2014). These materials possess a textile-like form, providing a soft and pliable surface. This quality makes these fungal materials particularly interesting for applications requiring direct contact with the body. Additionally, their adaptable texture allows them to conform to various shapes and surfaces, an attribute that can be exploited to create products with customized fits and enhanced comfort. A notable characteristic of these materials is their memory foam-like quality, which can be heat-pressed to achieve both cushion-like softness (padding quality) and selective firmness within the same piece. This versatility offers significant advantages, particularly in the creation of items that require both cushioning for comfort and a firm surface for support; one could also imagine a combination with even stiffer composite fungal materials. However, some

Table 2. Takeaways from fungal textile explorations with blended fruiting bodies

Types of Fungus White Button, Snow Fungus, Bracket Fungi	White button has better cohesion and a pleasant color. Snow fungus has a cushion like quality, but breaks apart. Bracket fungi has a cotton-pad like quality and cohesion
Processes Vacuum forming, Freeze-dry, Bleaching	Vacuum forming creates thin, even material, but difficult to remove from drying surface. Freeze-drying creates memory-foam like material with less bounce back. Freezing at colder temperatures beforehand yields less sticky material, probably because of more sublimated glycerol content. Bleaching with NaOH changes the deep brown color to more taupe-yellow
Oven Dry - Glycerol Concentrations 2%, 3%, 6%, 9%, 16%	2%, 3% and 6% lead to plastic-like material. 9% feels a bit more moist but does not feel like it'll crack when folded. 16% too moist
Oven Dry - Weight/Thickness 10 mL to 100 mL, 4 mm to 14 mm	Thicker layers more prone to cracking and noticeable stiffness difference with lower percentages of glycerol. More weight also leads to increased deformation in the oven drying process
Oven Dry - Molds Varying shape and texture	Grease needs to be applied to aid removal from the larger molds. Varying texture difficult with uneven shrinkage
Freeze Dry - Thickness 1.1 cm, 1.25 cm, 1.5 cm, 2 cm	Thicker can lead to material breakage during pressing. Similar results appeared for denser mixtures by weight
Freeze Dry - Molds Varying height, material	Need to be taller than material thickness to account for any bubbling over that occurs. No difference whether plastic or glass. Similar % glycerol conclusions as above
Heat Press Pattern, Infill, Temp, Duration, Coating	Too intricate of a pattern is difficult to remove. Lower infill leads to crosshatch pattern, 36% smoother. Need grease coating to aid material removal. Higher temp and longer duration leads to more color change. Beeswax coating helps with retaining surface finish

challenges also need to be addressed. The stickiness of the material from the fruiting body (white button) can complicate handling and finishing processes, potentially limiting its use in applications where a smooth, non-tacky finish is desirable. Moreover, the fragility of fungal materials raises concerns about their durability and suitability for long-term use, prompting a need for further development or integration with existing industry processes to enhance their strength and resilience. Despite these challenges,

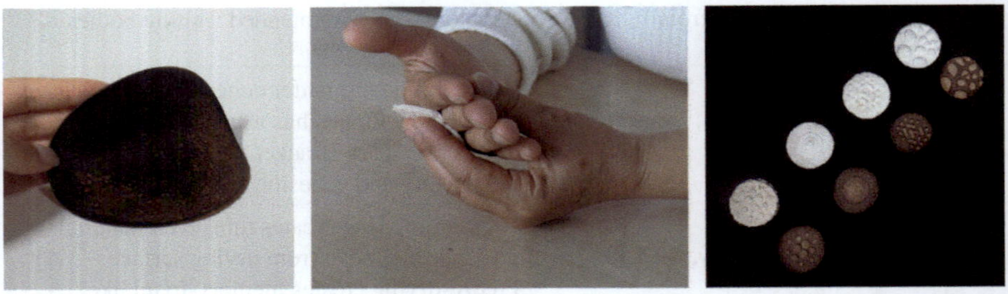

Fig. 6. Left: Flat sample of white button fruiting body textile, Middle: Interaction with flat sample of shelf fungi fruiting body textile, Right: Pressed textures on samples from both white button and bracket fungi fruiting bodies

the unique combination of comfort, adaptability, and versatile surface properties makes these fungal materials promising candidates for applications in wellbeing, particularly in the development of assistive wearables and other products that enhance user comfort and support. By harnessing the unique properties of fungal textiles, we can develop sustainable solutions that benefit both human and planetary health, addressing an increasingly urgent concern within textile research and industries (see, for an overview, McQuillian & Karana 2023).

Potential categories of needs in patients suffering from dexterity affected diseases have been highlighted in previous research interviews and could include temporal needs (with swelling and pain following a circular loop from morning to night), differing localization of pain points per person and joint, and the changing nature of those throughout

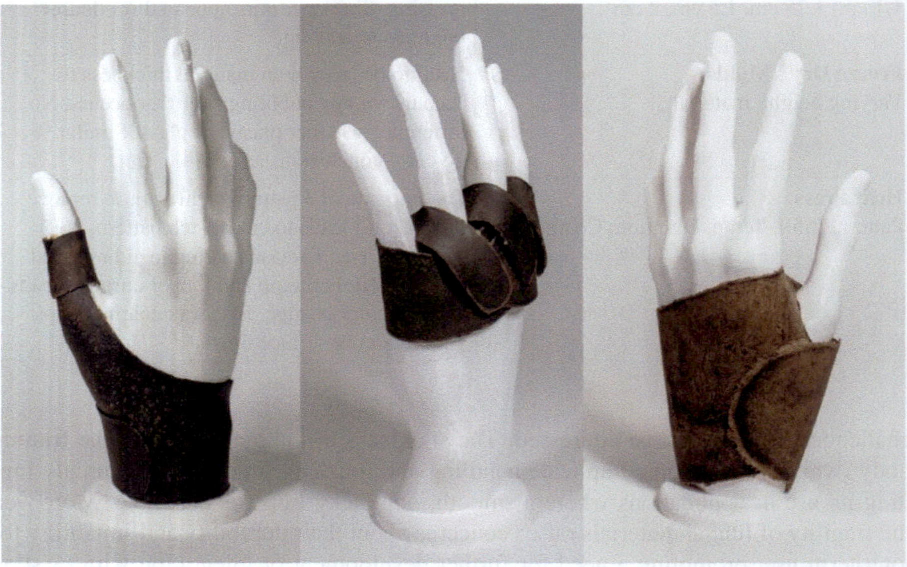

Fig. 7. Prototypes of fungal textiles applied to assistive wearable forms

time (months or years) (Fang 2020). Research into various fungal species (like Ganoderma taugae and Phellinus linteus) for wound care and healing (Suarato et al. 2018; Su et al. 2005) as well as to atopic dermatitis (Hwang et al. 2012) show potential of skin-to-skin contact of certain compounds in fungal mycelia for anti-inflammation, antioxidant, and anti-tumor effects (see examples in Manan et al. 2021). For these reasons, our fungal textiles hold great potential for further exploration in assistive wearables, providing localized cushioning and support, along with additional potential therapeutic qualities. We propose this as an exciting direction for future study. Figure 7 showcases our initial prototypes in this promising area.

5 Conclusion

In conclusion, this paper has documented the exploration and development journey of a fungal textile material utilizing two distinctive methods: growing from culture and blending fruiting bodies. By critically comparing and reflecting on these approaches and their outcomes, we have identified and discussed the ability to vary padding and flexibility qualities within the material. Drawing inspiration from these qualities, we speculated on and prototyped potential applications for this material in the field of human wellbeing, particularly within assistive wearables. By leveraging the unique properties of fungal textiles, we can create sustainable solutions that promote both human and planetary health, aligning material life cycles with the needs of their users and the environment.

Acknowledgements. Thanks to Richard Groen and the team at the Science Center's Biolab; Joren Wierenga and the team at IDE's Biodesign Lab; Mascha Slingerland and the team at IDE's Applied Labs; the Center of Design Research for Regenerative Material Ecologies and the Materials Experience Lab at TU Delft, for advice and invaluable host facilities. The project was possible through support from Stichting Fulbright Commission the Netherlands and the Netherland-American Foundation.

References

Adamatzky, A., Gandia, A., Chiolerio, A.: Towards fungal sensing skin. Fungal Biol. Biotechnol. **8**(1) (2021). https://doi.org/10.1186/s40694-021-00113-8

Air Mycelium. Ecovative (2024). https://www.ecovative.com/airmycelium/

Amobonye, A., Lalung, J., Awasthi, M.K., Pillai, S.: Fungal mycelium as Leather Alternative: a sustainable biogenic material for the Fashion Industry. Sustain. Mater. Technol. **38** (2023). https://doi.org/10.1016/j.susmat.2023.e00724

Appels, F.V.W., Wösten, H.A.B.: Mycelium Mater. Encycl. Mycol. **2**, 710–718 (2021). https://doi.org/10.1016/B978-0-12-809633-8.21131-X

Appels, F.V.W., et al.: Fabrication factors influencing mechanical, moisture- and water-related properties of mycelium-based composites. Mater. Des. **161**, 64–71 (2019). https://doi.org/10.1016/j.matdes.2018.11.027

Appels, F.V.W., van den Brandhof, J.G., Dijksterhuis, J., de Kort, G.W., Wösten, H.A.B.: Fungal mycelium classified in different material families based on glycerol treatment. Commun. Biol. **3**, 334 (2020). https://doi.org/10.1038/s42003-020-1064-4

Bae, B., Kim, M., Kim, S., Ro, H.S.: Growth characteristics of polyporales mushrooms for the mycelial mat formation. Mycobiology **49**(3), 280–284 (2021). https://doi.org/10.1080/12298093.2021.1911401

Camere, S., Karana, E.: Experiential characterization of materials: toward a toolkit. In: Proceedings of DRS 2018, Limerick, Ireland (2018). https://doi.org/10.21606/drs.2018.508

Chang, J., Chan, P.L., Xie, Y., Lee, K., MA, Cheung, M.K., Kwan, H.S.: Modified recipe to inhibit fruiting body formation for living fungal biomaterial manufacture. PloS One **14**(5), e0209812 (2019). https://doi.org/10.1371/journal.pone.0209812

Danninger, D., Pruckner, R., Holzinger, L., Koeppe, R., Kalternbrunner, M.: MycelioTronics: fungal mycelium skin for sustainable electronics. Sci. Adv. **8**, eadd7118 (2022). https://doi.org/10.1126/sciadv.add7118

EDEN power corp. (n.d.). EDEN power corp. EDEN Power Corp. https://edenpowercorp.com/

Ekiz, E., et al.: Exploring the potential medicinal benefits of ganoderma lucidum: from metabolic disorders to coronavirus infections. Foods **12**(7), 1512 (2023). https://doi.org/10.3390/foods12071512

Elsacker, E., Vandelook, S., Brancart, J., Peeters, E., De Laet, L.: Mechanical, physical and chemical characterisation of mycelium-based composites with different types of lignocellulosic substrates. PLoS ONE **14**(7), e0213954 (2019). https://doi.org/10.1371/journal.pone.0213954

Elsacker, E., Vandelook, S., Van Wylick, A., Ruytinx, J., De Laet, L., Peeters, E.: A comprehensive framework for the production of mycelium-based lignocellulosic composites. Sci. Total Environ. **725**, 138431 (2020). https://doi.org/10.1016/j.scitotenv.2020.138431

Elsacker, E., Zhang, M., Dade-Robertson, M.: Fungal engineered living materials: the viability of pure mycelium materials with self-healing functionalities. Adv. Funct. Mater. **33**, 2301875 (2023). https://doi.org/10.1002/adfm.202301875

Elsacker, E., Vandelook, S., Peeters, E.: Recent technological innovations in mycelium materials as leather substitutes: a patent review. Front. Bioeng. Biotechnol. **11** (2023). https://doi.org/10.3389/fbioe.2023.1204861

Fang, J.: Responsive wearables for rheumatoid arthritis. MIT (2020). https://hdl.handle.net/1721.1/127855

Fraxinea. Home. (n.d.). https://www.fraxinea.com/

Gandia, A., van den Brandhof, J.G., Appels, F.V.W., Jones, M.P.: Flexible fungal materials: shaping the future. Trends Biotechnol. **39**(12), 1321–1331 (2021). https://doi.org/10.1016/j.tibtech.2021.03.002

Gantenbein, S., et al.: Three-dimensional printing of mycelium hydrogels into living complex materials. Nat. Mater. **22**, 128–134 (2023). https://doi.org/10.1038/s41563-022-01429-5

Giaccardi, E., Karana, E.: Foundations of materials experience: An approach for HCI. In: Proceedings of the 33rd Conference on Human Factors in Computing Systems, pp. 2447–2456. ACM, New York, NY (2015). https://doi.org/10.1145/2702123.2702337

Haneef, M., Ceseracciu, L., Canale, C., Bayer, I.S., Heredia-Guerrero, J.A., Athanassiou, A.: Advanced materials from fungal mycelium: Fabrication and tuning of physical properties. Sci. Reports **7**(1) (2017). https://doi.org/10.1038/srep41292

Hoitink, A.: Mycotex by Neffa: Home. MYCOTEX (2022). https://www.mycotex.nl/

Home Mogu. Mogu (2024). https://mogu.bio/

Hwang, J.S., Kwon, H.-K., Kim, J.-E., Rho, J., Im, S.-H.: Immunomodulatory effect of water soluble extract separated from mycelium of Phellinus linteus on experimental atopic dermatitis. BMC Complement. Altern. Med. **12**, 159 (2012). https://doi.org/10.1186/1472-6882-12-159

Jones, M., Mautner, A., Luenco, S., Bismarck, A., John, S.: Engineered mycelium composite construction materials from fungal biorefineries: a critical review. Mater. Des. **187**, 108397 (2020). https://doi.org/10.1016/j.matdes.2019.108397

Jones, M., Gandia, A., John, S., Bismarck, A.: Leather-like material biofabrication using fungi. Nat. Sustain. **4**, 9–16 (2021). https://doi.org/10.1038/s41893-020-00606-1

Karana, E., Pedgley, O., Rognoli, V.: Materials experience: fundamentals of materials and design, 1st edn. Elsevier, Butterworth-Heinemann, UK (2014)
Karana, E., Barati, B., Rognoli, V., Van Der Laan, A.Z.: Material driven design (mdd): a method to design for material experiences. Int. J. Des. **9**(2), 35–54 (2015). https://www.ijdesign.org/index.php/IJDesign/article/view/1965/693
Karana, E., Blauwhoff, D., Hultink, E.-J., Camere, S.: When the material grows: A case study on designing (with) mycelium-based materials. Int. J. Des. **12**(2), 119–136 (2018). https://www.ijdesign.org/index.php/IJDesign/article/view/2918/823
Karana, E., McQuillan, H., Rognoli, V., Giaccardi, E.: Living artefacts for regenerative ecologies. Res. Direct. Biotechnol. Des. **1** (2023). https://doi.org/10.1017/btd.2023.10
Lululemon athletica. (2021). lululemon-mylo. https://corporate.lululemon.com/media/our-stories/2021/lululemon-mylo
Manan, S., Ullah, M.W., Ul-Islam, M., Atta, O.M., Yang, G.: Synthesis and applications of fungal mycelium-based advanced functional materials. J. Bioresources Bioproducts **6**, 1 (2021). https://doi.org/10.1016/j.jobab.2021.01.001
McQuillan, H., Karana, E.: Conformal, seamless, sustainable: multimorphic textile-forms as a material-driven design approach for HCI. In: Proceedings of CHI '23, pp. 1–19. ACM, New York, NY, USA, Article 727 (2023). https://doi.org/10.1145/3544548.3581156
Meet myloTM. Mylo. (n.d.). https://mylo-unleather.com/
Nakauchi, H., Amano, Y., Tagawa, S.: Preparation of mycelium pulp from mushroom fruiting bodies. ACS Sustain. Chem. Eng. **11**(44), 15789–15794 (2023). https://doi.org/10.1021/acs suschemeng.3c04795
Parisi, S., Rognoli V., Sonneveld, M.H.: Material Tinkering. An inspirational approach for experiential learning and envisioning in product design education. Des. J. **20**(sup1), S1167–S1184 (2017). https://doi.org/10.1080/14606925.2017.1353059
Pegler, D.: Useful fungi of the world: Amadou and Chaga. Mycologist **15**(4), 153–154 (2001). https://doi.org/10.1016/s0269-915x(01)80004-5
Raman, J., Kim, D., Kim, H., Oh, D., Shin, H.: Mycofabrication of mycelium-based leather from brown-rot fungi. J. Fungi **8**(3), 317 (2022). https://doi.org/10.3390/jof8030317
SQIM. (2024). Home - SQIM. https://www.sqim.bio/
Su, C.H., et al.: Development of fungal mycelia as a skin substitute: characterization of keratinocyte proliferation and matrix metalloproteinase expression during improvement in the wound-healing process. J. Biomed. Mater. Res. **72A**, 220–227 (2005). https://doi.org/10.1002/jbm.a.30235
Suarato, G., Bertorelli, R., Athanassiou, A.: Borrowing from nature: biopolymers and biocomposites as smart wound care materials. Front. Bioeng. Biotechnol. **6** (2018). https://doi.org/10.3389/fbioe.2018.00137
Sustainable textile production from mycelium. Mylium (2023). https://www.mylium.nl/
Vandelook, S., Elsacker, E., Van Wylick, A., De Laet, L., Peeters, E.: Current state and future prospects of pure mycelium materials. Fungal Biol. Biotechnol. **8**(1) (2021). https://doi.org/10.1186/s40694-021-00128-1

Building-By-Growing with Textile Logic - Speculative Morphologies of Biohybrid Textile Architecture

Eliza Biala[1(✉)], Anne-Kathrin Kühner[2], and Martin Ostermann[1]

[1] Institute for Building Construction IBK, University of Stuttgart, Stuttgart, Germany
eliza.biala@ibk2.uni-stuttgart.de
[2] Independent Material and Textile Designer, Stuttgart, Germany

Keywords: Mycelium-based composites · Mbc · Biofabrication · Biodesign · Biomaterials

1 Introduction

How could the intersection of textile design, biofabrication and construction interweave in delivering scalable solutions for regenerative architecture in Anthropocene? How hands-on biodesign and making experiments can fuel the grassroots development of new construction systems?

1.1 Extended Abstract

The fundamental rethinking of how architecture is materialised is key for reducing the embodied carbon of the newly erected structures. Substantial research endeavours in the past 10 years have identified mycelium-based composites (MBCs) as a possible low-carbon alternative for future architecture (Elsacker et al. 2020, Jones et al. 2020; Agravi-ador et al. 2022). These biomaterials result from the lignocellulosic fibres bounded by the growth of the three-dimensional mycelial network. Unlike other biomaterials, which rely predominantly on pressing or rigid moulding in the production process, working with MBCs also allows for developing alternative fabrication systems (like 3D printing and soft sacrificial moulding) that offer easily achievable geometric complexity and optimisation. One of the particular cases involves using permeable textile tubes for the creation of upscaled MBC-textile-hybrid filaments, which are subsequently aggregated by textile techniques assemblies (knitting, weaving, breading, or knotting). These explorations were performed on a very limited scale due to laborious hand-made characters or high contamination rates (Biala 2022, Kühner 2023).

This study aims at advancing the aforementioned method by semi-automating filament production and therefore unfolding the possibility of design-research explorations for upscaled-textile morphologies for architectural spaces.

A. Mateus et al. (Eds.): WAD 2024, What's Around Design?, pp. 58–66, 2026.
https://doi.org/10.1007/978-3-032-08163-6_5

Two main investigation stages can be distinguished. Firstly, off-the-shelf textile tube products are tested for interoperability with the electrical sausage-making machine for the semi-automated stuffing process. The MBC paste recipe is developed to allow for both extrudability with the possessed machine and workability with the available textile tubes. Subsequently, produced filaments are functionalised as yarns to produce architectural morphologies as solidified textiles. This stage was performed as two one-week workshops conducted in a framework of elective seminars during the Master's program in Architecture at the University of Stuttgart. They operated within the same filament production protocol, but each attempted to focus respectively on knitting or weaving as the key facilitator of the morphology-generation method. Gravity-formed archetypical architectural elements (wall, column) and shells are targeted. The process involves knitting or weaving with the upscaled yarn on the custom-made knitting beds- or looms, subsequently hanging the soft forms to develop their final gravity-optimised shapes. The hanging scaffolds are sealed into small cultivation chambers, creating the optimal conditions for mycelium growth and skin formation. The fully grown structures are air-dried and investigated as free-standing elements and spaces.

The paste formulation, utilising guar gum and psyllium husk as plasticisers, allows for archiving viscosity suitable for handling with the machine. Two tube types (Nobanetz and Tg, size 0), manufactured as tubular bandages, proved to function well in the fabrication scenario, allowing for diameter regulation between 2–4 cm. All the demonstrator structures grew without significant contaminations, regardless of being handled in non-sterile conditions and reached self-supporting stability. All highly transparent demonstrators, knitted or loosely weaved, present distinctive light-transmitting properties and geometric complexity that are hard to replicate with other materiality. However, their structural stability is solely dependent on the bio-welding of knots, which requires further studies. Thick weave, with flax strings functioning as warp, allows for easy production of surface forms due to continuous bio-welding along the yarns - not only in the knots. The cylindrical dens-weave demonstrator withstand 80kg of load without damage. The tacit knowledge of material behaviour was a decisive aspect of effectively fabricating the structures. Tackling the parameters experimentally and inscribing them into the trans-scalar feedback loops is crucial for advancing this biohybrid construction system (Figs. 1, 2, 3, 4, 5, 6, 7 and 8).

Fig. 1. Semi-automated filament production

Fig. 2. Hanging of the circularly knitted column demonstrator

Fig. 3. Iterations of the knitted columns

Fig. 4. Knitted knot with extensive growth of bio-welded skin

Fig. 5. Knot with distorted material filing within the tube

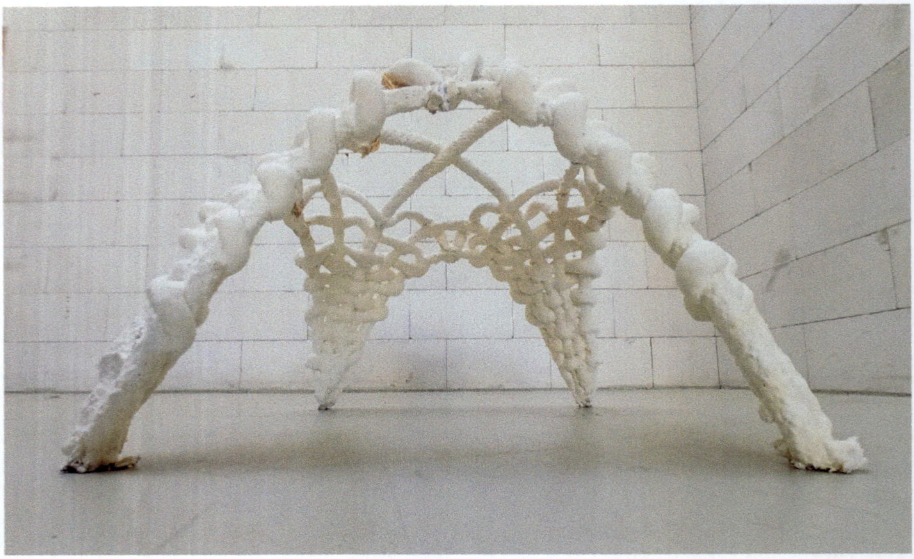

Fig. 6. Weaved shell

Fig. 7. Weaved and layered column demonstrators

Fig. 8. Assembly of the workshops outcomes

References

Agraviador, A., et al.: BioKnit: The coordination of computation with material investigation in the design of biohybrid textiles towards architectural integration. In: Proceedings of ACADIA 2022: Hybrids & Haecceities (2022)

Biala, E., Ostermann, M.: Mycostructures—growth-driven fabrication processes for architectural elements from mycelium composites. Architect. Struct. Construct. **2**(3), 509–519 (2022)

Elsacker, E., Vandelook, S., Van Wylick, A., Ruytinx, J., De Laet, L., Peeters, E.: A comprehensive framework for the production of mycelium-based lignocellulosic composites. Sci. Total Environ. **725**, Article 138431 (2020)

Jones, M., Mautner, A., Luenco, S., Bismarck, A., John, S.: Engineered mycelium composite construction materials from fungal biorefineries: a critical review. Mater. Des. **187**, Article 108397 (2020)

Kühner, A.-K.: Knitted mycelium: transforming mycelium hybrids into solidifying and lightweight textile structures. Cambridge Open Engage (2023)

MYO: Mycelium Technology for Kids

Elena Albergati(✉), Valentina Rognoli, and Michela Grisa

Politecnico di Milano, Milan, Italy
elena.albergati@polimi.it

Abstract. In exploring future sustainable alternatives in product design, this study emphasizes the importance of biobased materials, focusing particularly on integrating mycelium, the vegetative part of fungi, into children's toys. Despite growing environmental awareness, consumer acknowledgment and acceptance of these materials are still consolidating. This project born thanks to the collaboration between Politecnico di Milano and Mogu S.r.l., an interior design company leading in the creation of interior design products through fungal bio-manufacturing. In this paper we'll present the development of a toy for kids incorporating mycelium, merging touchscreen technology with biomaterials. The aim is to create a play experience that is both educational and enjoyable, fostering an early appreciation for biobased materials in children. The technological component has been thoughtfully integrated, ensuring that its use enriches the child's sensory and interactive experience, rather than diverting them from the material and tangible world.

Keywords: Mycelium · Biofabrication · Educational Toy · Interaction Design · Game Design · Materials Perception

1 Introduction

The end of the 20th century presented a big challenge in design. We've realized that our modern society consumes and pollutes more than what's sustainable. A strategy that's getting more attention, and might be key to a greener future, is the incorporation of both living and non-living biological systems into material science, nanotechnology, and subsequently into design (Solanki 2018). This process aims to obtain materials from biological resources, developed in a controlled and sustainable manner. This initiative aligns with the contemporary imperative to innovate and employ materials that have minimal environmental impact (Acaroglu 2018). Issues arising from synthetic plastics and the depletion of fossil fuels underscore the importance of transitioning to renewable material alternatives. Today, numerous studies aim to develop materials from natural sources, such as cellulose, lignin, pectin, plants, fungi, algae, and bacteria (Groll et al. 2016). All these materials are sustainable, biocompatible, and biodegradable with a wide range of properties. However, the development of such materials typically requires complex processing methods that can be costly, time-consuming, and with low production yield. For this reason, even though these materials could address various environmental issues, they are still underutilized and with very limited applications (Haneef et al. 2017).

A. Mateus et al. (Eds.): WAD 2024, What's Around Design?, pp. 67–83, 2026.
https://doi.org/10.1007/978-3-032-08163-6_6

One strategy to overcome these challenges might be the development of composite biomaterials with controlled and adjustable properties during their growth, ready for use without the need for overly sophisticated processing methods. The perfect example of this strategy is mycelium (Tsing 2015).

2 Mycelium-Based Materials

Mycelium represents the vegetative apparatus of fungi. It is the part of the fungus found in the soil and can be considered its root. It is a crucial organ, capable of transforming and fragmenting both organic and inorganic materials in the subsoil to derive nourishment. Diving deeper, mycelium is a eukaryotic cell that creates a network composed of individual hyphae, which grow from the inoculation of a fungal spore into a cell-based material, consuming carbon and nitrogen to build an increasingly intricate net. Mycelium exhibits a rapid growth rate and will continue to grow indefinitely if provided with ample nutrients (Sydor et al. 2021).

When introducing a specific variety of mycelium into a compound made from select raw materials such as cotton and hemp residues, it has been observed that the fungal organism assimilates this substrate, subsequently converting it into a new matrix. Indeed, through mycelium, various waste materials are transformed into nutrients, thereby regaining value and function. The manufacture of composite materials with mycelium would fit into a reuse cycle, leading to reduced waste and offering new economic possibilities. (Alemu et al. 2022) Another notable feature is that, once the product has completed its life cycle, it can be broken down and composted, further serving as a fertilizer for crop fields. This completes what could become a virtuous cycle, with great environmental and economic benefits. From a broader perspective, mycelium-based materials have the potential to catalyze the transition towards a biobased economy. The application of mycelium as a decomposer and a bio binder has been employed in various fields, from architecture, design, and art to engineering and medicine (Sheldrake 2021). Many of these projects explore and reflect upon how we might rethink the production of numerous materials using biological resources, through natural regenerative, self-sustaining, and self-feeding processes. The incorporation of mycelium in the realm of interactive toys is primarily motivated by its numerous inherent qualities that make it an ideal candidate for the fabrication of cutting-edge interactive play items.

2.1 Biodegradability

The significance of biodegradability in the realm of toys cannot be overstated, and mycelium technology plays a pivotal role in this transformative shift. Thanks to the innovative process of mycelium, all mycelium-based products can be unequivocally classified as biodegradable. In the toy industry, the concept of Circular Economy takes on paramount importance. While the global toy market's value soared to over $90 billion in 2019, the grim reality is that a staggering 80% of these toys, once they reach the end of their short lifespan, find themselves condemned to landfills, incinerators, or even worse, the vast expanse of our oceans (Yale Environment Review 2023). This translates into a colossal loss of both economic and environmental value. In this alarming scenario, the advent of mycelium technology offers an unprecedented opportunity. By harnessing mycelium's potential, we can craft toys that are not just eco-friendly but 100% biodegradable. This represents a great opportunity to transition towards an increasingly circular economy (The Ellen McArthur Foundation 2013), where toys, once cherished and now discarded, can seamlessly return to nature without harming the environment.

2.2 Low Density

Low density is a critical factor in enhancing the competitiveness of mycelium-based materials, particularly when applied to the realm of children's toys. The importance of lightweight toys for children's safety cannot be overstated. Lighter toys are essential for preventing injuries, enabling easy handling, reducing the risk of fatigue, and ultimately enhancing the overall safety and enjoyment of a child's playtime. In mycelium-based products, the density is primarily determined by the porosity of the selected substrate. A study (Sivaprasad 2021) reveals a density range of 0.03–0.05 g/cm^3 for a novel biomaterial composed solely of compressed mycelium. Notably, this falls within the same range as Expanded Polystyrene (EPS), which typically has a density of 0.022–0.030 g/cm^3. This underscores the potential of mycelium-based materials to attain an impressively low density while remaining competitive with the frequently employed plastic materials within the realm of toy manufacturing.

2.3 Sensoriality

Another notable characteristic of mycelium pertains to its surface texture. Notably, it exhibits a velvety and soft texture, inducing intriguing and engaging tactile sensations for users. Numerous research findings have underscored that soft, velvety textures, as opposed to rough surfaces, elicit positive and soothing sensory experiences in individuals (Camere and Karana 2018). This aspect of mycelium shows great potential for advancing the creation of materials that are both tactile-friendly and sensorially captivating, especially when it comes to designing products specifically intended for children.

2.4 Hypoallergenic

Materials based on mycelium possess hypoallergenic attributes for various compelling reasons. Firstly, their natural composition warrants consideration. Mycelium, the vegetative component of fungi, shares an analogy with plant roots in terms of growth. These mycelium-based materials predominantly comprise mycelium and natural organic substrates such as straw or sawdust. Crucially, they are devoid of chemicals or recognized allergenic agents, thereby minimizing the likelihood of provoking allergic reactions. Secondly, their hypoallergenic nature is intricately tied to the controlled growth process (Camere and Karana 2018). The production of mycelium-based materials entails a meticulously managed cultivation process conducted under stringent monitoring conditions. This method serves to forestall contamination from common allergens, preserving a sterile environment throughout the growth phase. In addition, Mycelium-based materials consistently demonstrate a tendency to emit very small or almost negligible amounts of VOCs, which are often identified as a primary cause of allergies and respiratory irritations. This quality makes them especially suitable for use indoors (Bianca et al. 2023).

2.5 Malleability

Undoubtedly, modeling is one of the favorite activities among children (Skedung et al. 2013). Moreover, handling any material contributes to the enrichment of the child's fine motor skills (Montessori 1999). In the case of mycelium, the creative possibilities it offers are boundless. This biobased material initially presents itself as easily moldable and malleable, allowing children to spend long hours while unleashing their creativity and imagination.

3 Biobased Material Perception

Despite the growing awareness of environmental issues, there are significant challenges in introducing biomaterials to the market, primarily due to their underestimated perception by the public. Many designers indeed admit to having difficulties in finding the right strategy to meet consumer expectations, lacking a clear understanding of how customers' inclination towards sustainability is evolving. For a transition to a "bioeconomy" to take place, it is essential that, the material, is not only technologically feasible and economically sustainable but also socially desirable (Langaveld 2010). Consequently, this transition depends not only on the effort and enthusiasm of professionals but also on the acceptance and involvement of consumers. It is, therefore, important to explore what kind of image, expectations, or concerns consumers have towards biobased materials. In the article by M. Petersen and S. Brockhaus, a study (Petersen and Brockhaus 2017), is presented that compares the perception of two types of headphones and two different models of trash bags. The results reveal that the "green" aspect can influence both the material and design perception of products, making them appear sustainable and environmentally friendly. Consumers are drawn not only to color but also to surface textures and cues related to the material's nature. Subsequently, when the consumer becomes aware of the material used in the product, their choice definitively leans towards the sustainable option, as there arises a "sense of guilt" in choosing the fossil alternative. However, for certain product categories, from the first impression, the sustainable option is perceived as having lower performance. This perception does not apply to food packaging, where sustainability is a very important factor that communicates naturalness and, therefore, higher quality. The inconsistent and complex perception society holds regarding biomaterials has also been demonstrated through a study conducted by the Monterey Regional Waste Management District Office Building and the EPA's Regional Headquarters in Philadelphia (Monterey Regional Waste Management District 2003). This document highlights three points that seem to summarize the image people have acquired of these materials:

- "Sustainable products are of lower quality than traditional materials." Some products might indeed have certain application issues, such as installation challenges or durability, but these issues are rarely due to the sustainability features of the product.

- "Sustainable products have a different aesthetic appearance than traditional materials." This point relates to the notion of the "perfect" material that our society has developed over the years through industrial production. However, it's also accurate that many eco-friendly products on the market deliberately conceal their sustainable nature.
- "Sustainable products are overpriced." Indeed, some "green" products cost more than their unsustainable counterparts available in the market. However, this additional cost is often offset by the product's enhanced sustainability, reduced maintenance costs, and benefits to the user (e.g., being free from toxic substances). Moreover, nowadays, some of these products are also available at affordable prices, competitive with standard options.

These three considerations should prompt designers to ensure that their sustainably made products are not mistakenly perceived in a negative light.

3.1 Acceptance and Acquisition of Identity of the Materials

The understanding of customer acceptance of a new technology has often been overlooked in marketing research, yet it is crucial for successful innovation. Companies need to understand why customers accept or reject a product in order to improve their efficiency and competitiveness (Dunphy and Herbig 2002). For success, a material or service must meet specific needs, demonstrate value, and resonate with the customer's mindset. Regarding materials, perception is vital. Like wood, which has been experimented with in countless ways over time, its use and properties have become an integral part of culture, making it a "known" material with a distinctive identity. The same applies to other traditional materials like steel and gold. This familiarity is based on the predictability of the material's behavior and the associated cultural values, which become key points in communication. In a traditional context, once a material becomes "known," it becomes a benchmark in its category. Stone signifies endurance, wood the passage of time, wool intimacy, steel strength, and gold power. From these associations, objects gain cultural depth and significance. The identity of these traditional materials has evolved organically due to a limited variety of materials, production technologies, and a gradual progression that fostered stable relationships between materials, forms, and symbolic meanings. According to Manzini, the challenge is tied to our cognitive structures: to truly understand and recognize entities, we need time to repeatedly experience them (Manzini 1986). Yet, what used to unfold over extensive periods, such as the discovery of a new material or processing method, has now been condensed into fleeting moments. People no longer have time to develop a relationship with the material, to become familiar with its colors, odors, or tactile sensations. The rapid pace of technological discoveries often results in them appearing mute and anonymous, contributing to growing disorientation. Antonio Petrillo also emphasizes that there's a lack of "time for things" to evolve and solidify their identity. Thus, it's crucial for humans to create meanings for these materials, ensuring they can be integrated stably into our society (Manzini e Petrillo 1991) (Fig. 1).

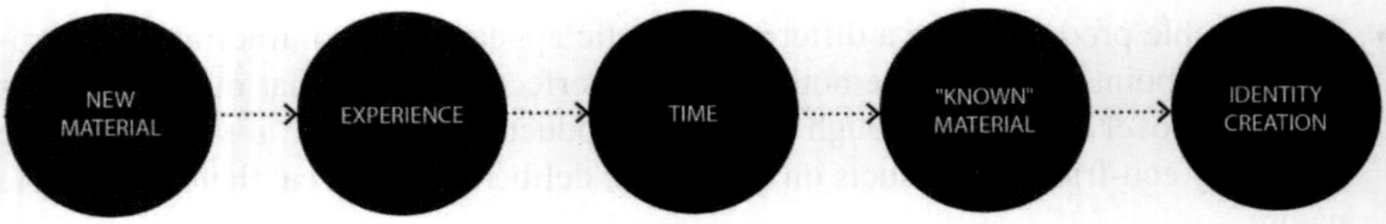

Fig. 1. material's identity creation path - Credits: Michela Grisa

3.2 A New Identity Through Children

"A child plays with a ball: he throws it against a wall, the ball follows a trajectory, bounces, and with a swift motion, the child catches it mid-air. This simple, everyday sequence of movements implies a vast personal and collective experience each time it occurs. Through play, the child acquires knowledge regarding the ball's appearance, its vibrant colors, tactile elasticity, and smell. All this embeds deeply in the child's memory, intertwining scent, color, elasticity, and childhood." (Manzini 1986)

This is how Manzini begins in "The Matter of Invention." Materials have a unique potential to guide diverse learning paths in young children (Penfold 2019). Playing with wooden blocks, for instance, can foster understanding of arrangement, balance, and height through stacking and construction. Conversely, large paper sheets can teach kids about gravity, weight, and shape. Children learn the physical properties of materials and their potential applications in the world as they play, understanding how materials can be rolled, stacked, or moved. The sensory and aesthetic qualities of a material might change when a child interacts with it, leading to further experimentation. These transformations allow children to broaden and complicate their learning over time. Moreover, this direct interaction lets a child attribute a distinct meaning to the material itself. Andrea Branzi speaks of the meaning that, over time, humans tend to assign to their surroundings, to everyday objects (Branzi 1999). This meaning is "internalized" primarily through memory and experiences with the world. As previously mentioned, the significance of past experiences and memory in establishing material identity is evident. Wilson points out that children, whose first experiences come from toy stores like Toys 'R' Us, might develop different material values compared to adults raised among wood, stone, and metal (Wilson 1988). Individual pasts, childhoods, could be crucial in fostering a new set of values, aiding the future in forming novel aesthetic standards for sustainable development. In the case of mycelium, due to its recent introduction, its identity remains undefined. Thus, a pressing question arises: in our rapidly changing world, can a recognizable identity be carved out for this material? Can we nurture appreciation and acceptance for mycelium?

"Today's children are tomorrow's adults," Munari often remarked, asserting that his most vital work was children's workshops. In his words, "Unable to change adults, I chose to work on children so that better ones might grow. It's a revolutionary strategy to work on and with children as future men." (Munari 2017).

This project aims to work with children to develop in them a sensitivity and appreciation for biomaterials, especially mycelium. How? According to Munari, the most potent and effective means of children's learning is an activity that is often underestimated: play (Munari 2017). Is it possible, then, to cultivate respect and deep understanding of mycelium in children's minds through play, so that in the future this material is valued and accepted?

4 MYO Project

Starting from the current perception of biomaterials, we wanted to develop a product that could promote their integration into society and help the mycelium material acquire a clear identity through direct experience with it, making it more accessible to people. We, therefore, decided to focus on a specific target group, children, specifically those aged between 4 and 10 years. We chose this age group because it is precisely this period when children develop their perspective, made up of symbols and meanings (Montessori 1999). Myo project aims to create a toy made of mycelium, merging two seemingly opposite worlds: biomaterials and technology. Technological components in a product, especially in a toy, are vital today both for desirability and to enhance and enrich a child's play experience (D'Alessandro 2018). However, one aspect often missing in a 100% digital environment is the physicality of the product, which is essential for ensuring the child's proper sensory and physiological development (Greco 2019). So, the intention was to integrate into the two-dimensional world of touchscreen devices the more tactile and sensory realm inherently belonging to a particular biomaterial, mycelium (Grisa M. 2021) (Fig. 2).

Fig. 2. Myo Project - Credits: Michela Grisa

4.1 Concept

The idea was to merge the digital and physical worlds within a single playful product. The analog-digital relationship was rethought to introduce a toy proposal that could be stimulating and, at the same time, support the child's intellectual and motor growth, allowing them to use their senses (Bradt 2010). The starting idea was to develop a product that combined three very different material categories: wood, a symbol of tradition; digital devices; and a highly innovative biobased material, mycelium. This unique combination was also utilized to overturn the cold and sterile aesthetics of digital devices, making them warmer, friendlier, and more natural. The result is Myo, a hybrid toy that merges the sensory world with the digital one through the intrinsic materiality of mycelium and its byproduct chitin (Grisa M. 2021). The packaging has been specially designed using rigid recycled cardboard, allowing for the boxes to be reused. The colors are bright and vibrant to attract the child's attention. Given the fragility of some components, the packaging has been designed with specific compartments to prevent the objects from moving and colliding with each other. The name "Myo" was conceived starting from the English word "mycelium." To emphasize the child's ability for personalization and creativity, we settled on the term "Myo," which immediately evokes the possessive pronoun "my." The underlying idea was to come up with a name that is easily pronounceable and recognizable for children (Grisa M. 2021).

4.2 Components

The Myo Standard Kit is primarily made up of four interconnected elements: the Wood Box, the Chitin Filter, the Smartphone, and the Mycelium Bricks. Four components from contrasting worlds that, through the project, come together to form a single identity. Each individual object has been designed in line with the target audience. Therefore, components that are simple to install were developed, with effective affordance, so that through its physical properties, it suggests the appropriate actions for the user to handle. As highlighted by the exploded view below, the elements that make up the project are essentially the following:

- **The Mycelium Bricks**, mycelium blocks of various shapes designed to stimulate the child's creativity.
- **The magnets**, which were previously inserted into the Mycelium Bricks and at the bottom of the Box.
- **The Wood Box**, a wooden box that is the connecting element between all the other components. Inside, it houses the smartphone, chitin, and on the external surface, the Mycelium Bricks.
- **The wooden layers**, which allow the child to personalize their creation to the fullest.
- **The Chitin Filter**, which makes the sterile surface of the digital device more tactile and material.
- **Wooden closures** to secure all elements inside the box (Fig. 3).

Fig. 3. Myo Packaging - Credits: Michela Grisa

4.3 Materials Selection

The choice of materials in the project was of utmost importance. The criteria for their selection were: Materiality, Safety of use, Environmental sustainability, Analog and natural appearance, Ease of use, Innovation, Natural vs technological aesthetics. Thanks to the collaboration with Mogu, it was possible to use and incorporate mycelium and its derivative, chitin, into the project. In addition to these biobased materials, wood was also chosen. It has always been recognized as an ecological and natural material that perfectly fits a plastic-free dimension. Another very important goal is the desire to disguise technology. The solution that we choose is to completely transform its often sterile and cold aesthetics, making it more natural. To achieve this, the analog aspect was emphasized, without, however, altering technological performance. Moreover, according to many researchers, one of the trends to engage children in play is to create a synergistic union between technology and nature (Ernst and Burcak 2017). In this way, children can establish a deep relationship with materials that represent the future for a sustainable world. As illustrated on the graphics below, there are three variants available for each material. For the wood, three types were selected: Pine, Fir, and Oak. As for the mycelium, depending on the growth period duration, it is possible to obtain endless results and textures. Lastly, for the chitin, specific fungal species were selected to achieve warm and natural colors (Fig. 4).

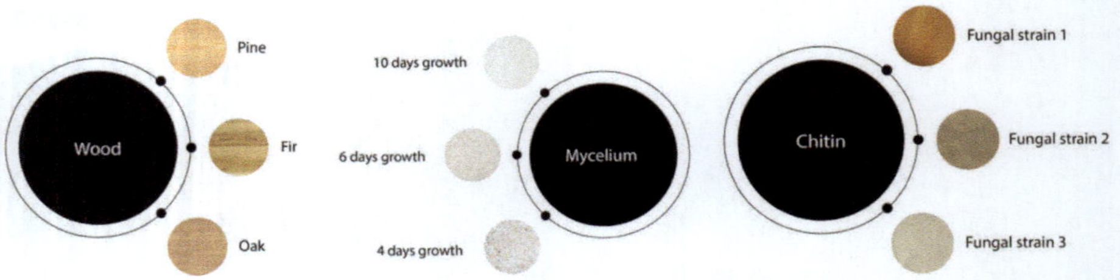

Fig. 4. Material's variations; Wood essences, Mycelium and Chitin - Credits: Michela Grisa

4.4 Analog Components

The Wood Box serves as the central unifying element among the various components. Its minimalistic and linear design ensures versatility and ease of handling, with dimensions tailored to commonly used smartphones for manageability. While the primary choice of wood is lightweight and easily workable spruce, options for aesthetic preferences include pine and oak. The box's lightweight nature is also due to its internal cutouts: an upper cavity for the smartphone and a hollow bottom for the magnets, essential for attaching the Mycelium Bricks externally. Specific compartments ensure the smartphone remains undamaged by magnets, with both components kept safely apart. The magnets' strength was deliberately chosen to be weak to avoid potential issues. Lastly, the kit includes various wooden layers, providing children with added customization options for their projects. These layers, with sufficiently large cutouts to maintain visibility, offer creative design solutions (Grisa M. 2021). The chitin filter was designed to make the touchscreen experience more tactile and textured, which normally, due to its surface, is not very stimulating from a sensory point of view. Another interesting feature of chitin is its visual texture. In fact, when combined with a touchscreen device, the digital image takes on a more analog and distinctive appearance. In addition to the aesthetic characteristics mentioned, this material also significantly reduces some eye issues caused by blue light (Grisa M. 2021). The heart of the project is undoubtedly represented by the Mycelium Bricks. Made entirely of mycelium, these magnetic bricks can be assembled to design infinite creations. The bricks have a naturally white appearance and are surprisingly soft and velvety to the touch. This last aspect, thanks to the production process, has been emphasized to allow the child to have an even more intriguing and unique experience. The Mycelium Bricks are made through a sustainable and circular production process (Grisa M. 2021). Once initially alive, they are dried to stop fungal growth. Their velvety appearance and inherent properties encourage user interaction. Unlike cold materials like aluminum, the warmth of mycelium invites physical touch, especially drawing children's curiosity (Vatavu et al. 2014). As detailed before, the material was chosen for its qualities suitable for children: lightness, engaging texture, hypoallergenic nature, moldability, and biodegradability. Each brick contains two magnets, enhancing play by broadening assembly options. Designed for compatibility, the magnets inside the bricks are oppositely polarized, allowing limitless combinations. These bricks can be used alone or with the magnetic Wood Box. The kit offers 32 bricks of different shapes, supporting easy assembly and play diversity. Over time, the bricks' natural composition causes them to evolve in color and texture, highlighting product uniqueness and personalization, strengthening the bond with the user. At the end of their lifecycle, after removing the internal magnets, the bricks can be fragmented and composted as soil fertilizer (Grisa M. 2021) (Fig. 5).

Fig. 5. Chitin filter - Credits: Michela Grisa

4.5 Digital Components

The Myo app has been specifically designed for children, catering to their needs and abilities to provide a unique gaming experience that the user can navigate entirely on their own. Considering the starting point - the ubiquitous presence of technology in daily life and the designer's role as a bridge between the analog and digital worlds - it's crucial for the project to harness the strengths of both while mitigating their challenges (D'Alessandro 2018). The digital realm can offer numerous gaming dynamics and variations, given the ability to continually download and update the system. Children quickly grow bored of constrained games that don't evolve over time. Having an app to complement the analog gameplay becomes almost essential, greatly enhancing the interaction and variability of the user experience. The digital nature primarily considered three factors: storytelling, the opportunity to create an immersive gaming experience, and the ability to consistently update and diversify the app's uses (Grisa M. 2021) (Fig. 6).

Fig. 6. Myo App - Credits: Michela Grisa

One of the steps required by the child is to select their preferred setting. In this interface, a whimsical scenario has been created, integrating various contexts. The child is simply asked to choose one of these by touching the display, allowing the user to subsequently interact with the specially crafted 360° video for each setting. Each scenario features animations and sounds that further enhance the gaming interaction. There are essentially four designed settings: Space, City and home interiors, Urban streets, Sea and ocean. The 360-degree video, or immersive video, is an emerging medium that offers an exploratory experience within spherical footage. The novelty of these videos is that the scene unfolds not just in front of the viewer but all around, in a full 360-degree field, prompting viewers to move around to see what's happening. In the Myo project, various backdrops were envisioned, one of which was designed and prototyped (Sea and ocean). These settings allow the child to move and manipulate their creation of her, made through analog components, to explore the space and discover what surrounds them. The 360° video solution was designed to encourage and further promote the child's motor skills. In this way, the user is "forced" to physically move to fully appreciate the experience. Each video contains fun and appealing animations and sound effects to capture children's attention (Fig. 7).

Fig. 7. 360° environment - Credits: Michela Grisa

5 Prototype

Fig. 8. Myo Project prototype - Credits: Michela Grisa

5.1 Mycelium Bricks

The Mycelium Bricks were produced following a standard and 100% circular manufacturing process. First, a designated bag (with an air filter) was filled with materials useful for nourishing and promoting the growth of the mycelium (cotton and various waste from the food and textile industries). Fungus spores were added to the bag to colonize the substrate. Subsequently, it was necessary to wait a few days for the mycelium to propagate through the material until it displayed a completely white and uniform surface. Afterward, the required material for the next step was disinfected with alcohol: the placement process. In this step, the material is broken down into small pieces to be more easily inserted into the mold. Once the mold is filled, it is covered with kitchen film to maintain the correct internal humidity and a consistent temperature. At this point, it's crucial to wait a few more days until the mycelium has grown back. Once the desired result is achieved, the block is removed from the mold. Lastly, to stop the fungal growth, it must be dried using specific dryers or otherwise using any low-temperature oven or exposing it to the sun (Grisa M. 2021). An intriguing aspect of mycelium is its metamorphosis during growth. Indeed, it's astonishing to see how a 100% natural material can expand and develop so rapidly and with unexpected uniformity. Initially, it appears heterogeneous, crumbly, and brownish. After just two or three days, one can already observe the initial changes: the hyphae branch out into the material, turning it white but still fibrous and inconsistent. After about seven or eight days, a soft, velvety surface texture is achieved, which makes the product not only aesthetically pleasing but also high-performing and durable. For the creation of the Mycelium Bricks, a single mold divided into different compartments was prepared to best grow the mycelium. The mold, made of cardboard, was lined with film to prevent any contamination of the material. Magnets were inserted into the mixture to make the bricks truly magnetic. The bricks have retained their original white color and their velvety texture, delicate and soft to the touch (Fig. 8).

5.2 Chitin

The material chitin, which was analyzed in more detail in Chapter I, is used in the project as a layer to coat the smartphone's touchscreen screen. Its primary function is to make technological interaction once again more tactile and less "cold" through 100% biobased materials. To obtain the different chitin layers, several experiments were carried out with the help of the SQIM laboratory. The specific process for creating the material cannot be described due to company privacy reasons. Prototyping clarified which properties of chitin were best suited to the project. Several sheets were developed with different colors, based on the fungal species used, and with various thicknesses (Grisa M. 2021). For the final prototype, selections were made following two different criteria: Display color alteration and Fragility. For the first factor, the chitin with a more intense color was discarded as it altered the smartphone display too intensely. In the second case, the chitin containing less biomass proved too fragile and was therefore also discarded. The photographs in this section highlight the main qualities of chitin, such as its texture and its natural, warm colors.

5.3 Wood Box

For the creation of the box, two different materials were used: Pressed cardboard (3 mm) and Oak veneer. To provide rigidity and structure to the box, cardboard was used, which was then entirely covered with the wood sheets. In this way, an aesthetic closely resembling the desired final product could be achieved. The lower compartment of the box was sealed only after inserting magnets inside to allow the correct use of the Mycelium Bricks. Lastly, in the upper opening, the chitin filter was attached. Additionally, to make it more realistic, one of the possible wood layers was created.

5.4 App and 360° Video

Regarding the creation of the Myo App, the Unity program was used entirely. This tool allowed the realization of a true 360° video. For the content, specific illustrations were designed and created using Adobe Illustrator. To animate and enable the actual functioning of the app, and the 360° videos, several tutorials were followed to learn the basics of C# programming. C# is an object-oriented programming language developed by Microsoft as part of the.NET initiative. The syntax and structure of C# are inspired by various previously established languages, particularly Delphi, C++, Java, and Visual Basic. In this way, it was possible to prototype the application, focusing mainly on the creation of the 360° video. Finally, the Unity project was exported to support Android systems and subsequently tested.

6 Conclusions

The project presented so far aims to be a starting point to encourage the use and appreciation of mycelium, and more generally, biobased materials in society. We need to rethink our values, including aesthetic ones, which are often too much in opposition to nature and its principles. It has been observed that working with children can be a possible solution to more easily achieve this purpose, educating them from an early age to new ideals in line with environmental sustainability. At this point, the steps that could be developed in the future are:

- **User testing:** This crucial phase will involve organizing various playful workshops to introduce children to the world of Biodesign and to test the acceptance of the material and the project by users.
- **Perception of mycelium:** Indeed, despite the research carried out, there are still few published studies, and it would be useful and interesting to delve deeper into this topic.
- **App development:** Creating more customizable content that aligns with the different needs and personalities of each child.

References

Acaroglu, L.: A manifesto for design-led systems change (2018). https://medium.com/disruptive-design/a-manifesto-for-design-led-systems-change-28ac240db6dd

Alemu, D., Tafesse, M., Mondal, A.K.: Mycelium-based composite: the future sustainable biomaterial. Int. J. Biomater. **2022**, 8401528 (2022). https://doi.org/10.1155/2022/8401528

Bianca, I., Vette, J., Bontiff, N.: Exploring the potential of mycelium composites as natural board materials (2023). https://doi.org/10.1007/978-3-031-33465-8_50

Bradt, S.: How touch can influence judgments. The Harvard Gazette, Massachusetts (2010). https://news.harvard.edu/gazette/story/2010/06/how-touch-can-influence-judgments/

Branzi, A.: The hot stone: Neolite, p. 83. Milano (1999)

Camere, S., Karana, E.: Fabricating materials from living organisms: an emerging design practice. J. Clean. Prod. **186** (2018). https://doi.org/10.1016/j.jclepro.2018.03.081

D'Alessandro, E.: Growing between natural rhythms and digital technologies. ScienzainRete (2018). https://www.scienzainrete.it/articolo/crescere-tra-ritmi-naturali-e-tecnologie-digitali/elena-dalessandro/2019-11-25

Dunphy, S., Herbig, P.: Acceptance of innovations: the customer is the key! vol. 6, issue 2, pp 193–209 (2002). https://www.sciencedirect.com/science/article/abs/pii/1047831095900144

Ernst, J., Burcak, F.: Young children's contributions to sustainability: the influence of nature play on curiosity, executive function skills, creative thinking, and resilience. Sustainability **11**(15), 4212 (2019). https://doi.org/10.3390/su11154212

Greco, A.: The function of sensory material (2019). https://www.boboto.it/it/blog/montessori/145-la-funzione-del-materiale-sensoriale

Groll, J., et al.: Biofabrication: reappraising the definition of an evolving field. Biofabrication **8**(1), 013001 (2016). https://doi.org/10.1088/1758-5090/8/1/013001

Haneef, M., Ceseracciu, L., et al.: Advanced materials from fungal mycelium: fabrication and tuning of physical properties. Nature, Article number: 41292 (2017). https://doi.org/10.1038/srep41292

Langeveld, H., Sanders, J., Onna, M.J.G.: The biobased economy: biofuels, materials, and chemicals in the post oil era (2010)

Manzini E.: The Matter of invention - Materials and design. Arcadia, Milano (1986)

Manzini, E., Petrillo, A.: Neolite: Metamorphosis of plastics. Domus Academy, Milano (1991)

Monterey Regional Waste Management District. Case Studies Sustainable Materials. Monterey Regional Waste Management District Office Building, EPA's Regional Headquarters. Philadelphia (2003)

Montessori, M.: The child's mind. Absorbent mind. Garzanti, Milano (1999)

Munari, B.: One thing leads to another. Notes for a design methodology. Ediz. illustrata. Mi: Laterza (2017)

Penfold, L., et al.: Material Matters in Children's Creative Learning. (s.l.): JoDS (2019).https://jods.mitpress.mit.edu/pub/bwp6cysy/release/1

Petersen, M., Brockhaus, S.: Dancing in the dark: Challenges for product developers to improve and communicate product sustainability. J. Clean. Prod. **161**, 345–354 (2017). https://doi.org/10.1016/j.jclepro.2017.05.127

Sheldrake M.: entangled life: how fungi make our worlds, change our minds and shape our futures. Vintage. First Edition (2021)

Sivaprasad, S., Byju, S., Prajith, C., Shaju, J., Rejeesh, C.:Development of a novel mycelium bio-composite material to substitute for polystyrene in packaging applications. Mater. Today: Proc. **47** (2021).. https://doi.org/10.1016/j.matpr.2021.04.622

Solanki, S.: Why Materials Matter: Responsible Design for a Better World. (s.l.): Prestel Publishing (2018)

Sydor, M., Bonenberg, A., Doczekalska, B., Cofta, G.: Mycelium-based composites in art, architecture, and interior design: a review. Polymers **14**(1), 145 (2021). https://doi.org/10.3390/polym14010145

Skedung, L., Arvidsson, M.: Feeling Small: Exploring the Tactile Perception Limits. Scientific Reports volume 3, Article number: 2617 (2013)

The Ellen McArthur Foundation. Towards the Circular Economy. Economic and Business Rationale for an Accelerated Transition (2013). Disponibile su https://www.ellenmacarthurfoundation.org/assets/downloads/publications/Ellen-MacArthur-Foundation-Towards-the-Circular-Economy-vol.1.pdf

Vatavu , R.D., et al.: Touch interaction for children aged 3 to 6 years: experimental findings and relationship to motor skills. Int. J. Hum.-Comput. Stud. **74** (2014). https://doi.org/10.1016/j.ijhcs.2014.10.007

Wilson, F.: Plastics, Past and Future. London: Architecture 77 April 108(1988)

Yale Environment Review. Most materials are recyclable, so why can't children's toys be sustainable? (2023) https://environment-review.yale.edu/most-materials-are-recyclable-so-why-cant-childrens-toys-be-sustainable

Circular Territories Inspiration

Prologue 1: How to Design the Change We Need?

By Luca Simeone - Aalborg University, Denmark

Parodying a famous scene in the movie Apocalypse Now (1979), the Italian semiotician and writer Umberto Eco once said: “I love the smell of book ink in the morning.” When I was asked to write a foreword for this book, I wondered whether a similar motif was at play here. We live in a world saturated with content. Already more than 30 years ago, way before the advent of Internet-based massively distributed content creation, the architect and graphic designer Richard Wurman (1989) noted that one copy of the daily edition of The New York Times contained more information than a normal person in 17th-century England would encounter in an entire lifetime. Today, this phenomenon has acquired an even higher relevance given the proliferation of social media and content relentlessly produced by generative AI, which, according to various estimates (Europol Innovation Lab, 2021; Thompson et al. 2024), might soon end up accounting for up to 90% of the content we encounter on the Internet. As the management scholar Roberto Verganti put it: “We live in a world awash with ideas” (2016). In all this, we might wonder, what is the point of producing one more book and, thus, contributing even more to today’s endless proliferation of content?

Yet, a book, particularly this type of book, belongs to a different category of content. It is content that was produced with care. Design scholars and practitioners from various walks of life initially crafted their conference contributions by reflecting on their previous studies or practical experiences. They subsequently refined their reflections through a peer-reviewing process and further rewriting. Afterwards, they presented at a conference where these pieces of writing became parts of the inspiring threads of a passionate 4-day conversation, which gave the authors the opportunity to be exposed to a variety of perspectives and, thus, advance their reflections even more. As such, this book results from a slower and more profound way of thinking – a way of thinking that is broader and multifaceted.

It is precisely this way of thinking that is needed to support our transition towards more circular and sustainable societies. According to the systems thinker Donella Meadows (1999), changing people’s mindsets is the most effective leverage point to improve our societies. To understand the complexity required to transition our societies towards different and more sustainable economic, social and cultural systems, we need to sharpen our thinking and understand the multilayered nature of change processes (Simeone et al. 2023). Along these lines, one could see how thinking could become an essential object of design processes and projects (Neuhoff et al. 2022), i.e. changing people’s default way of thinking from a default state riddled with planet-destroying and ego-centered biases towards a state where openness and compassion towards the needs and wants of other beings become the norm (Noera et al. 2024). For this to happen, we need to embrace relational and systemic approaches and see the world not simply as an empty stage where we (humans) are at liberty to perform our comedies and dramas without considering other beings (Akama 2015).

In this perspective, we should celebrate these two volumes and the tireless work of the editors and the conference organisers as powerful instruments that invite us to change our default thinking. Approaches such as Biodesign, Circular Territories, Speculative Biofutures, Nature-based Solutions, and Strategic Design for Green Transition are essential for all of us to learn how to inhabit the world more humbly and respectfully.

References

Akama, Y. (2015). Being awake to Ma: Designing in between-ness as a way of becoming with. CoDesign, 11(3–4), 262–274. https://doi.org/10.1080/15710882.2015.1081243

Europol Innovation Lab (2021). Facing Reality: Law Enforcement and the Challenge of Deepfakes. https://www.europol.europa.eu/cms/sites/default/files/documents/Europol_Innovation_Lab_Facing_Reality_Law_Enforcement_And_The_Challenge_Of_Deepfakes.pdf

Meadows, D. (1999). Leverage Points. Places to Intervene in a System. The Sustainability Institute.

Neuhoff, R., Simeone, L., & Holst Laursen, L. (2022). The potential of design-driven futuring to support strategising for sustainable futures. The Design Journal, 25(6), 955–975. https://doi.org/10.1080/14606925.2022.2121057

Noera, F., Simeone, L., and Neuhoff, R. (2024) Using contemplative approaches in education as a pathway to cultivating hope in design-driven futuring, in Gray, C., Ciliotta Chehade, E., Hekkert, P., Forlano, L., Ciuccarelli, P., Lloyd, P. (eds.), DRS2024: Boston, 23–28 June, Boston, USA. https://doi.org/10.21606/drs.2024.317

Simeone, L., Drabble, D., Morelli, N., & De Götzen, A. (Eds.). (2023). Strategic Thinking, Design and the Theory of Change: A Framework for Designing Impactful and Transformational Social Interventions. Edward Elgar.

Thompson, B., Dhaliwal, M. P., Frisch, P., Domhan, T., Federico, M. (2024) A Shocking Amount of the Web is Machine Translated: Insights from Multi-Way Parallelism. arXiv.doi: 10.48550/arXiv.2401.05749

Verganti, R. (2016). Overcrowded: Designing Meaningful Products in a World Awash With Ideas. MIT Press.

Wurman, R. S. (1989). Information Anxiety. Doubleday.

Introduction to Part 2: Around Design and the Circular Territories

By Gabriel Patrocinio, ISMAT Portimão and IFHT/UERJ (Conference track-chair)

Starting from the late sixties, a generation of thinkers – economists, architects, designers and others – established de foundations of what we know today as the circular economy.

The economist Kenneth Boulding proposed, in 1966, that society should be seen as a closed system, not borderless and inexhaustible as it might have been seen in the past. According to Boulding, "Primitive men (…) imagined themselves to be living on a virtually illimitable plane. (…) there was always some place else to go when things got too difficult." (Boulding 1966, p. 3–4). He explains that a closed system has not external inputs or outputs. For a closed system, there is no 'outside'. This perception of the harms of unconditional development of the economy led to the creation, in 1968, of the Club of Rome, to discuss the dilemmas of humanity's uncontrolled growth. The organization's ideas originated the book The Limits to Growth (Meadows, Randers & Meadows 1972). In the seventies, the Romanian economist Nicholas Georgescu-Roegen (1971) proposed a bio- or ecological economy, advocating the importance of considering raw materials – and the prospect of their collapse. His ideas were a starting call-out to review an unsustainable process of growth.

Leaving aside the economists, the field of design has also taken part and interacted in this exchange of ideas. Buckminster Fuller with his Operating Manual for Spaceship Earth (1969), or Victor Papanek's Design for the Real World (1971), added to this debate of what became known as ecological economics. But it was the Swiss architect Walter Stahel and his concept of a closed loop economy (1982) that was one of the most resounding voices in the development of the idea of a circular economy. Stahel advocated for a longer product life, recycle, and reuse – concepts treasured by circular economy.

In 2002, Cradle to Cradle (Braungart & McDonough 2002) warned us that we have a huge design problem to fix, reminding that we can't just throw out things, because there is no "out" – we live in a closed system, as Boulding had stated in 1966. The authors worked on a closed loop conceptual model, with two variants: with biological and technological closed systems, where product cycles went from being extracted from nature (at minimum levels and less harm), to manufacturing, use, and then, instead of being thrown away, are recycled, reused, and remanufactured, and ideally with no discard. Their circularity models are the basis for a circular economy.

Currently, the Ellen MacArthur Foundation is a beacon for circular economy since its creation in 2010. Their website appoints design as a paramount tool for circularity, introducing a framework with six leverage points that designers can take to embrace circularity – and help unlock the potential of circular design: "[A] Observe and interpret the system; [B] Envision circular futures; [C] Create the conditions for collaboration; [D] Build circular design capabilities; [E] Rewrite the rules; [F] Develop tools to design and evaluate" (EMF 2023). These are some cues to designers willing to commit to change companies or territories, supporting a more sustainable society.

This historical viewpoint serves as an introduction to the following contributions by some enthusiastic authors, and also reminds us of What's Around Design in the field of circularity.

References

Boulding, K. E. (2013). The economics of the coming spaceship earth. In Environmental quality in a growing economy (pp. 3–14). Washington: RFF Press.

Braungart, M.; Mcdonough, W. (2002). Cradle to cradle: Remaking the Way We Make Things. Albany, California: North Point Press.

Buckminster Fuller, R. (1969). Operating Manual for the Spaceship Earth. Carbondale: Southern Illinois University Press.

EMF (Ellen MacArthur Foundation) (2023). The six levers designers can use to unlock the potential of circular design. Available at: [https://ellenmacarthurfoundation.org/news/the-six-levers-designers-can-use-to-unlock-the-potential-of-circular-design]. Accessed: 04.02.2025.

Georgescu-Roegen, N. (1971). The Entropy Law and the Economic Process. Cambridge: Harvard University Press.

Meadows, D. H., Randers, J., & Meadows, D. L. (1972). The limits to growth. New York: Universe Books.

Papanek, V. Design for the Real World: Human Ecology and Social Change. New York: Pantheon, 1971.

Stahel, W. R. (1982). The product-life factor. An inquiry into the nature of sustainable societies: The role of the private sector (Series: 1982 Mitchell Prize Papers), NARC. Available at: [https://www.quebeccirculaire.org/data/sources/users/4/32217.pdf]. Accessed 04.02.2025.

Repurposing Marine Waste: A Circular Design Approach

Isabel Farinha[1(✉)], Miguel Pina[3], Sara Gancho[2], and Carlos A. M. Duarte[2]

[1] Life Quality Research Centre (LQRC-CIEQV)/Polytechnic Institute of Setúbal (IPS), Setúbal, Portugal
isabel.farinha@estsetubal.ips.pt
[2] UNIDCOM/IADE - Universidade Europeia, Lisbon, Portugal
[3] IADE, Universidade Europeia, Lisbon, Portugal

Abstract. This research article explores the potential of applying circular design strategies to marine plastics to improve the health of coastal and ocean ecosystems within the Cascais captaincy. Drawing on evidence from literature reviews, document analyses, two case studies, and thirteen semi-structured interviews with fishermen from Cascais and Ericeira—focused on their working conditions and tools, such as fishing nets, while they performed tasks at the port—the study reveals a promising opportunity for circular design in marine plastics. This opportunity is not only viable for the future but could also yield environmental, social, and economic benefits for the planet, oceans, and nearby communities.

To arrive at this conclusion, we followed a structured and scientifically sound methodology. Our findings suggest that further research is necessary, particularly on redesigning fishing gear. Additionally, we developed a diagram and documented the evolution of the Blue Circular Postbranding Project (BCpb) as a potential template for implementing circular design strategies in addressing marine plastics.

Keywords: Marine Plastics · Fishing Nets · Circular Economy · Ocean Pollution · Design Methodologies

1 Introduction

The aim of this research is to apply design methodologies to improve the coastal and maritime health of the Cascais captaincy ports (Portugal), with the broader goal of making a small but positive impact on the planet. Although this research focuses on a small section of the Portuguese coast, the fact that 60% of Portugal's total borders consist of coastline (CIA, April 7, 2022) suggests that this case study could serve as a valuable model for creating procedures and templates applicable to similar contexts.

The study is centered around the Blue Circular Postbranding Project (bluecircular.org), which originated in Ericeira and was later expanded to Cascais under the Captaincy of the Port of Cascais—a local branch of the Maritime Authority General-Directorate responsible for regulating maritime spaces within its jurisdiction, which extends from Foz do Rio Sizandro (Torres Vedras) to Forte de S. Julião da Barra (Cascais). The project encourages local fishermen to recover marine plastics and fishing gear

A. Mateus et al. (Eds.): WAD 2024, What's Around Design?, pp. 91–105, 2026.
https://doi.org/10.1007/978-3-032-08163-6_7

caught in their nets and deposit them in the organization's marine recycling bins rather than discarding them back into the sea. Following this, the research investigates methods to reintroduce these materials into the economy through circular strategies.

Plastic debris is a matter of significant societal concern, as it has been identified as an unnatural stressor to a wide range of organisms and ecosystems. It is also considered an aesthetic nuisance and an unethical addition to nature. Cleaning contaminated areas requires substantial effort and resources, which has economic implications. The visibility of plastic debris may also contribute to public concern (Koelmans et al., 2017).

In any given context, understanding every small system in place and their complex interconnections is not a simple task; a strong line of thought on this matter is that a holistic approach lends itself as the better tool to resolve problems in complex contexts. Epstein (2020) calls these contexts the Wicked World. Epstein's book Range, a strong influence in this study's approach, is an exploration of the advantages and disadvantages of people considered specialists and generalists lie in; the idea presented is that specialists excel at kind learning environments which he refers to as the Kind World which are defined by having a static set of rules such as a chess master studying and practicing chess; patterns repeat over and over, and feedback is extremely accurate and usually very rapid. On the other hand, he presents us with the generalists and late specializers who sample a variety of disciplines before truly committing to one (if ever), who excel at problem-solving in the Wicked World – the rapidly changing reality which requires conceptual reasoning skills that can connect new ideas and work across contexts, and for which you'd need the ability to apply knowledge broadly, thus requiring broad training. This line of thinking is not exclusive to this author, Science Advice for Policy by European Academies (SAPEA) states that close interdisciplinary collaboration between the natural, social and behavioral and regulatory sciences is the way forward for addressing the complex issue of plastic waste and pollution (SAPEA, 2019, January 15). Despite plastic's very important role in our economy and crucial role in many industry sectors due to its low cost and high functionality, the short-term one-time applications these materials are often used in are a very inefficient way of using resources – even though on the surface they might seem cheap, and so we would like to explore an alternative way of thinking about plastic consumption while at the same time reducing the already existing amount floating in the ocean, which in 2014 was measured to exceed 250,000 tons (Eriksen et al, 2014).

According to Moir (2022, April 20) the European Court of Auditors (2021, July 5) and our own observation during this investigation, there are three main challenges in reusing plastics recovered from the ocean:

- Cost/ Availability: Throughout the course of this investigation, we found that most of the industries that have the recycling capabilities for these plastics are already allotted to recycling domestic waste and have little room for expansion, or, have no financial incentive in opting to recycle national plastics which is more expensive in comparison to the alternative plastics from international seas that get purchased much cheaper.
- Greenwashing: The alternatives available to these transformation industries include buying new plastics for a much cheaper cost than that of the whole recycling process or, in case their product's marketing speech is environmentally aligned, importing the already separated washed and extruded plastics from Asian countries such as

Singapore, which further pollutes the planet but gives them the ability to market their products as recycled ocean plastics. In the present, most of the financial incentive to get companies on board with circular economy ideas comes from the ability they must associate their brand with these initiatives, boosting their bottom lines because of being seen as greener – effectively exploiting this loophole is what we call greenwashing. Thus, we can say that greenwashing is the act of making false or misleading claims about the environmental benefits of products or services (Moir, 2022, April 20).

- Enforcement of Legislations: With the Asian plastics loophole being so commonly exploited, another course of action would be to create and enforce legislations that prevent greenwashing from happening and further incentivize the recycling of products even at a theoretical financial loss. Unfortunately, this represents yet an-other challenge as even though there exist regulations already in place for this purpose, they don't seem to be properly adjusted and/or enforced. In 2021 the European Court of Auditors published a Special Report titled "The Polluter Pays Principle: Inconsistent application across EU environmental policies and actions" outlining the flaws of application of these environmental policies (2021, July 5).

The evidence suggests that the focus should be on circular economy approaches and away from linear processes and end-of-life cleanup (SAPEA, 2019, January 15). In this study we will work under this assumption raised by SAPEA that the answer lies within a societal/ design-shift, away from linear economy processes.

The general objective of this study was to determine whether there is a circular way to reuse the recovered discarded fishing nets mainly composed of Polyamide 6 (PA 6) nationally, and to design a model/template that represents the developed circular operation for ease of implementation in similar contexts.

The specific objectives which we designed to help us achieve our main goal are: (i) involvement of the community in the virtual supply chain through the design of a circular strategy that exchanges some kind of reward for their cooperation; (ii) the growth of ocean pollution awareness within the fishermen community; (iii) research of other attempts at tackling the environmental problem of ocean pollution and determining their usefulness as an example; and (iv) determining by order of proximity which of the industries from the plastics transformation sector have interest in forming meaningful partnerships that work towards the betterment of the Cascais captaincy coastal hygiene.

2 Circular Economy: The Plausible Evolution of Consumption

A Circular Economy is "a model of production and consumption involving sharing, reusing, repairing, refurbishing, and recycling materials and products for as long as possible" (European Parliament, 2015). Unlike the traditional "take, make, waste" linear model, a circular economy aims to minimize resource use, waste, pollution, and emissions by closing the loop through reuse and recycling (Geissdoerfer et al., 2017).

For the past 150 years, the linear model has dominated industrial evolution, driving prosperity but at a high cost, based on the flawed assumptions that Earth's resources and regenerative capacity are infinite (Wautelet, 2018). We are now at a critical juncture, or "chasm," in the adoption lifecycle (Moore, 2002), where only a few projects will succeed in transitioning to a circular model.

However, signs of progress include laws limiting single-use plastics and preventing the rejection of recovered materials for profit (Presidência do Conselho de Ministros, 2021), as well as the Mediterranean's goal of "No Plastics in Nature by 2030" (WWF, 2015). Though full transition will be slow and incremental, it remains a promising path for the future.

While the Circular Economy model has challenges, and a 100% circular system is unlikely, these limitations are beyond the scope of this research. Mentink (2014) provides further detail on these challenges in his work on circular business model innovation.

2.1 Microplastics in Nature and Society

Microplastics are tiny plastic particles that result from both commercial product development and the breakdown of larger plastics (Liitschwager, 2019, July 1) that can represent harm to the environment and animal health. The issue with microplastics is that because they are plastic, they do not readily break down into harmless molecules – and can take hundreds or thousands of years to decompose. On beaches, tiny multicolored plastic bits can often be seen in the sand, while in the oceans, these particles are frequently consumed by marine animals (Liitschwager, 2019, July 1)."

In January 2019, SAPEA's (Science Advice for Policy by European Academies) researchers couldn't conclude specific consequences of ocean pollution (Galgani et al, 2019, April 26) although they left the topic on a high-alert tone, with attention to the ocean exploration growing at an unprecedented pace (SAPEA, 2019, January 15). But in 2022, a study shows that microplastic pollution has been detected in human blood in almost 80% of the studied population and raises concerns that they can travel around the body, cause damage to human cells and may lodge in organs and cause millions of early deaths a year (Carrington, 2022, March 25). This confirms that microplastics have infiltrated the food chain, which is a major source of concern not only as an immediate danger to us humans but to the whole ecosystem.

2.2 Maritime Activities and Sources of Marine Litter

Plastics in the ocean can either be sea-based or land-based but what is important is to define an outcome we intend to achieve (e.g. pollution mitigation) and carefully design the mechanism of action that will take us to the desired outcome (e.g. introduction of new regulations). Judging from the information described in Table 1, we can immediately see that education is, unsurprisingly, the most flexible and 'low-hanging fruit' mechanism for both prevention and mitigations outcomes across the board – and while it might not be the one with the strongest impact on this list, its strength lies in its straightforward, cross-board effect. New fishing regulations seem to favor finetuning of Extended Producer Responsibility (EPR) schemes and as for product innovation, the most promising line of thought is to find smart ways of designing products such as ropes and nets from scratch with their specific recycling purpose in mind right from the get-go (European Commission, 2020, August 19).

Table 1. Scope of the Plastics Treaty (Tessnow, 2019, p.48).

Source	Category	Mechanism	Outcome
	(mainly) Lost & abandoned fishing gear	Fishing Industry Regulations	Pollution Mitigation
		Product design innovation	Pollution Prevention
		Education	Pollution Mitigation/ Prevention
	Hard Plastics + Soft Plastics	Improvement of Collection and Recycling Systems	Leakage Prevention
		Circular Economy	Pollution Mitigation/ Prevention
		Product Design Innovation	Pollution Mitigation/ Prevention
		Bans	Pollution Prevention
		Economic Mechanisms	Pollution Mitigation/ Prevention
		Education	Pollution Mitigation/ Prevention
	Microplastics	Bans	Pollution Prevention
		Product Design Innovation	Pollution Mitigation/ Prevention
		Technology development	Pollution Prevention
		Education	Pollution Mitigation/ Prevention

3 Qualitative Research – Semi-structured Interviews

The semi-structured interviews followed a focus group methodology with guiding questions, conducted in the two locations selected by the BCpb project. This method, often used in social sciences and user research, facilitates group debates moderated by the interviewer, aiming to extract insights from the group dynamic (George, 2021).

To explore challenges and solutions for developing a circular design process involving community stakeholders, we developed a categorization table and a detailed interview plan. The interview objectives were to: understand cooperation dynamics with the fishermen (Circular cooperation); assess the impact of Docapesca's awareness intervention at the Cascais docks (Social action); gauge fishermen's awareness of ocean pollution and their role in it (Environmental impact); and understand how fishermen perceive themselves and BCpb (Community engagement).

1. The seven interviewees in the port of Cascais (between 56 and 25 years old) were happy to talk about their profession and comment on a few questions regarding the fishing nets while they were performing work tasks at the port. Given their eagerness to discuss certain aspects of their profession, work conditions and tools such as the

fishing nets, the interview quickly developed beyond the pre-defined questions and became a very rich source of information on the community, which fortunately was in line with what we looked for and the semi-structured format we prepared. Our main content analysis from these interviews were (Table 2):

Table 2. Content analysis of the fishing community of Cascais.

Circular cooperation	Due to the price of new nets (~50€) being superior to sending them to a reparation professional in a different port to fully repair (~20€), fishermen are reluctant to throw them away, and when they do it's after thoroughly confirming they can no longer be repaired which happens in land – making the containers placed in the port for the recovery of the fishing nets highly convenient in terms of location and usefulness, which results in the containers being used very often. Apart from very few older aged outliers, fishermen themselves don't repair their own nets in the port of Cascais and instead employ someone to do them for them in a different port. Reparation of own nets seems to be a fading trend as the younger generation no longer learns it
Social action	Although the practicality of the Docapesca recycling plastic bags, the intervention didn't reach the small sample we interviewed, nevertheless it does seem that it wasn't totally useless, as the topic of ocean pollution and fishing net disposal seemed to be very fresh on fishermen minds
Environmental impact	The kind of litter they find in the sea and in nets and octopus cages are almost always not prevenient from fishing activities, instead the report that disposable hygiene products such as masks and menstrual pads are the most common, along with bottles, plastic and glass alike
Community engagement	The BCpb project, and its net containers are mostly welcome since they provide a comfortable way of disposal for them

2. The six interviewees at the port of Ericeira were also eager to share their opinions and expertise. Among them was the treasurer of the fishermen's association. Ericeira sees less fishing activity compared to Cascais, which is reflected in the number of active fishermen. Five of the six interviewees were retired veterans (all over 60) who spent most of their lives at sea and now visit the port daily to assist with equipment repairs and preparations. The sixth participant was a sport fisherman. Key insights from these interviews included (Table 3):

4 Research Results

The conducted research allowed us to make a few conclusions:

- The fishing nets retrieval operation put in place by the BCpb is successful and well accepted by its users, mainly due to not giving the fishermen work that they already wouldn't do and satisfying their need of having any container placed in proximity of their workstations so that they can dispose of the nets with ease.

Table 3. Content analysis of the fishing community of Ericeira

Circular cooperation	Similarly, to Cascais port, bringing the broken nets to the shore is standard procedure that every fisherman abides by. The fishermen at the port of Ericeira report the same price for nets (~50€) and a slightly lower price for full repairs (15€) making it even more worth for these men to properly maintain their nets. There are very few remaining fishermen that repair their own nets (<2) as most are not willing or can't afford the time cost of doing it themselves (between two and three full days of work for one net), hence it's not something that is even taught on by the previous generations
Social action	Due to the nature of the interviewee's current occupations (five retired veterans and one sport fisherman), it wasn't possible to measure the impact of the Docapesca intervention. What we can say is that according to these seniors and the treasurer of the fishermen association all are in agreement that every single boat is equipped with trash containers that they religiously make use of – "nowadays there is no fisherman that throws garbage to the sea" – and even without the bags they claim that waste is separated and taken back to the port – which might suggest that the introduction of the recycled plastic bags from the initiative could be redundant
Environmental impact	The type of waste present in the ocean vastly differs between the two locations. We can hypothesize that this is due to the proximity of Cascais to the Tejo River bringing about more urban waste to the sea
Community engagement	Regarding the BCpb project, all agree that bringing the nets back is something they do anyway and it's nice to be rewarded for it, while #4 and #5 added that the fishermen association is a very small entity that survives through tiny helpings from different places, adding the BCpb to those is welcome as it diversifies sources of help

- Although environmental practices are not the fishermen's first concern, the fishermen crowd are generally very aware of the subject and can report with a degree of specificity what kind of waste they do or do not find while at sea.
- Previously *alcatruz* (Fig. 1) weren't made of plastic but of clay. The replacement of the devices for their plastic versions (Fig. 2) enables fishermen to fish the highly demanded octopus for the entirety of the year, which represents most of their income in comparison to the fishing of other species.
- There may be something to be done regarding pollution originating from *alcatruz* fishing in the future; although the damage done is invisible to the naked eye it might be responsible for the endangerment of species in the future due to the rate at which they accumulate at the sea floor without ever having the chance of being removed.
- The floaters used to keep the GPS marked signaling flags afloat are simply a block of cheap Styrofoam, which constantly gets replaced since it breaks down and floats away to the sea.

- Lastly, in regards to rubbers and plastics being utilized that potentially contribute to the microplastic catastrophe in the ocean is that corvina, a type of fish fished with the *aparelho* (Fig. 3) device which is basically a collection of hooks and lines, is fished not with live bait but with rubber bait (Fig. 4), that is cheap "durable" (in comparison with live fish) and that this particular type of fish will still swallow, allowing them to fish out this specific fish from say a very big school of sardines.
- The costs of purchasing and repairing equipment, boat maintenance, transportation, and time lost due to poor conditions quickly add up, affecting profits. This partly explains the lack of young fishermen, as it's a highly specialized profession with tough hours and physical labor, paying crew members between €800 and €1500, depending on the catch, for six days of work per week—conditions that are not very attractive. Profits are distributed after the catch is sold, with each crew member receiving a share, while the boat master takes two shares to cover equipment and boat ownership costs. This thin profit margin forces them to extend the life of equipment, especially fishing nets, which are repaired repeatedly before being discarded.

Fig. 1. Alcatruz fishing device

Fig. 2. Covo fishing device

Fig. 3. The aparelho fishing device

Fig. 4. Rubber bait used in aparelho

5 Contribution

Our contribution for the improvement of ocean health and future research is a diagram that shows the current circular process of the BCpb that we tuned and developed in parallel with our research and the Design methodologies discussed in the introduction and literary review. We hope to contribute to this field of research by providing a solid case study for future researchers and an actionable template usable by others in similar circumstances. Figure 5 shows the conceptual starting point for the BCpb prior to this research.

Let us consider five different circular objectives: longevity/maintenance, reuse, refurbishment, remanufacture and/or recycling (Table 4). The first challenge when coming up with a diagram that frames the circular value chain of the BCpb project is examining each of these objectives and using them as a platform to inspire a solution.

Fig. 5. The BCpb starting point

5.1 Summary: Diagrams

To achieve the BCpb ambitions of creating a simple and replicable circular model in the form of a diagram a few public-private-community partnerships and relationships had to be built with other organizations. We based ourselves on the circular business model canvas (Ellen McArthur Foundation & IDEO, 2017, January 23) to lay the groundwork of processes, relationships and interactions needed. With the key partnerships secured and the loop closed, we came up with the circular diagram below in Fig. 6 as the output for this research.

The diagram above represents the start to finish circular journey of going from point A: waste recovery from the ocean; to point B: giving it a purpose, benefiting the community and the planet and infinitely looping the system.

This loop starts (and ends) with the fishermen, who for their own financial motivations do their utmost best to extract the biggest lifespan out of each fishing net and deposit them into the BCpb's containers once they finally reach their limit of usability so they can be collected by us.

To this effect a contact was made between the BCpb and Docapesca – a state-owned company that provides public service on mainland Portugal by organizing the first sale of fish and supervising the fishing and fishing ports sector – resulting in a partnership Docapesca/APLM under their sustainability project "Fishing for a Sea Without Litter"

Table 4. Circular objectives and observations (Based on Franconi et al.2022, pp 5–6)

Circular Objective	Description	Observations
Longevity/ maintenance	Design for maintenance/longevity is an approach for extending the life of a product by incorporating maintenance considerations. Frequently, this circular objective is associated with the product–service system. Design for maintenance/longevity is close in meaning to design for slowing resource loops, design for long-lasting products, design for obsolescence resilience, design for durable products, design for extending life cycle, or design for product-life extension	Fishing nets are already cared for to ensure they reach their full durability potential There is nothing to be done in this aspect since it's not realistic within the scope of this research to come up with an infinite lifespan type of fishing net
Reuse	Design for reuse implies creating the conditions for value to be reused in a new product life cycle as-is, with minimal rework. As with the other circular objectives, planning for reuse should begin prior to the beginning of the product lifetime. The design for reuse method has received less attention compared with other circular objectives. Design for reuse is different to design for creative upcycling and design for repurposing	Fishing nets are not a one-use device, and since they are made of a single type of plastic (PA6) and thus are already at their maximum purity level for recyclability, at first glance there doesn't seem to be much else that can be done in the manufacturing process to affect reusability
Refurbishing	Design for refurbishing is defined as extending the product's lifetime by restoring its full functionality and/or aesthetics by reworking only what is compromised. Only portions of the product that have failed or are badly worn can be disassembled and rebuilt with old and/or new components. Consumer acceptability of refurbished products has recently become a study focus. Design for refurbishing is close in meaning to design for reconditioning	When fishing nets reach the "soft" limit of their repairability, the harder structures within the nets get used as the base for "new" refurbished nets From our empirical research we found this practice to be widespread, thus we consider this objective is already put into practice

(*continued*)

Table 4. (*continued*)

Circular Objective	Description	Observations
Remanufacturing	Design for remanufacturing extends product lifetime by restoring used products to like-new or better-than-new condition. To clean, restore, and replace components during remanufacturing, total disassembly is required. What sets remanufactured products apart from refurbishes is their condition, performance and warranty. Some scholars consider remanufacturing to be the most promising circular objective in the CE	Remanufacturing is possible and indeed very promising, but since in the specific case of fishing nets that would always require them to be extruded first, it seems logical to first analyze the recycling logistics
Recycling	When there are no other options to extend the product lifetime except recapturing the value of materials, recycling can be considered as the main circular objective. Design for recycling ensures proper material selection, separation, and reprocessing for new material flow. Design for recycling is close in meaning to design for upcycling	PA6 wasn't always the easiest to recycle, but with companies such as Aquafil and PlastiX paving the future and putting forward powerful recycling solutions for the material, the outcomes seem exciting since it allows to outsource a part of the process

which in turn opened up the contact opportunity with their close associates – the municipality, local authorities, and the fishermen association – which we used to introduce our interference in the fishing net life cycle: two big white containers to be placed in the docks of each location for the specific purpose of fishing net disposal. During the projecting, the idea of using bigbags personalized with logos inside of the containers came up to make the transportation of the fishing nets an easier and cleaner process.

Whenever the containers reach full capacity, the collected fishing nets are then collected by official waste management entities at the service of the municipality who bring them to Tratolixo's facilities to be temporarily stored before it gets transported to Ambiberica's facilities to be sorted and treated. There the waste gets packed and sent to their partner Aquafil who turns the waste materials into profit by recycling them into new nylon fibers that cut up to 90% of the pollution used in refining brand new fibers and sells them as a raw material for others to transform. Ambibérica then proceeds to donate to and support the fishermen who retrieve the waste through the donation of professional grade fishing gear.

As previously noted, the project is highly dependent on establishing strong partnerships and keeping them engaged. We believe that the key to the project's health and

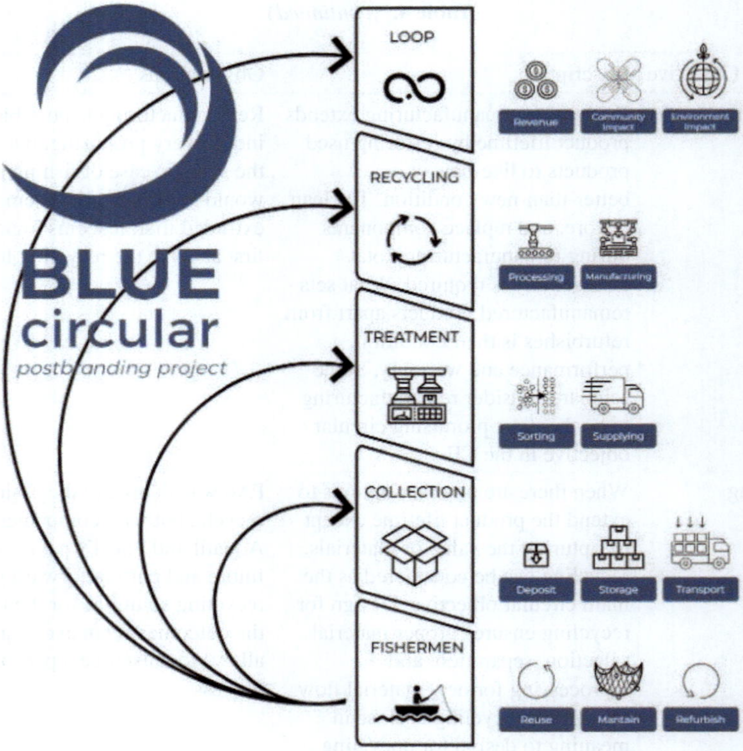

Fig. 6. Diagram of the BCpb circular model

longevity is that each relationship formed within partners is in one way or another mutually beneficial and never just one sided, so something that could be a takeaway for future projects is to make sure that everyone in the supply chain gets involved with some depth.

6 Closing Remarks

The main goal of this research was to determine whether circular economy strategies could improve the coastal health of the selected ports. Not only did we find it possible, but we also designed and implemented a model that could inspire long-term environmental solutions. We discovered that not all recycling methods are equally effective for plastics from end-of-life fishing gear, but advancements in Nylon recycling have removed past limitations, allowing for ambitious projects to reuse discarded fishing nets.

Our first objective—to involve the community in the virtual supply chain—was achieved through a mutually beneficial cooperation strategy. The second objective—raising awareness of ocean pollution—was also met, as fishermen already demonstrated high awareness, and our ongoing interactions will continue to reinforce these values.

However, fishermen's financial viability remains a concern. We believe redesigning certain fishing tools could improve sustainability without increasing costs, making ocean health a higher priority for them. Our third objective—investigating companies and projects addressing ocean pollution—was achieved through case studies and literature. We found that although the industry for Nylon recycling is in its early stages, it already offers real-world potential for creative circular models.

Finally, our last objective—assessing the practicality of the project's location—was confirmed through document analysis. We found that operations could stay within reasonable proximity, and partnerships with plastics transformation industries could be established, often supported by their interest in social responsibility.

6.1 Limitations

The sample size of interviewees was proportionate to the two fishing locations but not representative of the national picture, potentially limiting the accuracy of insights for other regions. Additionally, it is unclear whether the fishermen's actions at sea align with their claims. To maximize the reliability of responses, we avoided rigid views on environmental practices, quickly acknowledged when interviewees admitted to harmful actions, and kept the dialogue relaxed and open-ended during the interviews.

Redesigning equipment is beyond the scope of this research but aligns with our goals of improving ocean health and community awareness. The exact amount and type of fishing gear stuck on the seafloor remain unknown without further research and advanced screening technology. The BCpb addressed this by focusing on materials that matched our resources while still positively impacting the environment.

6.2 Further Work

As companies strive to meet environmental regulations and replicate successes like Aquafil and Plastix, future research could explore scaling this model to other waste materials. By consolidating efforts from various organizations, marine waste recovered by fishermen and volunteers could be gathered at a central point, reducing transportation costs to industries in the north. Competition from rival companies would also stimulate innovation and research.

Further investigation is needed to assess whether modern fishermen, using standard technology, can accurately distinguish between sand and rock on the seafloor and what new methods could improve this accuracy.

Acknowledgment. The Mar 2020 programme funded the Blue Circular PostBranding Project (BCpb).

The funding was operationalised through A2S (Association for the Sustainable Development of the Saloia Region).

References

Aquafil: Environment and Sustainability – Aquafil. https://www.aquafil.com/environment/

Blue Circular Economy: Products from Waste Fishing Nets. Accessories, Clothing, Footwear, Home Ware, Recreation, March 2022. https://keep.eu/api/project-attachment/730/get_file/

Carrington, D.: Microplastics found in human blood for first time. The Guardian, 25 March 2022. https://www.theguardian.com/environment/2022/mar/24/microplastics-found-in-human-blood-for-first-time

CIA: Photos of Portugal - The World Factbook. CIA, 7 April 2022. https://www.cia.gov/the-world-factbook/countries/portugal/#geography

Ellen McArthur Foundation & IDEO: Resources. The Circular Design Guide, 23 January 2017. https://www.circulardesignguide.com/resources

Ellen McArthur Foundation: Circular Design. Explore the Circular Economy, 15 December 2021. https://archive.ellenmacarthurfoundation.org/explore/circulardesign

Epstein, D.: Range: Why Generalists Triumph in a Specialized World. Penguin LCC US, pp. 53–54 (2020)

Eriksen, M., et al.: Plastic pollution in the world's oceans: more than 5 trillion plastic pieces weighing over 250,000 tons afloat at sea. PLoS ONE **9**(12) (2014). https://doi.org/10.1371/journal.pone.0111913

European Commission: Study on circular design of the fishing gear for reduction of environmental impacts. Publications Office of the EU, 19 August 2020. https://op.europa.eu/en/publication-detail/-/publication/c8292148-e357-11ea-ad25-01aa75ed71a1/language-en/format-PDF/source-147995096

European Court of Auditors: Special Report 12/2021: The Polluter Pays Principle: Inconsistent application across EU environmental policies and actions. Special Report 12/2021: The Polluter Pays Principle: Inconsistent application across EU environmental policies and actions, 5 July 2021. https://www.eca.europa.eu/Lists/ECADocuments/SR21_12/SR_polluter_pays_principle_EN.pdf

European Parliament: Circular economy: definition, importance and benefits | News | European Parliament. European Parliament, 2 December 2015. https://www.europarl.europa.eu/news/en/headlines/economy/20151201STO05603/circulareconomy-definition-importance-and-benefits

Franconi, A., Cezschin, F., Peck, D.: Structuring circular objectives and design strategies for the circular economy: a multi-hierarchical theoretical framework. Sustainability **14**(15), 9298 (2022). https://doi.org/10.3390/su14159298

George, T.: What Is a Focus Group? | Step-by-Step Guide & Examples. Scribbr, 10 December 2021. https://www.scribbr.com/methodology/focus-group/

Galgani, L., Beiras, R., Galgani, F., Panti, C., Borja, A.: Editorial: impacts of marine litter. Front. Mar. Sci., 26 April 2019. https://doi.org/10.3389/fmars.2019.00208

Geissdoerfer, M., Savaget, P., Bocket, N., Hultink, E.J.: The circular economy-a new sustainability paradigm? J. Cleaner Prod. **143**(1), 757–768 (2017). https://www.repository.cam.ac.uk/handle/1810/261957. https://doi.org/10.1016/j.jclepro.2016.12.048

Koelmans, A.A., et al.: Risks of plastic debris: unravelling fact, opinion, perception, and belief. Environ. Sci. Technol. **51**(20), 11513–11519 (2017)

Liitschwager, D.: Microplastics. National Geographic Society, 1 July 2019. https://www.nationalgeographic.org/encyclopedia/microplastics/

Mentink, B.: Circular business model innovation: a process framework and a tool for business model innovation in a circular economy. TU Delft repositor, May 2014. https://repository.tudelft.nl/islandora/object/uuid:c2554c91-8aaf-4fdd-91b7-4ca08e8ea621?collection=education

Moir, A.: Greenwashing in Singapore. Lexology, 20 April 2022. https://www.lexology.com/library/detail.aspx?g=3dc68517-d679-4b5f-be92-82ef288b8ac4
Moore, G.: Crossing the Chasm: Marketing and Selling Disruptive Products to Mainstream Customers. HarperCollins (2002)
Presidência do Conselho de Ministros: Publicação: Diário da República n.º 187/2021, Série I. Diário da República Eletrónico (2021). https://dre.pt/dre/detalhe/decreto-lei/78-2021-171871496
SAPEA: A scientific perspective on microplastics in nature and society – SAPEA, 15 January 2019. https://sapea.info/topic/microplastics/
Tessnow - von Wysocki, I.: International Cooperation for the Protection of Global Public Goods: Towards a Global Plastics Treaty. University Alliance for Sustainability (UAS) Working Papers, p. 2 (2019). https://refubium.fu-berlin.de/bitstream/handle/fub188/24616/TessnowWP-GlobalPlasticsTreaty_OnlineVersion.pdf?sequence=5&isAllowed=y
Wautelet, T.: Exploring the role of independent retailers in the circular economy: a case study approach, February 2018. https://doi.org/10.13140/RG.2.2.17085.15847
World Wildlife Fund (WWF): Principles for a Sustainable BLUE ECONOMY. UNCTAD, 28 May 2015. https://wwfint.awsassets.panda.org/downloads/15_1471_blue_economy_6_pages_final.pdf

Be Circular. The Push of the City

Giovanna Tagliasco(✉)

University of Genoa, Genoa, Italy
giovanna.tagliasco@edu.unige.it

Abstract. The cities are places where a lot of pollution is produced due to the higher concentration of the population. However, they are the places where it is possible to intercept actions for change. To respond to sustainability issues, the idea of circular cities has developed and spread widely, especially in large European urban areas. This paper examines the city of Genoa, as a circular city, within the C City project. The project actions will be analysed looking at the contribution of design, in particular Service Design, through the analysis of projects and interviews, to verify the adaptation to the circularity criteria. Within the project process, co-design actions were implemented with the partnership and stakeholders, according to the principles of Service Design. In another context, touchpoints were designed to define the identity of the individual parts of the project such as to be coherent within the larger system.

The results of the analysis describe an early-stage process in which a general vision characterised by a cross-sectoral approach emerges. This has highlighted parts of supply chains, such as food, that are accessible for concrete initiatives and actions by the city and its citizens. In conclusion, the paper defines the actions that the city of Genoa has planned to become a circular city and the contribution of Service Design and Design in the process.

Keywords: Circular City · Circular Economy · Service Design · Identity · Event Design

1 Introduction

The cities are responsible for about 78% of the world's energy consumption and produce more than 70% of global greenhouse gas emissions (Barth et al., 2022).

One model that addresses these issues is, for example, 'The 15-Minute City' (Moreno, 2024), which describes a city designed so that distances between places of living, work or services can be covered by foot or bicycle (or zero-emission vehicles) in 15 minutes. This model is based, for example, on the valorisation of living time, but also on energy saving and emission reduction.

Another example is 'Collaborative Cities' (Manzini, 2018, tda, p. 170) cities driven by bottom-up initiatives, characterised by collaboration between citizens, associations and organisations in the area. To address the issues of environmental, social and economic sustainability and urban regeneration, projects on cities start from the real needs

A. Mateus et al. (Eds.): WAD 2024, What's Around Design?, pp. 106–113, 2026.
https://doi.org/10.1007/978-3-032-08163-6_8

of inhabitants and their experiences. Another approach concerns Circular Cities developed all over the world in recent years, defined in these terms by the Ellen MacArthur Foundation:

A circular city has embedded the principles of the circular economy across the entire urban area. Everything is operating within an interconnected network of systems that are designed to eliminate waste and pollution, circulate products and materials, and regenerate nature. There's collaboration between citizens, government, research facilities, and businesses. And the entire city is powered by renewable energy (Ellen MacArthur Foundation, n.d.).

The Circular City is a city that through various actions is committed to managing resources for the well-being of its citizens and the environment. This paper presents an analysis of Genoa as a Circular City, in particular with a focus on the C-City project. The project was financed through the PON METRO Cities 2014-2020, integrated with the Recovery Assistance for Cohesion and the Territories of Europe (REACT EU) – Asse 6 – Ripresa verde, digitale, Azione 6.1.4. "Qualità dell'ambiente e adattamento ai cambiamenti climatici". The project, also, won the 1st Prize for Sustainable Development 2023 - circular economy sector at Ecomondo (Rimini, 7–10 November 2023)

The aim of the project is to launch activities, information and a network of stakeholders to implement the circular economy within the city. The city of Genoa extends along a narrow strip of land enclosed between the sea and the mountains, following the outline of a gulf. This strongly linear geographical feature seems to contrast with the forms of circularity that are developing.

As argued by the digital and ecological transition manager of AMIU (Azienda Multiservizi di Igiene Urbana), this morphological characteristic of the city of being long and narrow, for example, forces the logistics of waste collection transport to move along this extended strip of land. This entails a high cost and a high environmental impact.

Moreover, Genoa stretches along the coast, which means an additional responsibility for the protection and preservation of the sea. In fact, continuous and constant monitoring of the accidental and non-accidental dropping of waste into the water is necessary (Digital and ecological transition manager of AMIU).

2 Methodology

At the methodological level, it was decided to follow the model adopted in the paper "Circular cities: Mapping six cities in transition", in which, through Document review, Project mapping and Semi-structured interviews, is analysed the circularity of six different cities (Prendeville et al, 2018). The approach taken by the authors was to evaluate projects and initiatives through the criteria of ReSOLVE (EMF, 2015). The ReSOLVE framework is a tool to facilitate governments' and companies' adoption of circular strategies. This framework consists of six macro-areas (Regenerate, Share, Optimize, Loop, Virtualize, Exchange) that feed into each other (EMF, 2015). In addition to the ReSOLVE criteria, again following the methodological approach of the paper by Predenville et al. (2018), it was decided to also identify the types of strategies that policymakers put in place to implement circular economy policies. The strategies identified, started from the EMF's

"Toolkit for Policymakers" (2020), are: "education, information and awareness (knowledge development), business support schemes, collaboration platforms, public procurement and infrastructure, regulatory frameworks and fiscal frameworks" (Predenville et al., 2018, p. 175).

The analysis on the city of Genoa focuses mainly on the actions implemented by the C-City Genoa Circular City project.

In contrast to the methodological approach taken as a reference (Prendeville et al, 2018), the paper focused on a single city and a specific project, however characterised by several actions. All this was evaluated through the contribution of Design as a co-activator of services to implement circularity within the city context.

3 Genoa

Regarding the Document review, the action plan "Genoa 2050. Action Plan for a Lighthouse City"(Comune di Genova, 2021) and "Circular Economy Action Plan – For a cleaner and more competitive Europe" (European Commission, 2020) were examineted to understand the vision and the strategy of the city of Genoa in terms

of circular economy. Also Genoa is a signatory of the 'Circular Cities Declaration' and appears with its examples in the report 'Circular Cities Declaration. Report 2024. Insights on implementation, measurement, and nature' published by ICLEI and edited by experts from ICLEI Europe, Ellen MacArthur Foundation and Circle Economy.

Concerning the Project mapping the ReSOLVE criteria were attributed through field experience, desk research on the project website, and re-discussed with interviewees.

The Semi-structured interviews were conducted both to gain a general overview of the city of Genoa and its potential to be circular, and to collect specific feedback on the projects most managed by the interviewees (Table 1).

Table 1. Interviews for selected projects

Role	Type and duration of interview	Project
Director Job Centre srl- Genoa Blue District	30 min, face-to-face	C-city, Circular Hub, Circular Desk
Administrative and Reporting Area Projects Job Centre Srl - Genoa Blue District	30 min, face-to-face	C-city, Circular Hub, Circular Desk
Responsible press office and social channels (AMIU)	30 min, (telephone)	C-City, Surpluse
Digital and ecological transition area manager (AMIU)	30 min, (online)	C-City, Surpluse
Sustainability & Resilience Manager – City of Genoa	60 min, (online)	C-City

Within the C-City Genoa Circular City project, the following project actions were analysed as tools to implement the strategy: Circular Hub, Circular Desk, and the Reuse and the Repair Centre Surpluse in Via Bologna.

The general strategy of the C-City project was defined by the Municipality of Genoa and the University of Genoa, specifically with the Department of Economics (DIEC) through 3 macro-processes: eco-design, long use, reuse.

Otherwise, the Blu District(www.genovabluedistrict.com) with Job Centre was identified as the implementing entity .

The aim of the Circular Hub is to increase the awareness and activating circular solutions in different sectors: food, blue economy, textiles, energy, logistics, buildings, materials and products. The Circular Hub project is realised in a specific physical space, within the Blu District site. This allowed the project to be included in an ecosystem already active on sustainability issues, both in terms of active networks and recognition of values.

According to the interviewees, this project was the start of a broader discourse on the cross-sectoral circular economy. This broader discussion made it possible to visualise different scenarios and to discover sectors where concrete action can be taken, such as the food supply chain. In some ways, the Circular Hub's main function is to be a "relationship densifier". (Director Job Centre srl- Genoa Blue District). This is also a space to give a voice to the territory to provide a showcase for realities working on circular economy issues, in fact as a city you can create conditions to facilitate and optimise the supply chain, create awareness, inform about the concept of waste, and bring out the contradictions of the mechanisms of large-scale distribution (Administrative and Reporting Area, Projects Job Centre Srl - Genoa Blue District).

It is therefore necessary to work on territoriality to bring out local realities (Sustainability & Resilience Manager – City of Genoa).

The Circular Desk was imagined as a service to accelerate and facilitate the meeting between supply and demand. A space, also physical, for small and medium-sized enterprises to meet, with the aim of sharing good practices and suggesting new solutions.

For both the Circular Hub and the Circular Desk, in agreement with the interviewees, the correspondence with the ReSOLVE criteria does not concern the concreteness of actions, but the relationship with the topics. In fact, the project foresees the realisation of meetings and information exchange events that have brought out critical issues and the initiation of possible actions. Given the nature of these two projects, it was decided to include them within the "Knowledge development" and "Collaborative platform" strategies. Table 2.

Another key partner of C-city is Amiu (Azienda Multiservizi di Igiene Urbana), indeed the Digital and Ecological Transition Manager of AMIU said that the company wanted to change its cultural approach and vision by giving a new message to citizens.

The paradigm shift involves a new waste management, which considers waste no longer as a waste, but as a valuable resource. This vision contains the project of transforming materials, incentivising reuse, facilitating separate collection processes, etc., which also creates new jobs. To take this vision forward, the first Reuse and Repair Centre Surpluse was opened in 2020 in Coronata, district of Genoa. With the C-city project in 2023, the one in Via Bologna was opened with a collaboration between the municipality of Genoa, Amiu, the DAD and local realities. Surpluse is a network of reuse and repair centres spread over the territory (Fagnoni et al., 2022; Olivastri & Tagliasco, 2024).

In this centre, it is possible to exchange objects, such as books, small furniture components, to give them a new chance. In addition, this centre is designed as a meeting space for workshops or seminars on the circular economy. The aim is also to activate knowledge exchanges on reuse and repair practices. In fact, the centre has been registered as a Repair Café. The Repair Cafè, a place for sharing knowledge and skills for repair and up-cycling, was conceived by Martine Postma in 2007 in the Netherlands (Repair Cafè, n.d.).

As pointed out during the interview with AMIU's press office and social channels manager, citizens have a responsibility to behave in a sustainable manner, for example by sorting waste correctly, but the city and therefore also the company must provide the tools to help people behave appropriately.

In this sense, the communication and awareness-raising part, including social management and flyer distribution, but also the organisation of events and meetings with the local community, plays a fundamental role in involving the community. (Press office and social channels manager, AMIU). Regarding the ReSOLVE criteria, the categories Share, Optimise and Loop were chosen because, as the Reuse and Repair Centre is a space of exchange and reuse, the criteria were representative. Finally, both Knowledge Development and Collaboration Platforms were chosen for the strategies, because it is both an educational and information place, but also a space for real collaboration between different stakeholders in the territory.

Table 2. Initiatives Identified in Genoa

	Re	S	O	L	V	E	
Ge 1 Circular-HUB	x	x	x				Knowledge Development
Ge 2 Circular-Desk	x					x	Collaboration Platforms
Ge 3 Surpluse		x	x	x			Knowledge Development, Collaboration Platforms

4 The Role of Design

As mentioned, Circular Desk was imagined as a service. Service Design tools were used to facilitate co-design among the partners (Job Centre, TICASS, DLTM, Il Rastrello, DAD) and the practice of focus groups with the various stakeholders. Following the Service Design approach, it was decided together with the partnership to submit a questionnaire to small and medium-sized companies in the area. The questionnaire served to identify the positioning of the companies in relation to the circular economy. A mapping of actions already activated and to be implemented, together with needs and solutions, was highlighted. The data collected from the questionnaires were visualised and shared with partners and companies to open a discussion table. In fact, the aim of the Circular

Desk was to identify new solutions through meetings with companies. The other aim of the Desk is to activate networking and provide information on the topic, which was achieved by organising a discussion meeting to verify potential solutions and analyse the needs that emerged with a co-creation approach (Huang & Villari, 2021).

At the end of these initial activities, inputs were provided for the implementation of the website in line with the service infrastructure. The tools of Service Design, whose prerogative is to design new connections, to visualise and represent the dynamics of relationships, can facilitate the realisation of complex processes that have to consider as valuable resources many elements to be interconnected to the environment on which they will act (Morelli, et al. 2021). The system must contain elements of a different nature, such as knowledge, ideologies, institutions, norms, ethics, economic interests and others, which in the interactions they produce make the action possible and situated. From this develops the representation of a user within an ecosystem that reconfigures needs and behaviour compatible with the protection of the environment, which is the first resource of value

To summarise, the principles of circularity can help regenerate tools with a view to greater design complexity, which considers how to activate new connections that act on the life cycle of products, so that they become valuable resources again. The system composed of the factors of the service and the context of reference, acquires value depending on how it interacts with the wider ecosystem. Differently, within the Reuse and Repair Centre, the role of design, in collaboration with Amiu, was to conceive the set-up of the space as a touchpoint of the service itself (De Rosa, 2022; Olivastri & Tagliasco 2024). The different phases were: the choice of furniture and its arrangement in the space, communication according to the guidelines of the Surpluse brand (Fagnoni et al., 2022)

Design is also developing its identity in the conception of events (Richards, et al. 2014). Indeed, the value perception of the relationship between product, service and event is increasing (Vannicola, 2017). This is why design has also been involved in the realisation of small events to start animating the centre. In November and December 2023, moments of sharing on the themes of repair and reuse were organised to give a concrete start to the Repair Café. The activities were designed to start having a draft shared calendar, where all those who are interested, work or invest in the world of reuse, repair and recycling can find space for initiatives (Table 3).

Table 3. Design Activities

	Activity description
Ge 1 Circular-Desk	Service facilitation and co-design activities
Ge 2 Circular-Hub	/
Ge 3 Surpluse	The set-up and communication design, defining event activities for the repair café

5 Future Scenario

The c-city project is moving forward with the "C-City Genoa circular city: Sustainable textiles". The idea is to make "vertical" insights by developing the different sectors identified: Energy, Food, Sea, Mobility, Texile, Building, Material & Product. Even though Genoa is not a textile-producing city, it is important to work on awareness of responsible consumption.

The topic of textiles brings several points to the attention, such as social and environmental justice, even in more immaterial terms (Administrative and Reporting Area, Projects Job Centre Srl - Genoa Blue District).

It is not always possible and sufficient to work concretely on the infrastructure of the supply chain; therefore, it is important to intervene on the ethical and cultural factors linked to these production and disposal systems.

Another possible focus of actions, identified by both the municipality and AMIU, and especially by the Blu District, which is already working on it, is that related to water and the sea, boats and the port area, also using strategies from the blue economy (Pauli, 2015).

Regarding the reuse and repair centre, it is possible to implement the service with a digital space, which not only showcases the objects but also illustrates the skills of possible stakeholders who can animate the space. In this sense, the relationship with the territory and the people who inhabit it becomes fundamental. This is also because the objects in the centre do not only have an economic value but are also full of stories that can be intertwined, told and passed on. Design in this process can contribute by finding new solutions to represent, narrate and visualise the stories also to give new value to the objects. Another step could be to define a sort of 'sustainability passport' to systematise all actions within the city to reward and/or incentivise citizens' actions. Defining a sustainability identity also means providing tools for practice and awareness.

Finally, a possible future scenario proposed by the Sustainability & Resilience Manager (City of Genoa) is to distribute and decentralise small systems and activities in the territory, as is already being done for the Reuse and Repair Centres Surpluse (Fagnoni et al., 2022), possibly characterising and specialising neighbourhoods differently.

6 Circular Conclusion

This project, as emphasised by the interviewees, is a first step to trigger various actions in terms of the circular economy within the city. As pointed out in the text, the city has the power to act on some but not all aspects.

For example, the limitation may concern certain rules, regulations or the management of complex supply chains, but on the other hand it has an urban community on which to act in terms of awareness and manageable policies in the city context. The fragmented and generalist vision that characterised the start of the process, on the one hand, may have slowed down the implementation of immediate concrete policies, on the other hand, it allowed a vision of supply chains and areas of intervention that otherwise would not have emerged. The design contributed by reinforcing the identity of the project, co-designing a coherent infrastructure in line with the principles of circularity. The theoretical framework is certainly implementable, the theoretical approach (Prendeville et al, 2018) followed allowed the case study of genoa to be compared with other realities.

As the Sustainability & Resilience Manager (City of Genoa) said, the 'Genoese are circular by nature, they don't throw anything away', so it is necessary for citizens to communicate and share their sustainability choices, and these projects can provide the tools to share this type of action in an innovative way.

References

Barth, B., et al.: Urban climate action. the urban content of the NDCs: global review 2022, pp. 1–59 (2023)

C-CITY: Il progetto C-City (2023). https://c-citygenovacircolare.eu

Comune di Genova: Genoa 2050. Action Plan for a Lighthouse City (2021). https://www.investingenova.com/sites/default/files/archivio/download/Action-Plan-Def-digitale.pdf

De Rosa, A.: S+S – Spatial Design + Service Design. Maggioli (2022)

Ellen MacArthur Foundation: Delivering the circular economy: a toolkit for policymakers (2015). https://www.ellenmacarthurfoundation.org/publications

Ellen MacArthur Foundation: Towards a circular economy: business rationale for an accelerated transition (2015). https://www.ellenmacarthurfoundation.org/towards-a-circular-economy-business-rationale-for-an-accelerated-transition

Ellen MacArthur Foundation: Cities in the circular economy: an initial exploration (2017). https://www.ellenmacarthurfoundation.org/cities-in-the-circular-economy-an-initial-exploration

Ellen MacArthur Foundation (n.d.): Circular cities: thriving, liveable, resilient. https://www.ellenmacarthurfoundation.org/topics/cities/overview

European Commission: Circular economy action plan – for a cleaner and more competitive Europe. Brussels (2020)

Fagnoni, R., Ferrari Tumay, X., Olivastri, C.: Surpluse – Centri del riuso e del riparo. In: Ferrara, C., Germak, C., Imbesi, L., Trapani, V. (eds.) Design per Connettere | Atti Assemblea Annuale SID, 25–26 February 2021, Palermo, pp. 714–727 (2022)

Huang, L.-T., Villari, B.: Co-creation in circular cities: a design perspective. DISCERN Int. J. Des. Soc. Change Sustain. Innov. Entrepreneurship **2**(1), 69–88 (2021)

ICLEI Europe: Circular Cities Declaration Report 2024. Insights on Implementation, Measurement, and Nature (2024). https://circularcitiesdeclaration.eu/fileadmin/user_upload/Resources/CCD-Report-2024.pdf

Manzini, E.: Design, When Everybody Designs: An Introduction to Design for Social Innovation. MIT Press (2015)

Manzini, E.: Politiche del quotidiano: progetti di vita che cambiano il mondo. Edizioni di comunità (2018)

Manzini, E.: Abitare la prossimità. Idee per la città dei 15 minuti. Egea (2021)

Morelli, N., De Götzen, A., Simeone, L.: Service Design Capabilities. Springer Nature (2021)

Moreno, C.: La città dei 15 minuti. Per una cultura urbana democratica. Add Editore (2024)

Olivastri, C., Tagliasco, G.: Servizi per il riuso e il riparo: L'allestimento tra touchpoints e infrastrutture relazionali. Agathon Int. J. Archit. Art Des., 15 (2024)

Pauli, G.: Blue economy 2.0: 200 progetti implementati, 4 miliardi di dollari investiti, 3 milioni di nuovi posti di lavoro creati. Edizione Ambiente (2015)

Prendeville, S., Cherim, E., Bocken, N.: Circular cities: mapping six cities in transition. Environ. Innov. Soc. Trans. **26**, 171–194 (2018)

Repair Cafè (n.d). What is Repair Cafè. https://www.repaircafe.org/en/about/

Richards, G., Marques, L., Mein, K.: Introduction: designing events, events as a design strategy. In: Event design, pp. 1–13. Routledge (2014)

Vannicola, C.: La prosev strategy. Il design del prodotto servizio evento. Forma (2017)

Co-design as an Enabling Method in Systemic Design Methodology for Circular and Sustainable Ecosystem

Martina Spinelli(✉), Davide Montaquila, Davide Nonis, and Amina Pereno

Politecnico di Torino, Torino, Italy

Abstract. This contribution explores the potential role of co-design as an enabling method in systemic methodology to promote circular ecosystems within the circular economy. By analysing two ongoing projects, we reflect on the integration of co-design into systemic design to develop sustainable solutions that effectively address contemporary environmental challenges. Co-design actively involves various stakeholders in the design process, facilitating relationships and creating resilient systems that optimize resource use and reduce waste. The systemic methodology, with its holistic approach, enables the analysis and understanding of the interconnections between different elements of an ecosystem, promoting synergies between economy, society, and environment. The first project focuses on using co-design to connect the diverse professional backgrounds within the working group. This approach aims to develop a training program tailored to a varied user base with different backgrounds and needs. By involving stakeholders such as universities of design, health professions and management in the co design process, the project ensures that the training materials are relevant, comprehensive, and accessible to all participants which arrive from different academic backgrounds. The second project utilizes co-design to facilitate industrial symbiosis and promote circular economy practices. Through co-design, different industries collaborate to identify opportunities for resource sharing and waste minimization. By engaging stakeholders from various sectors, the project creates a network of businesses that can exchange materials, energy, and by-products, thus closing the loop and reducing environmental impact. This industrial symbiosis exemplifies how co-design can drive circular economy initiatives by fostering collaboration and innovation across industries.

Keywords: Systemic Design · Co-design · Circular Economy · Circular Eco-System · Paticipatory Action Research

1 Introduction

The circular economy (CE) has emerged as a pivotal framework for advancing sustainability across multiple industrial sectors, garnering significant support from the European Union (Ghisellini, Cialani e Ulgiati, 2016). At its core, the CE paradigm emphasizes resource efficiency, waste reduction, and sustainable development through closed-loop

A. Mateus et al. (Eds.): WAD 2024, What's Around Design?, pp. 114–126, 2026.
https://doi.org/10.1007/978-3-032-08163-6_9

systems (Nogueira A. et al., 2020). Establishing effective circular ecosystems requires engaging multiple stakeholders working synergistically towards shared sustainability goals.

This research explores the impact of co-design and co-creation on the specified process, particularly emphasizing the roles and interactions of various stakeholders in driving change. It examines two ongoing projects: one on sustainable healthcare systems and the other on digital industrial symbiosis, both utilizing systemic design methodology. The aim of this paper is to offer an empirical reflection on the role of co-design and co-creation in the specified framework.

This study builds on the foundational work of numerous scholars to gain a deeper understanding of the practices and theories that have established co-creation as a key mode of participatory engagement in design. Over the past decade, the promise of participatory design has become embedded in mainstream practice as design co-creation (Jones, 2018). Jones also highlights the evolution of co-creation over the past 50 years. Originally rooted in social systems and democratic practice theory, co-creation has evolved into a standard mode of participatory engagement in design, encompassing ideation, creative problem-solving, and decision making.

The case studies presented are based on a systemic design methodology, which is characterised by its holistic approach, provides a framework for analysing and understanding the complex interconnections between different elements of an ecosystem (Battistoni, Giraldo Nohra, & Barbero 2019). This methodology promotes synergies between economy, society, and environment by recognizing the interdependencies and potential feedback loops within a system. In the context of the circular economy, systemic design aims to create closed-loop systems where waste is minimized, and resources are continuously cycled back into the production process (Pereno, Amina & Barbero, Silvia, 2020). This approach aligns with the principles of sustainability, aiming to reduce environmental impact while promoting economic viability and social well-being in the territory.

Systemic design is evolving and incorporating more and more co-design approaches, but it is still an emerging and developing field. As Jones (2018) observes, there is a notable absence of comprehensive studies and a paucity of literature that specifically addresses co-creation within the context of systemic design. Nevertheless, several tools have been developed to facilitate co-design dynamics (De Kooning et al., 2016) within the systemic design methodology, including the systemic design toolkit (Systemic Design Toolkit, 2021). These tools are pivotal in advancing the co-design approach, supporting the creation of sustainable systems through collaborative practices.

The potential of co-design for fostering collaborative ecosystems is considerable. When we talk about circular ecosystems (Felicitas Pietrulla, 2022), we mean an interconnected network of actors who collaborate to promote resource efficiency, waste reduction, and sustainability practices. In both the cases we analyse, this involves designing systems and processes that prioritize the reuse and recycling of materials, reduce environmental impacts, and enhance overall efficiency. As Pedersen and Clausen (2019) observe, this approach combines design, value chain considerations, and circular economy mindsets to inform negotiations and address concerns among actors in the value chain. The integration of co-design into circular ecosystems facilitates the collaborative and participatory engagement of diverse stakeholders, thereby enhancing the comprehensive, context-specific, and widely accepted solutions that are the hallmark of co design. These solutions address the complex challenges of sustainability, aligning with the goals of circular ecosystems, which are driven by collective expertise and perspectives that drive innovation and ecological transitions.

The key actors in the circular economy ecosystem include academia, industry, policy-makers, and communities. Academia comprises researchers and educators who provide the fundamental knowledge base and training that is necessary for the development and implementation of CE principles. Industry encompasses a diverse range of entities, including healthcare providers, medical product manufacturers, and technology companies that apply CE practices in real-world settings. Policymakers, including government and regulatory bodies, are responsible for creating the policies and regulations that support CE initiatives. Last but not least, communities, consisting of citizens, benefit from and actively participate in sustainable practices.

In conclusion, the foundation of the concepts discussed thus far will be illustrated through the presentation of two case studies. These studies will elucidate the dynamics of co-design application, outlining the objectives and outcomes for each case. Despite their distinct aspirations, backgrounds, goals, and purposes, both cases leverage co-design as a facilitator and incubator for innovative strategies. They share the overarching aim of developing virtuous systems that advance towards increasingly circular and sustainable production practices. Both cases involve a high number of stakeholders or diverse figures, highlighting the importance of co-design.

2 Case Studies

The methodology framework that establishes common ground for the two proposed case studies is based on the systemic design methodology pursued at Politecnico di Torino. This approach comprises four principal phases, which provide guidance on the process of developing a new system.

The initial phase, entitled 'Understanding Complexity', represents the point of departure for the process, with a particular focus on the collection and analysis of data, as well as the examination of the intricate relationships that exist within the system (Battistoni, C., Giraldo Nohra, C., & Barbero, S., 2019). The second phase, entitled "Tackling Challenges," entails the identification and resolution of pivotal issues that emerge from the analysis of the system in question. The third phase is concerned with the design of the new system. This involves proposing integration with the existing system, as well as a future direction and vision based on the insights gained during the preceding phases. The final phase, evaluating the proposed system, assesses the system's viability and effectiveness.

The utilisation of co-design is of paramount importance throughout the various stages of this methodology, as it serves to facilitate interaction and collaboration between stakeholders. This participatory approach is particularly relevant when considering the Quadruple Helix Model, which addresses the complexity of relationships among government, civil society, academic research, and business. The inclusion of society in these processes has become increasingly important due to the systemic challenges facing contemporary times and the necessity for a more holistic approach to address them. (Sevaldson, B. & Jones, P. 2020)

The first case study demonstrates the use of co-design to promote a design process that integrates heterogeneous backgrounds. In contrast, the second case study presents a circular ecosystem where the establishment of a common language is crucial to its creation, an objective that could be effectively achieved through co-design. Although co-design has not yet been implemented in the second case, the diversity of the entities involved has underlined the need for a common language, which co-design could facilitate. These examples underline the crucial role of co-design in creating resilient, adaptable and sustainable ecosystems within the circular economy.

2.1 DesHealth

DesHealth, an Erasmus+ project, aims to create training programs that teach management and sustainability practices to health professionals, students, and industry practitioners. The goal is to raise awareness and enable informed decision-making to drive targeted changes where needed.

The training is designed for those involved in the design of medical equipment or services, for those whose task it is to manage resources or waste flow, and for those whose task it is to decide on the quantities and qualities of hospital supplies to be purchased, such as medical managers responsible for placing orders. The project targets both current and future professionals, such as students; and aims to enhance their awareness in order to facilitate more precise decision-making and targeted modifications where necessary. Additionally, medical practitioners may derive benefit from exerting influence over the provision of services or participating in the redesign of existing services. One example of this is the rethinking of glove dispensers developed by TuDelft design project (Van den Berg L., 2022; Van den Berg L., Albayrak A., Hunfeld N., & Diehl J. C., 2024).

In order to achieve this, it was necessary to establish a collaborative partnership between professionals from a range of disciplines, including the health sector, the business and management sector, and the design sector. In order to develop an optimal training design, the research team for the DesHealth project was constituted with a diverse range of European partners, representing different specialisms and expertise. The project partners include universities specialising in systemic design, with some focusing on services and impacts and others on sustainable product design. The university partners include other academic institutions, some of which specialise in medicine, while others focus on economics. In addition to these university partners, the consortium includes corporate entities such as a research and development company for ecological transition and a consortium of health realities. It is evident that collaboration between partners of diverse natures is fundamental to the project itself.

It is evident that the diverse specialisations of the partners are important in order to gain an insight into the specific requirements of the target groups that the project is seeking to address. These requirements are not straightforward and emerge from the various fields in which the partners operate, health design and management, and the health professions themselves. During the design phase, the project team benefited from a heterogeneous composition, which proved advantageous in addressing the diverse needs of the target users. To achieve this, the team adopted a multi-faceted approach that included not only the identification of user requirements, but also the development of effective communication strategies tailored to the specific drivers perceived by each user.

In order to integrate the disparate knowledge bases into a unified design for the training programme, a workshop was organized among the partners of the Deshealth project; the objective was facilitating a collaborative design process. At the time of this activity, an extensive study had already been conducted on the potential of systemic product and service design and management in the context of sustainable healthcare. However, this was not the only area of investigation. Prior to this activity, a questionnaire was distributed to partners who were in close relation with the target groups, namely students and professionals in the afore mentioned fields.

The first activity of the workshop was structured around the assignment of key area tackles, as determined by a previously conducted questionnaire (e.g., management/sustainability strategy, waste management, sustainable procurement), to the various groups that had been formed among the participants for the occasion. Each group employed a supporting board to identify the most pertinent topics and to deliberate upon the requisite knowledge and abilities necessary to excel in that field. Furthermore, visual cards illustrating sustainable healthcare best practices were distributed to all groups, showcasing positive actions within the relevant field, with the intention of inspiring the selection of optimal skills. The concluding outcomes of this activity elucidated the indispensable skills and knowledge domains pertaining to sustainable healthcare. The key areas identified were management and sustainability strategies, energy efficiency, green public procurement, digitalisation, and waste management. The specific skills deemed essential were multi-horizon vision, carbon footprint measurement, and life cycle assessment. The importance of integrating digital tools and policies, fostering synergy among hospital units, and adopting life cycle design principles to minimise environmental impact from product development to disposal was emphasised. Overall, the activity underscored the importance of comprehensive and practical training for healthcare professionals. (Fig. 1)

The second activity involved the partners in the creation of four personas, with the objective of facilitating a more precise identification of the potential users of the platform. The participants utilised a simplified board to define personas, a design technique that represents different user types. This approach facilitated a deeper comprehension of the learners' needs, experiences, behaviours, and goals. The groups were divided according to their respective levels of familiarity with the target groups in order to facilitate the creation of reference personas. Three groups worked on personas from industry, hospitals, and higher education. These included a young health industry professional, a young hospital professional, a management student, and a design student. The last group, which worked with the higher education field, created two personas representing different typologies of students. Each persona included sections on sustainability goals, constraints to achieve these goals, and the necessary knowledge to pursue them (Fig. 2).

The third activity constituted a further development of the preceding two interactive sessions and was designed to define the modules of the training programme. A board was provided on which only the title and learning outcomes of each module were specified. The participants were permitted to engage in discussion, modification, and development of the outcomes and content of the modules using handwritten notes or Post-it notes.

Fig. 1. Through identifying good practices (case study cards), partners define the skills to be included in the training programme.

This approach permitted the input of collaborative and creative ideas, thereby ensuring that the training programme was comprehensive and tailored to the needs identified in previous activities. (Figure 3)

However, the planning process, up to the final form, was characterised by several challenges, including changes of direction and the introduction of new configurations of modules that were initially considered to be definitive. This was due to the fact that each participant assigned different levels of importance and a different hierarchical structure to the content related to their subject area. Following an extensive discussion that spanned several hours, the working group employed additional tools, including concept maps and timelines of the entire training programme, in order to arrive at a solution that encompassed the essential elements of all aspects under consideration. It became evident that, in order to develop an exhaustive and multidisciplinary training programme, it would not be feasible to examine every detail in great depth. Instead, a meticulous curation process was identified as being essential, with co-design playing a pivotal role in this endeavour. The objective was to guarantee a rigorous selection process, one that was mindful of the diverse needs of all stakeholders, the green skills required in this field and the pedagogical approaches that the various partners would have to adopt.

In conclusion, the case study demonstrates the value of co-design as a means of achieving a project that can meet the diverse objectives and consider the varying professional profiles of the participants. Once the project is fully implemented, this could result in a virtuous cycle of knowledge, enabling the target audience to gain a more

Fig. 2. Definition of the personas to whom the training course is directed.

comprehensive understanding of healthcare. This, in turn, could prompt professionals to move beyond their current compartmentalised approach to management and design.

2.2 Nodes

The project, entitled "NODES: Digital and Sustainable North-Western Italy", was developed under the National Recovery and Resilience Plan (PNRR) with the objective of fostering interdisciplinary collaboration and sustainable innovation within a localised geographical area. This initiative functions as an innovation ecosystem, wherein disparate research and industrial domains converge, driven by a collective vision to advance ecological transition practices and solutions. The project's objective is to synthesise the expertise of various actors into a comprehensive service for the circular economy and industrial symbiosis in the northwestern Italian region, which has the potential to make a significant impact in the fields of industrial symbiosis and circular economy.

NODES addresses the urgent environmental and societal challenges that require interdisciplinary collaboration (Ceschin, F., & Gaziulusoy, I. 2019), a principle that is wholeheartedly embraced by the project. The objective of NODES is to establish

a dynamic ecosystem for the exchange of knowledge and the emergence of innovative solutions, which will be achieved by bringing together experts from academia, industry and research centres.

A principal objective of NODES is to facilitate the convergence of research efforts among all participants onto a centralized digital platform. Although the objective may

Fig. 3. Partners collaboratively define the training programme, including circular strategies, necessary skills and different disciplinary sectors.

appear straightforward, the aggregation of research and expertise into a readily accessible and useful digital platform represents a significant challenge, given the inherent complexities involved in the process. The management of the diverse contents and knowledge from all partners necessitates the utilisation of tools that facilitate the various stages of the process towards the attainment of this final objective. The objective of this platform is to facilitate connection and collaboration among expertsfrom diverse disciplinary backgrounds. The integration of disparate research outputs into a unified portal is intended to enhance accessibility and foster a unified approach to the study of the circular economy and industrial symbiosis.

Nevertheless, the process has encountered certain obstacles, particularly in the domains of codesign and co-planning. Notwithstanding the considerable efforts that have been made, these collaborative approaches have not yet yielded the desired results. One of the challenges has been the unequal interaction between experts and the restricted exchange of information. In order to enhance the efficacy of NODES, it is necessary to implement strategies that facilitate more ejective communication and collaboration across disciplinary boundaries. Potential strategies for achieving this include the introduction of more robust facilitation techniques and the utilisation of digital tools designed to enhance collaborative workflows. By refining and rethinking the dynamics of these processes, NODES can more ejectively harness the collective expertise of its participants, thereby advancing its mission of sustainable innovation.

In order to achieve a comprehensive outcome that genuinely addresses the challenges at the core of the NODES project, it is essential that the contributions of each participant are aligned and synergistically integrated with those of others. In order to achieve this objective, it is desirable that relationships are not merely consolidated retrospectively, but rather emerge as inherent properties of the system itself (Capra, F. & Luisi, P. L., 2014). This necessitates the collaborative development of results through a shared endeavour, particularly the process of co creation. In the future, the project will focus on the development of a comprehensive, horizontally integrated digital platform for knowledge sharing across all participating departments. These improvements will facilitate not only the efficient exchange of research findings and proposals, but also reinforce the project's capacity to model the innovative digital ecosystem it aims to promote. By addressing these areas for improvement, the NODES project is well-positioned to make significant strides in aligning its technological infrastructure with the collaborative and interdisciplinary demands of complex sustainability projects. This approach will enable NODES to more ejectively utilise digital tools to facilitate the co-creation of knowledge and foster genuine interdisciplinary synergies.

3 Discussion

Co-design has the potential to significantly improve collaboration within circular economy ecosystems by fostering participatory engagement and integrating diverse perspectives. However, the success of co-design activities can vary considerably depending on the context and stakeholders involved. The two case studies we proposed, present a contrasting scenarios: a successful one, exemplified by the DesHealth project, and a more challenging one, represented by the NODES project.

The DesHealth project offers an illustrative example of the efficacy of co-design activities in a multidisciplinary context. In this project, stakeholders, including healthcare providers, medical product manufacturers, and technology companies, exhibited the capacity to align with the design practices. The partners involved were already conversant with the terminology, principles and associated methodologies of design. This common ground facilitated more ejective communication and collaboration, as the stakeholders were unified in their objective of developing sustainable healthcare systems.

The success of co-design in DesHealth can be attributed to a number of factors. Primarily, the stakeholders manage to reach an understanding of design principles, which facilitated engagement in co-design activities. Secondly, the focus on sustainable healthcare systems provided a clear and unifying objective that all partners could collectively pursue. Lastly, the utilisation of established co-design tools and methodologies ensured a systematic and inclusive process.

In contrast, the NODES project illustrates a scenario in which the implementation of co-design has proven to be more challenging. The NODES project brings together a diverse group of stakeholders from academic backgrounds, with the shared objective of promoting sustainable innovation and a local circular economy.

However, the project's multidisciplinary nature has made it more challenging to identify a shared vocabulary and achieve ejective co-design.

The NODES project has encountered several challenges. The stakeholders from a multitude of disciplines, utilising disparate terminologies, methodologies and priorities, which collectively impede the establishment of a unified understanding. The absence of a common language has impeded ejective communication and collaboration, resulting in misunderstandings and misalignments. Moreover, the comprehensive and ambitious objectives of promoting sustainable innovation across numerous sectors have rendered it challenging to identify precise and feasible goals for co-design activities.

Despite these challenges, the NODES project is actively seeking solutions to overcome these difficulties. One proposed approach is the development and implementation of specialised co-design tools tailored to the project's needs. These tools will facilitate improved communication and collaboration among the diverse stakeholders, thereby helping to bridge the gap between different disciplines and foster a more integrated approach to co-design.

The contrasting scenarios of DesHealth and NODES illustrate the varying degrees of success that can be achieved in co-design activities within circular economy ecosystems. The success of DesHealth demonstrates the significance of shared understanding, common goals, and ejective collaboration tools. In contrast, the challenges faced by NODES illustrate the complexities of multidisciplinary collaboration and the need for specialised tools to facilitate co-design.

4 Conclusions

This study aimed to examine the potential contribution of co-design within the systemic design methodology, emphasising its capacity to facilitate reflection and multifaceted design processes. Given the numerous tools developed to facilitate co-design processes within systemic design (De Kooning et al., 2016), including the Systemic Design Toolkit (Systemic Design Toolkit, 2021), the authors saw these ampirical cases as a opportunity to further explore a field that remains under researched and lacks sufficient literature, as noted by Jones (2018).

In the context of the evolving landscape of co-design approaches within the systemic design field, the case studies presented in this paper employ this methodology as a tool to optimise the design outcome of an interdisciplinary team, while maintaining the integrity and multidisciplinary nature of the project. As previously stated, Jones (2018) emphasises the scarcity of research and literature on co-creation in systemic design. This paper, in conjunction with the NODES and Deshealth case studies, serves to enhance awareness of the current state of the art.

The Nodes project identified two significant challenges associated with co-design approaches. Firstly, the considerable variety of contexts in which co-design is implemented makes it unfeasible to standardise practices and tools; instead, these must be tailored to each specific environment. Secondly, the hierarchical structure of participants presents a challenge, as the ideal ecosystem of actors should be as horizontally aligned as possible, even if this is only a temporary state during the activity. It is of the utmost importance to refrain from any form of bias or influence, whether social or hierarchical, and to ensure that power is distributed equally among participants.

In the Nodes project, it proved challenging to achieve valid outcomes due to an imbalance among the participants. The vertical organisational structure and the division of roles constrained decision-making to a single perspective. Consequently, the results were homogeneous and predetermined, despite the involvement of a diverse range of professionals. This emphasises the necessity for collaborative endeavours to not only encompass a diverse range of professional profiles but also to address any hierarchical imbalances that may exist within the group.

Furthermore, the Nodes project demonstrates the challenges that emerge when co-design is employed in more complex settings, characterised by interdisciplinary diversity and a lack of a shared language. These issues illustrate the vital importance of contextual factors in the success of co-design initiatives. Notwithstanding these challenges, the Nodes project persists in its efforts to surmount them through the development of bespoke co-design instruments. The objective of these tools is to facilitate enhanced communication and collaboration among stakeholders from disparate fields, thereby advancing the project's overarching objective of promoting sustainable innovation within the circular economy.

In the Deshealth project, a horizontal collaboration was established from the outset, as some of the participants had previously worked together, and all shared a common background in healthcare, despite representing diverse professional expertise.

And so this project provides an illustrative case study of how co-design, when implemented in a well-defined context with stakeholders who are already attuned to design principles, can facilitate the successful development of sustainable healthcare systems. The project demonstrated the efficacy of co-design in fostering interdisciplinary collaboration, resulting in the development of a novel multidisciplinary training programme that is both innovative and responsive to the evolving needs of future healthcare professionals. The alignment of objectives and the existence of a shared understanding among stakeholders facilitated a seamless collaboration, thereby demonstrating the effectiveness of co-design in achieving tangible, sustainable outcomes.

In the future, the advancement of systemic design and sustainability will be contingent upon our capacity to integrate these collaborative practices in a coherent and structured manner. The successful implementation of co-design within circular ecosystems will necessitate the ongoing refinement of methodologies and the development of context-specific tools that can bridge disciplinary divides and foster ejective communication. Such an approach will enhance the capacity of co-design to address global sustainability challenges, ensuring that circular economy initiatives not only succeed but also contribute to the broader goal of creating a more sustainable and resilient world.

In conclusion, while co-design has emerged as a pivotal methodology for advancing circular ecosystems and circular economy initiatives, its success is contingent upon the careful consideration of context and stakeholder readiness. The experiences of DesHealth and NODES illustrate the necessity of developing flexible co-design strategies capable of navigating the intricacies of multidisciplinary collaboration. As these methodologies are further refined, it is evident that co-design will assume an increasingly pivotal role in the evolution of sustainable innovation and systemic design, ultimately propelling the global transition towards a more sustainable and regenerative future.

References

Battistoni, C., Barbero, S.: Systemic design for territorial development: ecosystem to support autopoietic local economies. In: Proceedings of RSD7, Relating Systems Thinking and Design, vol. 7 (2018). http://openresearch.ocadu.ca/id/eprint/2710/

Battistoni, C., Giraldo Nohra, C., Barbero, S.: A systemic design method to approach future complex scenarios and research towards sustainability: a holistic diagnosis tool. Sustainability **11**(16), 4458 (2019). https://doi.org/10.3390/su11164458

Blomsma, F., Pigosso, D.C., McAloone, T.C.: A theoretical foundation for developing a prescriptive method for the co-design of circular economy value chains. Proc. Int. Conf. Eng. Des. **1**(1), 3141–3150 (2019). https://doi.org/10.1017/dsi.2019.321

Capra, F., Luisi, P.L.: The Systems View of Life: A Unifying Vision. Cambridge University Press (2014)

Ceschin, F., Gaziulusoy, I.: Evolution of design for sustainability: from product design to design for system innovations and transitions. Des. Stud. **47**, 118–163 (2016). https://doi.org/10.1016/j.destud.2016.09.002

Ceschin, F., Gaziulusoy, I.: Design for Sustainability: A Multi-level Framework from Products to Socio-technical Systems. Routledge (2019)

Chick, A.E.: Design for social innovation: emerging principles and approaches (2012)

De Koning, J., Crul, M., Wever, R.: Models of co-creation (2016)

De Salvo, V., Carraro, M., Bianchini, M., Majei, S.: Validating collaborative practices for urban circular economy: the design perspective. Techne, 122–130 (2021). https://doi.org/10.36253/techne-10580

Dokter, G., Boks, C., Rahe, U., Jansen, B.W., Hagejärd, S., Thuvander, L.: The role of prototyping and co-creation in circular economy-oriented innovation: a longitudinal case study in the kitchen industry. Sustain. Prod. Consumption **39**, 230–243 (2023). https://doi.org/10.1016/j.spc.2023.05.012

Jones, P.: Contexts of co-creation: designing with system stakeholders. In: Jones, P., Kijima, K. (eds.) Systemic Design. Translational Systems Sciences, vol. 8, pp. 3–52. Springer, Tokyo (2018). https://doi.org/10.1007/978-4-431-55639-8_1

Kjøde, S.G.: Navigating designerly systemic approaches for sustainability transitions. In: Proceedings of Relating Systems Thinking and Design, RSD11 (2022)

Nogueira, A., Ashton, W., Teixeira, C., Lyon, E., Pereira, J.: Infrastructuring the circular economy. Energies **13**(7), 1805 (2020). https://doi.org/10.3390/en13071805

Pedersen, S., Clausen, C.: Staging co-design for a circular economy. Proc. Int. Conf. Eng. Des. **1**(1), 3371–3380 (2019). https://doi.org/10.1017/dsi.2019.344

Pereno, A., Barbero, S.: Systemic design for territorial enhancement: an overview on design tools supporting socio-technical system innovation. Strateg. Des. Res. J. **13**, 113–136 (2020). https://doi.org/10.4013/sdrj.2020.132.02

Pietrulla, F.: Circular ecosystems: a review. Cleaner Circular Bioeconomy **3**, 100031 (2022). https://doi.org/10.1016/j.clcb.2022.100031

Pirinen, A.: The barriers and enablers of co-design for services. Int. J. Des. **10**(3), 27–42 (2016)

Sanders, E.B.-N., Stappers, P.J.: Co-creation and the new landscapes of design. CoDesign **4**(1), 5–18 (2008). https://doi.org/10.1080/15710880701875068

Systemic Design Toolkit: Seven steps to tackle complex issues. This is our methodology (2021). https://www.systemicdesigntoolkit.org/methodology

Van den Berg, L.: Reducing the environmental impact of gloves used in the intensive care unit: towards greener ICUs. Master's Thesis. Delft University of Technology (2022)

van den Berg, L., Albayrak, A., Hunfeld, N., Diehl, J.C.: Towards greener ICUs: redesigning the use of disposable gloves. In: Melles, M., Albayrak, A., Goossens, R.H. (eds.) Convergence:

Breaking Down Barriers Between Disciplines. HEPS 2022. Springer Series in Design and Innovation , vol. 30, pp. 213–223. Springer, Cham (2024). https://doi.org/10.1007/978-3-031-32198-6_20

Van der Bijl-Brouwer, M., Malcolm, B.: Systemic design principles in social innovation. She Ji J. Des. Econ. Innov. **6**(3), 350–371 (2020). https://doi.org/10.1016/j.sheji.2020.06.001

Vink, J., Hollingshead, M.: Now and then: co-designing systems smart enough for the future (2014)

Sevaldson, B.: What is systemic design? Practices beyond analyses and modelling. In: Relating Systems Thinking and Design (RSD8) 2019 Symposium, Chicago, USA, 13–15 October 2019. http://openresearch.ocadu.ca/id/eprint/3233

Battistoni, C., Nohra, C.G., Barbero, S.: A systemic design method to approach future complex scenarios and research towards sustainability: a holistic diagnosis tool. Sustainability **11**(16), 4458 (2019). https://doi.org/10.3390/su11164458

Sevaldson, B., Jones, P.: An Interdiscipline emerges: pathways to systemic design. She Ji J. Des. Econ. Innov. **5**, 75–84 (2020). https://doi.org/10.1016/j.sheji.2019.05.002

Aguirre-Ulloa, M., Paulsen, A.: Co-designing with relationships in mind. FormAkademisk **10**(1) (2017). https://doi.org/10.7577/formakademisk.1608

Ghisellini, P., Cialani, C., Ulgiati, S.: A review on circular economy: the expected transition to a balanced interplay of environmental and economic systems. J. Cleaner Prod. **114**, 11–32 (2016). https://doi.org/10.1016/j.jclepro.2015.09.007

In Search of Wool(s): Rethinking the Future of Materials and Craft Communities

Bilge Merve Aktaş(✉)

University of Miami, Coral Gables, USA
bilgemerveaktas@gmail.com

Abstract. Sustainable and circular production has been highlighting the importance of material extraction, processing, and disposal. In order to reduce the negative environmental and social impact of design products, relying on regenerative resources that are biodegradable and non-hazardous have become important assets. Developing novel materials with such properties is one of the strategies for minimizing material-based impact. This paper explores possibilities of reviving existing materials in industrial and local production for bettering the conditions. The paper introduces a project titled "In Search of Wool(s)" to identify the challenges of growing and using local wool, specifically in North Carolina, USA. The project examines the potential of existing natural materials to be reintroduced as sustainable solutions for overcoming environmental problems that stem from material generation and disposal. By building on the findings developed from interviews and a workshop with sheep farmers, artisans, and educators, the paper proposes developing and strengthening material communities as a method of reviving the use of existing natural materials. Findings show that expanding people and things involved in material communities is a crucial first step to bringing innovative thinking back to local practices. These initiatives can also help small communities develop in socially sustainable ways.

Keywords: Wool · Material Communities · Sustainable Crafts · Sustainable Development · Design Research

1 Introduction

Recent decades have witnessed an increase in research and development projects in emerging materials. Such initiatives have been exploring the potentials of organic matter and turning them into alternatives to fossil-based, nonbiodegradable, and, at times, poisonous materials, such as plastic. Novel material examples include fungi-based mycelium for the design of everyday products (Angelova, et al., 2021; Muiruri, et al., 2023), kombucha tea by-product for obtaining material films (Ramirez, et al., 2020), or other biomaterials based in for example agricultural waste (Maraveas, 2020; Donner, et al., 2021). These initiatives are promising to lower the overall environmental impact of materials by changing the models of raw material extraction, waste management, and product longevity. Thus, following a circular model that minimizes the environmental damage created by material-related systems.

A. Mateus et al. (Eds.): WAD 2024, What's Around Design?, pp. 127–141, 2026.
https://doi.org/10.1007/978-3-032-08163-6_10

Though these innovations are impactful and needed, a brief study shows that local and natural materials can also achieve some of the promises after a revision of existing systems. Properties like biodegradability, regenerative capacities, and scalability are true for some of the traditional materials, such as wool. Wool is a protein-based animal fibre that has traditionally been used for making textiles, insulation equipment, and farming fertilizers. Recent research has also focused on its properties of sound and heat insulation, absorbing spills of oil and dyes, and light upholstery for the transportation industry, as well as its substances, such as lanolin, for pharmacotherapy cosmetics and medical treatment (for a review, see Allafi, et al. 2020). Engineering projects focus on wool as a bio-composite for insulation and heat absorbance in the construction industry (Denes, et al., 2022). Wool's biofunctionality has also been explored as a method of tissue engineering and wound healing (Ranjit, et al., 2022). Despite these initiatives and such valuable capacities, its general use has still been in decline in recent decades. Several projects investigating its reintroduction have yet to reverse this trend.

This paper explores possibilities of reviving local communities and systems around wool as a material in order to re-propose it as a socially, environmentally, and economically profitable material within the design field. Studying existing materials for unknown futures is relevant and needed for a few reasons; firstly, these materials already have existing human and nonhuman communities, and their current and future status influences the well-being of those involved. Secondly, these communities also embody deep knowledge about ways of developing, working with, and learning from these materials. This knowledge can be transferred to changing needs while being translated into developing other similar materials. Lastly, the so-called traditional materials can already fulfil some of the current needs, and some of their properties can be enhanced or advanced by research and development projects to fulfil others.

Building on these ideas, this paper presents a residency project that took place in the summer of 2023 in Asheville, North Carolina, USA. The project, titled "In Search of Wool(s)", explored the current situation of wool as a craft material and the existing obstacles that lead to its decline. By studying these issues, the paper identifies key issues that creates challenges.

Next, the paper will briefly introduce wool's current situation in the world. By doing this empirical presentation, the section will introduce its material connections with systems, communities, and the future. Then, the "In Search of Wool(s)" project and study setting will be presented. The final section will share findings and suggestions based on the learnings from the five-week long project.

2 A Look into Material Connections Through the Lens of Wool

Crafting and interacting with materials are some of the most humane acts. Materials and humans co-create each other (Ingold, 2013; Malafouris, 2013), becoming part of the same ecology (Deleuze & Guattari, 1980/2019). Materials, or matter, never exist alone. They are always in connection to many other things that can be humans, nonhumans, or both; everchanging.

Philosophers Deleuze and Guattari (1980/2019) discuss the activeness of matter as "being in flux" that can only be followed by the artisans. Ingold (2012, p. 434) also refers to the same dynamism and suggests that material properties should not be seen as attributes, but rather as their histories that can show how it has been changing. In a similar line of thinking, Jane Bennett (2010) argues for vital materiality, where matter and things are part of ongoing forces and create ecologies of coexistence.

As a result of this kinship, the well-being of one of the elements in an ecology affects the well-being of the other elements. Looking at this idea from the material perspective shows that the industry, material origins, creative practice, and makers are complementary elements of ensuring functioning, productive, and sustainable futures. However, as a result of rapid industrialization and digitalization for several decades, both crafting and material interaction have been evolving in ways that are only sometimes beneficial to small and local communities. The increased optimization of some fields has caused a decline in some others that lack smooth adaptation to technological advancement or social changes. Wool and small-scale craft production with it are among such fields.

The history of wool production's decline follows a similar timeline as the growth of industrialization. Bowden's (1956) research discussed the change in wool supply happening already in 17th century England in reference to changing pastures and eventually changing wool qualities – as the temperature and living conditions of sheep heavily influence wool fineness. An overview of the wool supply in the 17th century concluded that it affected the trading model, import and export relationships, and the development of local industries (ibid. p. 58).

More recently, a statistical study from 2018 showed that natural fibres account for 35% of global fibre production, and only 3.3% of it is wool (Popescu et al., 2019). Although the study also showed that natural fibres were in demand by consumers, the use of wool was not among those favoured due to the decline in the number of sheep and the increase in meat farming (ibid.). Though the global demand is in decline, in some countries, wool production continues to grow. For example, during the 2010s, the decline in wool production in Australia was associated with the decline in consumer demand (Gleeson, 2013), whereas still a rise in production was predicted (Barrett, 2011). Similar to Australia, currently, in several parts of the world, the marketing and consumption of lamb and sheep is either in a stable situation or even increasing. Among these countries, an interdependency also exists. For instance, wool produced in Australia and other countries like New Zealand, USA, EU countries, Nepal, and Turkey, is imported by China's textile industry (OEC, 2024; Longworth, Brown & Waldron, 2005). Therefore, developments in these countries affect each other's advancement as well. Still, compared to the large producers, fleece and wool use is in decline in most parts of the world; for example, this decline has been happening in the U.S. since the 1940s (Agricultural Marketing, 2024) despite it being one of the big players in the industry (OEC, 2024).

Thus, currently, raw wool production centralizes in certain regions, like Australia, and manufacturing of woollen objects, such as garments, centralizes in other regions, like China. This dynamic deterritorializes agricultural, industrial, and consumer systems leading up to challenges where sheep produce wool; however, since the industry does not efficiently use it, this once-valuable material becomes waste material or unwanted by-product. Indeed, for several meat-farmers, disposing wool is not always straightforward, and it becomes a problem that needs to be get rid of. Fieldworks in sheep farms in

Finland, Turkey, and the U.S. show that this situation is a commonly shared one (Aktaş, 2023).

In their books about industrial ecology, Arnes and Arnes (1996, p. 1) provocatively put forward that "… every substance extracted from earth's crust, or harvested from a forest, a fishery, or from agriculture, is a potential waste. … it soon becomes an actual case in almost all cases with a delay of a few weeks to a few years at most." Though for textiles, disposal may be longer than a few years, they are among commonly discarded materials. U.S. data from 2018 shows that 5,8% of overall municipal waste consists of textiles (EPA, 2024), while in the E.U., 87% of used garments end up being incinerated or landfilled (EP, 2024).

In 1996, Arnes and Arnes suggested four strategies for raising the productivity of material resources: dematerialization, meaning more efficient use of material; subsidizing a hazardous material with a safer one; repairing, reusing, remanufacturing, recycling; and finally, waste mining (p. 8). Further, they propose technology and policy as driving forces to develop such strategies for specific materials. According to them, following such strategies could overcome pollution and problems with waste disposal. After 30 years, these strategies are still relevant to manage material-related issues. Indeed, they are very similar to principles and competencies for circular design, such as impact assessment, recovery, multiple use cycles, user engagement, collaboration, and communication (Sumter, et al., 2020). As a further strategy, in this paper, I propose building or repairing material ecologies and the connections they are part of.

A previous study in the case of felting identified that in small-scale production, material connections relate to seven other issues: collaboration, making process, design ideation, materials, marketplace, economic situation, maker's societal relations, and attachment to the practice (Aktaş et al., 2020). Studying materials not in isolated projects to enhance their properties but as part of their ecologies and connections can contribute to designing sustainability strategies as well. In their in-depth literature review for digital ecologies, Turnbull et al. (2023, p. 6) propose that working with ecologies increases the productivity of investigations as it diversifies social and material relations embedded in the study topic. Although their focus is on digital ecologies, the definition lends itself to nondigital and material ecologies itself.

Following a similar thinking, wool ecologies include multispecies connections that have been evolving for thousands of years. This journey became more important for humans with the domestication of sheep around 11 thousand years ago (Demirci, et al., 2013). Through selective breeding, the fineness of wool was increased (Dwyer, 2008), and it became a material for shelter building and garment making. When it became a craft material, wool also became part of communities. As sociologist Richard Sennett (2008, p. 66) characterizes, craft is an "anonymous, collective and continued practice" compared to other creative practices. Therefore, it naturally builds communities of makers, users, suppliers, traders, and learners around it.

Similarly, wool growers are part of ecologies that include multispecies beyond humans and sheep and change over time as a result of the systems they are part of (Aktaş, 2023). These systems are social, spatial, creative, behavioural, cultural, and political, and they are all entangled (ibid.). The initial observations show that the connection among these systems is currently broken. Although various stakeholders work on wool-related matters, they do work independently in their own circles. For this reason, this study examined wool in the local context by interviewing people involved in different types of local practices to identify potential ways of rebuilding the community. Next, I will introduce the study setting and analysis.

3 The "In Search of Wool(s)" Project and Study Setting

In the summer of 2023, I became a researcher-in-resident at the Center for Craft in Asheville, North Carolina, for five weeks. The aim was to explore how people were tangled with wool and what type of support and opportunities they needed. Prior to this residence project, I had been researching wool as a felting material for several years and have had field works conducted in Turkey and Finland within the context of craft and design research (see, for example, Aktaş, 2020, 2022, 2023). These field works have pointed out a shared problem among farmers from Turkey and Finland, which was the inability to compete against wool products coming from industry-dominating countries, like Australia, Iceland, New Zealand, and China. Thus, resulting in burning large amounts of wool or keeping them in warehouses. The previous studies laid the ground for connections and systems that wool is part of to identify steps toward rehabilitating some of those. As a continuation, in this residency project, the aim was to work with locals for a longer period to specify the initial concrete steps toward a revival.

Preliminary research about the craft scene in Asheville presented a relatively long history of craft production in the region. The Cherokee people, the Indigenous people of current-day North Carolina, were very skilful at basketweaving and pottery. The culture of handmade production was revisited during the late 19th and early 20th centuries as a means of income generation. In this period, later identified as the Craft Revival, people who were not necessarily craftspeople started making their own objects when the agricultural period was off and engaged with trades (the Craft Revival Project, 2024). During the revival, woodwork, textiles, and pottery became common practices among the residents of North Carolina. This culture led to the preservation of traditional skills, the foundation of folk schools, and the opening of cottage and textile industries (ibid., Williams, 2014). In the early to mid-20th century, craft schools also flourished across the state. Two significant schools are Penland School of Crafts, which is still operating with international recognition, and the Black Mountain College, which had Anni and Josef Albers, Elaine de Kooning, Buckminster Fuller, and Walter Gropius among their faculty. The strong craft presence in North Carolina promised a rich ground to understand craft's evolution in the 21st century in relation to materials and local community.

Additionally, the desktop research about craft culture across the USA, as well as the availability of sheep farms, pointed out Asheville as one of the rare locations with institutions, one of which was the Center for Craft, that still actively engage in reviving craft and supporting others to research and practice contemporary craft issues. For this reason, Asheville was selected as the study location.

To collect information and insights from a wide range of experiences and perspectives, I followed four paths: visits to farms and interviews with farmers (N: 2), interviews with artists, artisans, and educator (N: 4), a workshop with people who are involved with wool in one way (N: 12), and personal explorations by visiting the shops around the town and playing with wool and other fibres myself. These interactions are documented via extensive field notes, photographs, and videos. Additional real-time notes and descriptions were also documented in a research catalogue. In this section, I will briefly introduce each path and initial findings from those interactions before discussing the main findings and suggestions.

3.1 Interviews with Farmers

During the five weeks, I visited two sheep farms, one of which was for meat production and the other for fibre production. These two farms were selected as they were accessible and responsive. The experience level, daily practices, and knowledge of the farmers about wool differed since one of them focused on meat production only. Though both farmers picked this practice as a retirement hobby, their approach to sheep and wool was different from each other. A shared issue was regarding the use areas of wool. While the meat farmer essentially stored his wool without an agenda for purposing it, the fibre farmer also shared challenges with shipping the fleece for processing and spinning to other states.

Despite the existence of several sheep farms, Asheville, and North Carolina in general, does not have wool cleaning, carding or spinning facilities. Thus, if the farmers decide to work with their raw wool, they need to ship this material out of state. As a low-density material, shipping wool creates large transportation costs, something that most farmers are unable to cover. It also requires long-term planning, as turning fleece into yarn may take up to a year. As a result, usually, the wool in the meat farm remains unused, while the wool in the fibre farm experiences struggles not all farmers can deal with (Fig. 1).

Fig. 1. Left: Blue Leicester sheep from the fibre farm, Right: unwanted wool in the meat farm waiting to be repurposed or eventually discarded. Photo: Aktaş, 2023.

3.2 Interviews with Artists, Artisans, and Educators

To learn from artists and their experiences of working with wool, four interviews were arranged; two were artists, and the other two were artisans whose main job was to work at educational institutions, one at the craft and textile department of a college, the other one in K-12 system (Fig. 2).

M., the artisan who works with local wool, is an educator at the local Warren Wilson College. Though her specialty is in flax, as part of her curriculum, she designs courses where students grow plants for material generation, hand-dye and hand-spun their fibres, including wool, and weave these in collaboration with a local organization called "Local Cloth" to promote local resources, craft, and knowledge. M. works with local wool as an act of preservation and local development and teaches her students steps of taking care of this material. The other educator, N., is a textile designer and has established a collective where she mostly works with kudzu fibres and, at times, incorporates wool. A significant note by this interviewee was that to her, wool was already perfect, and there was no need to make it better. This became a critical comment for me to think about what wool's future might look like. In both interviews, it was clear that these organizations were flourishing due to the hard work and dedication of their leaders. They shared similar ambitions about sustainable local development, and to contribute to this, both of the educator-artisans were networking, participating in, and organizing events to increase visibility and accessibility to materials and crafted works.

The interviews with the two artists focused more on their practice compared to the two educators. These interviews with L. and N.G. clearly demonstrated that they had different styles. They described how they value first becoming a craftsperson and then building their artistic expressiveness over their expertise. In order to achieve this, they mentioned the importance of exploring materials, their capacities, and behaviours. L., a felt artist, shared that only after truly exploring most aspects of wool she was able to move from visual expressiveness to structural designs. N.G., who mainly weaves, on the other hand, mentioned he respects his material source, in his case mostly Alpaca hair, and wants to highlight the material in his works through visual expressions. In both interviews, it was clear that artists and artisans need to build this close kinship, and for this, they need material accessibility. Both artists mentioned they purchased their materials from abroad.

Interviews with two artists and two educators showed challenges of being a textile artist and the need for finding supportive means to gain a living. All four artists had either teaching jobs or another means of income generation to support their art making. They were all deeply informed about their materials.

3.3 Group Explorations: Workshop

As part of the project, a workshop was offered to those who find the topic of wool interesting, relevant, or important. The local organization "Local Cloth" opened their common area to host the event and shared an invitation in their newsletter. Thus, the participants were all interested in textiles or fibres and came to share their knowledge. At the end, the workshop hosted sheep farmers, artists, and activists.

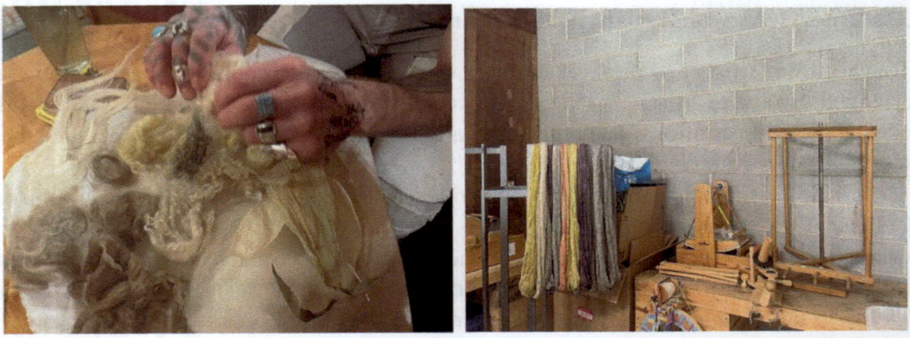

Fig. 2. Left: N.G., is playing with different kinds of local wool and sharing his tactile reflections during the interview. Right: From the interview with M. at her teaching studio at the college. Photos: Aktaş, 2023.

In the workshop, participants were invited to rethink the capacities of wool, its behaviours in creative making, and the values attributed to it. The workshop started with writing letters to wool. This exercise warmed up their perceptions and conceptions related to wool. Then, they worked with other natural fibres, such as flax, cotton, and moss (Fig. 3). This exercise aimed to explore if local wool could be paired with other fibres to simplify its processing line. The most fruitful part of the workshop was the last part of the round table discussion, where each participant shared their personal experiences of working with wool, some of the challenges they faced, and their suggestions for overcoming the barriers. Through speculation, we also envisioned new use areas for wool. The workshop ended with another writing exercise in which wool replied to the first letter. This exercise aimed at concretizing the needs from wool's perspective.

One significant takeaway from the workshop was regarding the reasons behind wool's invisibility. Although all participants expressed their awareness regarding wool's diminishing, they were also equally surprised why this was happening. Knowing its power, almost all of them expected this material to be more mainstream. Another insight from the workshop was about the consumers and their need for knowledge about the materials they use. Finally, when asked if they needed new policies or support from policymakers, univocally, all participants expressed their scepticism toward policymakers and their often unsustainable policies. The observations from the workshop showed that craftspeople and farmers in Asheville are very well informed about issues, and they are also in the process of strengthening their network. Yet, as the challenges are complicated, these initiatives require further efforts.

3.4 Personal Explorations

The strolls at the local museums, such as Biltmore Homespun Museum and Black Mountain College Museum, and craft shops, as well as working with local wool became tools for reflecting on the information gathered (Fig. 4, 5). Additionally, while walking around, I met people and shared my research with them. This also sparked new conversations and sometimes led to looking into new things, such as other textile fibres used in the

Fig. 3. During the workshop, participants shared their experiences and visions about wool while playing with it. Photo: Mellanee Goodman. 2023.

region, indigenous history, and local industries. I also participated in a workshop to make flax fibres from the harvest and gave a speech about thinking through textile making, followed by a group discussion at the Local Cloth with a different group of people. Throughout the residency period, these practices localized me and made me less of an outsider exploiting but a friend of this shared issue.

Fig. 4. Left: a view from the Biltmore Homespun Industries Museum, Centre: photo from the Black Mountain College Museum's library, Right: A craft shop in downtown Asheville. Photos: Aktaş, 2023.

In Asheville, the small shops and larger organizations identify crafts with different terminologies depending on their target groups, such as mountain crafts, Appalachian crafts, and crafts and fine arts. The crafted artefacts are mostly exhibited in high-end looking galleries. Local organizations like the Southern Highland Craft Guild and their Folk Art Center organize events and fairs throughout the year to promote and support craftspeople.

Fig. 5. Tests with wool and corn husk, flax, and coconut fibres. Photo: Aktaş. 2023

Overall, the observations showed that the craft scene was vibrant in Asheville. The city has a long history of textile production and there is still interest among the crafts community to continue doing things. However, two shortcomings were also prevalent. Most of these craft shops were for tourists coming to Asheville for their mountain vacation. This creates an invisible wall between relatively larger craft organizations and the local public. A second shortcoming, more specific to textiles, was that the textile visibility in the city was relatively low compared to, for example, ceramics and basketry, which are also local crafts historically done by Indigenous people of the Eastern Band of Cherokee. These observations point out the strong existence of a craft community that often remains in its bubble.

3.5 Wrapping up the Residency Project

After the interviews and workshop were concluded, a small installation was set up at the Engagement Studio of the Center for Craft (Fig. 6). The station included a brief overview of the project, its aims, stakeholders, and initial observations. It had local wool samples, images from the visits, and further reading.

As one of the key discussion points in the workshop and with the interviewees was that the locals and consumers are not well-informed about wool and its potential; for this installation, I also designed a pamphlet and left several of it on the desk to be picked up (Fig. 7). It introduced the wool, the reasons behind its decline, and some suggestions for the public to help reverse this trend. The installation was on at the gallery of the centre for about six months, from the end of 2023 till early 2024.

4 Discussion and Suggestions

Based on these observations and interactions, the identified reasons behind overlooking locally grown wool as a material for various industries can be summarised as following:

Fig. 6. The installation at the Engagement Studio. Photo: Mellanee Goodman, 2023.

Fig. 7. The pamphlet highlighted key issues related to wool communities and suggested simple steps to reverse the decline. Photo: Aktaş, 2023.

Processing wool from fleece to an easily controllable form, such as roving or yarns, is labour-intensive and costly. This situation demands expensive outcomes to ensure a profitable business. For many, working with wool or engaging with crafts or sheep farming is a passion-driven practice that requires lots of personal dedication. In return, these practices are challenging to fully support one's livelihood. Thus, the field lacks attraction for new investments and enthusiasts. The accessibility and production capacity of wool processing can sometimes be challenging regarding time management and efficiency. The service system of processing wool can become discouraging.

Consumers need to become better informed about the materials they interact with daily. Fast fashion and consumerism have created habits of over-consumption without building meaningful relationships with the owned objects. The disconnection from paying attention to surrounding materials creates a lack of empathy for interacting with materials, their qualities, and their origins. As an extension of this, users and consumers need to be better informed about wool and the value it offers.

An overarching finding is that **working with wool requires an interdisciplinary community. As such, the future of this material depends on creating and/or strengthening this community.** The data shows that the community has almost dissolved and, to a large degree, unified (Fig. 8). The interaction is active among members that are similar to each other but there needs to be communication among those that are from different backgrounds. For a well-functioning community, specifically for wool, diversity among the members is needed. However, the study shows that the kinship that makes individuals a community is disappearing for wool in Asheville. International studies about the decline in wool use confirm that this is a global trend. Currently, the small community is scattered and mainly consists of artisans and farmers. Although scientists and engineers are also working with wool, they are not in connection with those who actually grow this natural resource. This breaks the line of communication between different groups and also results in insufficient education and learning from the locals' perspective. The knowledge produced at the laboratories should be disseminated to the community who would benefit from it when it comes to implementing micro-scale innovations. Simultaneously, when local crafts are "elevated" to the gallery setting, its users and the makers become separated as they often do not share similar circles.

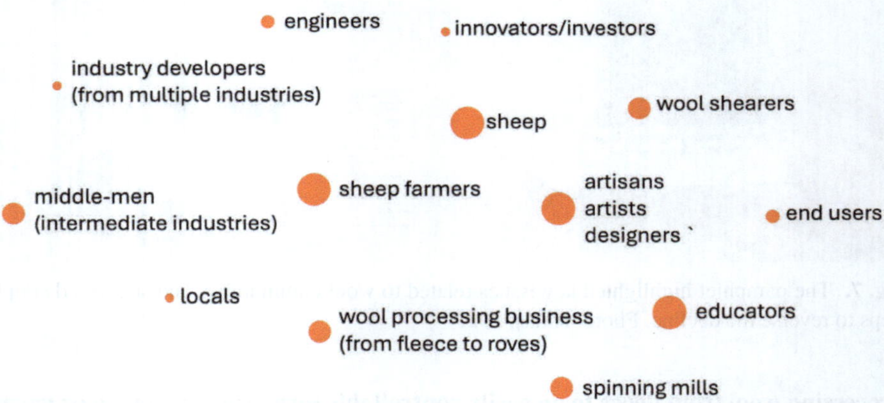

Fig. 8. The wool community is scattered and not equally involved.

Accordingly, one of the early steps should be to rehabilitate the community and bring together people from different industries, scales, and work contexts together. The data identified four main industries that can be easily included in the campaigning for wool. **The fashion and textile industry** can expedite and initiate local development projects to promote woollen artifacts as durable and eco-friendly alternatives to polyester. Projects already focus on wool recycling (Mu et al., 2023; Russell et al., 2016), but initiatives to change consumer perception are also needed. Taking care of woollen artifacts is relatively easy; when aired occasionally, cleaning is seldom needed. Wool can be easily repaired by darning, mending, or felting. These aspects can become points of departure for the reintroduction of wool. **The construction industry** has already developed new knowledge about wool's role in heat and sound regulation and insulation in large-scale buildings.

The knowledge can be made more accessible and transferred to DIY approaches to attract local growers and homeowners to adopt this behaviour. **The sustainable agriculture movement** has been exploring small-scale sheep breeding for sustainable meat and dairy produce. Additionally, using biomass produced by sheep has the potential to offset fossil fuels (Colley et al. 2020) and lead to regenerative agriculture. Wool and sheep waste can also be used as nonchemical/organic fertilizer; scaling this up can create new product lines for local sheep farmers and contribute to cleaner agricultural production. Finally, the observations show the need for simplifying the processing line. This requires collaborating with **Engineers, most likely Textile, Chemical and Mechanical**, to develop new machinery and tools for improving the cost-heavy and complicated process. Simplifying the processing could encourage new farmers and artisans.

These key findings indicate that both consumption and production cycles should be redesigned to attract more people to engage with wool. In order to accomplish this, stakeholders need to improve their network and start repairing the kinship to maintain sustainable futures for wool, sheep, farmers, artisans, users, industries, and the world.

5 Conclusions and Next Steps

A more sustainable future for wool can only be achieved through collective and cross-disciplinary endeavours. It requires campaigning for wool targeted at different stakeholders to ease their obstacles. Education, networking, and simplifying technology are essential first steps. Shared projects that bring together different stakeholders are crucially needed to disseminate different knowledge types about wool and scaling-up and down innovations. Additionally, a fundamental transition is needed towards more sustainable and value-driven consumption and production behaviours to revive the use of wool and implement similar studies to other natural fibres. From the consumer's side, such transition should promote demanding high-quality products and considerations for longevity, positive environmental impact, and local welfare as elemental aspects of consumption decisions. From the producer's side, this transition requires contacting innovators and investors to demand the development and distribution of new knowledge at the local, small-scale production level. This repaired connection can also support re-industrialization of rural areas. The study shows that the wool communities have been diminishing and becoming too consistent among mostly artisans and wool producers in different parts of the world. The homogeneity among community members may need more inspiration to spark innovation and new thinking. Diversifying people and places involved will increase the chances of reviving wool from the 21st-century perspective. From a similar perspective, exploring other material communities would be beneficial to learn from how other practitioners overcome obstacles while implementing emerging thinking in their fields. Considering that despite all the obstacles small scale wool growers and artisans have been persistent with maintaining their practices, working with local wool can be seen as an act of resistance and even silent activism.

Acknowledgements. I would like to express my deepest gratitude toward the Center for Craft for hosting the residency project, Local Cloth for hosting the workshop, and all the participants and interviewees for sharing their knowledge and time with me.

References

Aktaş, B.M.: Entangled Fibres-An Examination of Human-Material Interaction. Aalto University, Espoo (2020)

Aktas, B.M., Mäkelä, M., Laamanen, T.K.: Material connections in craft making: the case of felting. In: Design Research Society International Conference, pp. 2326–2343. Design Research Society, Brisbane, August 2020

Aktaş, B.M.: Entangling with materials: crafting as a way of repairing our relationship with the environment. In: Fredriksen, B.C., Groth, C. (eds.) Expanding Environmental Awareness in Education Through the Arts. Landscapes: The Arts, Aesthetics, and Education, vol 33. Springer, Singapore (2022)

Aktaş, B.M.: Following a material's journey to unravel becomings and systems in human-material interaction. In Relating Systems Thinking and Design, Washington DC, vol. 12, October 2023

Allafi, F., et al.: Advancements in applications of natural wool fiber. J. Nat. Fibers **19**(2), 497–512 (2020)

Angelova, G.V., Brazkova, M.S., Krastanov, A.I.: Renewable mycelium based composite–sustainable approach for lignocellulose waste recovery and alternative to synthetic materials–a review. Zeitschrift für Naturforschung C **76**(11–12), 431–442 (2021)

Ayres, R., Ayres, L.: Industrial Ecology: Towards Closing the Materials Cycle. Edward Elgar Publishing, Massachusetts (1996)

Barrett, D.: Wool. Agric. Commodities **1**(2), 71–75 (2011)

Bennett, J.: Vibrant Matter: A Political Ecology of Things. Duke University, London, UK (2010)

Bowden, P.J.: Wool supply and the woollen industry. Econ. Hist. Rev. **9**(1), 44–58 (1956)

Colley, T.A., Olsen, S.I., Birkved, M., Hauschild, M.Z.: Delta life cycle assessment of regenerative agriculture in a sheep farming system. Integr. Environ. Assess. Manag. **16**(2), 282–290 (2020)

Craft Revival Project. Western Carolina University, 2005–2010. https://tinyurl.com/mtb8aeve. Accessed 13 Aug 2024

Deleuze, G., Guattari, F.: A Thousand Plateaus: Capitalism and Schizophrenia. Bloomsbury, London (1980/2019)

Demirci, S., et al.: Mitochondrial DNA diversity of modern, ancient and wild sheep (Ovis gmelinii anatolica) from Turkey: new insights on the evolutionary history of sheep. PloSone **8**(12), e81952 (2013)

Dénes, T.O., et al.: Analysis of sheep wool-based composites for building insulation. Polymers **14**(10), 2109 (2022)

Donner, M., Verniquet, A., Broeze, J., Kayser, K., De Vries, H.: Critical success and risk factors for circular business models valorising agricultural waste and by-products. Resour. Conserv. Recycl. **165**, 105236 (2021)

Dwyer, C.: Environment and the Sheep. In: The Welfare of Sheep. Animal Welfare, vol 6. Springer, Dordrecht (2008). https://doi.org/10.1007/978-1-4020-8553-6_2

Environmental Protection Agency, USA (2023). https://tinyurl.com/58frb7jh. Accessed 13 Aug 2024

European Parliament: The impact of textile production and waste on the environment (infographics) (2024). https://tinyurl.com/5xun4wxb. Accessed 13 Aug 2024

Gleeson, T.: Wool. Agric. Commodities **3**(1), 111–118 (2013)

Ingold, T.: Toward an ecology of materials. Annu. Rev. Anthropol. **41**, 427–442 (2012)

Ingold, T.: Making: Anthropology, Archaeology, Art & Architecture. Routledge, London (2013)

Longworth, J., Brown, C., Waldron, S.: Features of the Wool Industry in China. China Agricultural Economics Group, University of Queensland, Brisbane (2005)

Malafouris, L.: How Things Shape the Mind: A Theory of Material Engagement. MIT, Cambridge, MA (2013)

Maraveas, C.: Production of sustainable and biodegradable polymers from agricultural waste. Polymers **12**(5), 1127 (2020)

Muiruri, J.K., Chuan Yeo, J.C., Zhu, Q., Ye, E., Loh, X.J., Li, Z.: Sustainable mycelium-bound biocomposites: design strategies, materials properties, and emerging applications. ACS Sustain. Chem. Eng. **11**(18), 6801–6821 (2023)

Mu, B., Yu, X., Yang, Y.: Sustainable and green process for recycling waste wool textiles into high-quality protein fibers on a pilot scale. Resour. Conserv. Recycl. **198**, 107190 (2023)

Popescu, A., Stoian, E., Şerban, V.: Trens in the world production of natural fibers of animal origin-silk and wool. Manage. J. **4**, 273–288 (2019). ISSN 2284-7995

Ramírez Tapias, Y.A., Peltzer, M.A., Delgado, J.F., Salvay, A.G.: Kombucha tea by-product as source of novel materials: formulation and characterization of films. Food Bioprocess Technol. **13**, 1166–1180 (2020)

Ranjit, E., Hamlet, S., George, R., Sharma, A., Love, R.M.: Biofunctional approaches of wool-based keratin for tissue engineering. J. Sci. Adv. Mater. Devices **7**(1), 100398 (2022)

Russell, S., Swan, P., Trebowicz, M., Ireland, A.: Review of wool recycling and reuse. In: Natural Fibres: Advances in Science and Technology Towards Industrial Applications: From Science to Market, pp. 415–428. Springer Netherlands (2016)

Sennett, R.: The Craftsman. Yale University Press (2008)

Sumter, D., de Koning, J., Bakker, C., Balkenende, R.: Circular economy competencies for design. Sustainability **12**(4), 1561 (2020)

Turnbull, J., et al.: Digital ecologies: materialities, encounters, governance. Prog. Environ. Geogr. **2**(1–2), 3–32 (2023)

Williams, E.M.: The craft revival: preserving culture, changing lives. In: Appalachian Studies Association Annual Conference. Marshall University, 28–30 March 2014

Wool in China: The Observatory of Economic Complexity (2024). https://oec.world/en/profile/bilateral-product/wool/reporter/chn. Accessed 20 Sep 2024

Wool Profile: Agricultural Marketing Resource Center (2021). https://tinyurl.com/vyxmmnn2. Accessed 13 Aug 2024

Speculative Biofutures Inspiration

Prologue 1: Speculating Around Design

By NORMALS - Design Studio Berlin

During his address at the What's Around Design conference, Ezio Manzini clarified the role of designers as being "to increase the number and variety of possibilities". This task, which speculative designers put at the center of their practice, is indeed crucial in bringing about conversations about the futures we want as well as their social, technical, and environmental implications. It initiates a process of challenging the status quo by prodding our collective imaginaries, deconstructing cultural heritages and contemporary power structures. Beyond highlighting new opportunities, the goal is to identify and discuss trade-offs. We must acknowledge that change never comes for free, and instead assess its costs and those most likely to shoulder them.

Many who work with futures also wrestle with the challenge to make their contribution truly useful to society. Speculative design can serve in multiple ways by expanding its field of inquiry beyond mere technical solutionism. It can help anticipate future problems, some of which derive from solutions implemented with little concern for their impact on the built and social environments. It can shed a new light on legacy systems and bring about constructive conversations between stakeholders that aren't just shareholders. It can highlight the impact of decisions on those—human or not, contemporary or futures—that aren't in a capacity to influence them.

In his third law of the future, Jim Dator quotes Marshall McLuhan's and John M. Culkin's claim that "we shape our tools and thereafter our tools shape us", which may be completed with Audre Lorde's statement that "the master's tool will never dismantle the master's house." A goal of speculative design is to provide new tools for informing and enacting change in new ways. It is a tightrope walker's challenge: to account for the multiple shapes futures can take, from the more evident and foreseeable consequences of contemporary decisions in familiar contexts to signals that are faint—either because they are yet to emerge, or because they have emerged in spaces we are unfamiliar with. While this might give the false impression that speculative designers must exhaustively address all possibles, the opposite is true: it is an exercise in curation and sense-making, a careful investigation picking its most meaningful clues. We aren't here to check boxes, but to imagine new ones.

Speculative biofutures is a field worthy of such investigation. To place genetic material at the center of production processes, to view the living as a partner rather than an obstacle, at the same time expands our horizons and invites us to reassess our place in the environment. Whether it is by recognising its microbial heritage, embracing biophilic design, or reviewing the metaphors behind our economic systems, it raises both technical and ethical questions, and provides elements of response through functional parables—narrative provocations that seamlessly merge the advisory and the cautionary. Ambivalence is key here, as the most successful futures aren't the ones that gather unanimous approval or defiance, but those that leave their audience pondering about the desirability of the presented path, and about their own moral compass.

Introduction to Part 3: The challenges of Speculative Biofutures

Susana Leonor, ISMAT, Portugal, and David Palma, ISMAT, Portugal (Conference Track Chairs)

We inhabit an era of rapid transformation, compelling heightened consideration of the consequences stemming from our collective actions. We are confronted with existential challenges that demand a fundamental reassessment of our relationship with the planet and the tools we create. The climate crisis accelerates, waste accumulates in alarming proportions – from ubiquitous plastics to mounting waves of electronic debris – and unsustainable consumption patterns strain the limits of Earth's ecosystems. Within this context of urgency, it is clear that conventional approaches are insufficient. We must introduce new ways of thinking that allow us to reimagine our living patterns, fostering a global consciousness and practices that promote regeneration and sustainable coexistence, looking towards technological innovation and deeper ecological integration.

Against this challenging backdrop, Speculative Biofutures emerge as a vital field of research, questioning, imagining, and projecting optimistic future scenarios. This avenue of thought does not merely seek to predict what might arise solely from advanced biotechnological interventions; it actively aims to project and speculate on the transformative potential of biological understanding and application in its broadest sense – encompassing high technology, biomimicry, ecosystem restoration, and the exploration of potential shifts in our integration with natural systems. It explores how these diverse approaches might revolutionize medicine or industry and the foundations of life, society, and our shared environment.

The speculative approach empowers communities – including designers, researchers, and citizens alike – to break free from the constraints of the present and the weight of the past, enabling the projection of alternative trajectories. Rather than succumbing to dystopian futures or naive techno-optimism, and in this context, Speculative Biofutures encourage critical and creative analysis. It prompts exploration into potential shifts in cultural and philosophical perspectives on life, identity, and the human condition. It considers how these might evolve in response to developments ranging from synthetic organisms and advanced interfaces to new ecological stewardship and community living forms. Such exploration is inseparable from rigorous ethical, legal, and social examination, anticipating dilemmas and seeking frameworks for responsible development and societal adaptation.

Recognizing the complexity of these issues, Speculative Biofutures thrives as an inherently interdisciplinary approach. It draws knowledge and methodologies from design, biology, ecology, engineering, ethics, sociology, anthropology, science fiction, and Indigenous knowledge systems, weaving together comprehensive, imaginative, and grounded visions of the future. The objective transcends mere forecasting; it is about cultivating a proactive mindset to shape preferable futures through conscious choices about technology, behavior, and our relationship with the biosphere.

Crucially, in the face of our current environmental crises, this speculative method offers a powerful resource for designing desirable, optimistic scenarios. Speculative Biofutures serves as an approach that empowers us to imagine futures where human activity actively harmonizes with natural systems. This includes contributing to the remediation of ecological damage, fostering biodiversity, creating regenerative material cycles, developing truly sustainable food and energy systems, and cultivating new habits and societal structures grounded in ecological awareness. It is an invitation to research communities, policymakers, and citizens to think beyond current limitations, catalyzing the necessary behavioral and systemic shifts, and fostering a global consciousness focused on environmental well-being, enduring sustainable practices, and fundamentally reconsidered ways of living that ensure a safer and more equitable world for all species.

A Theory of Speculation: Do Speculative Futures Really Relate to the Future?

Alakesh Dhibar(✉) and Fernando Oliveira

UNIDCOM/IADE - Unidade de Investigação em Design e Comunicação, Av. D. Carlos I, 4, 1200-649 Lisbon, Portugal
alakeshisavailable@gmail.com,
fernando.oliveira@universidadeeuropeia.pt

Abstract. The paper aims to argue that speculative futures are impacted by past and present experiences, in addition to future possibilities, by analyzing the Speculative Design methodology in the context of human cognition and cognitive biases. Drawing from contemporary literature and case studies on Speculative Design, the paper hypothesizes that despite the potential of Speculative Design methodologies to foster novel perspectives, their present applications may need a more comprehensive approach due to the pervasive influence of cognitive biases. To gain an in-depth understanding of the problem, the study utilizes a qualitative theory triangulation methodology grounded on a literature review. It explores and analyzes theories from diverse fields of knowledge, such as divergent thinking, decision-making in the design process, and micro trends in human cognition, to establish conceptual similarities with Speculative Design. Furthermore, it investigates the relationship between cognitive biases and Speculative Design. Finally, the paper proposes a more holistic way of thinking in Speculative Design as a theory of speculation to address the impact of cognitive biases effectively.

Keywords: Speculative Design · Cognitive Biases · Divergent Thinking · Microtrends · Futures

1 Introduction

The study aims to analyze the position of Speculative Design by evaluating its origin, methodologies, and current applications in the field of Design. It further argues that speculative futures are not solely focused on future scenarios but are also influenced by past and present experiences. This critique is supported by examining human cognition and imagination through the lens of cognitive psychology. For example, there is evidence of cognitive biases within the process speculation supported by imagination (Engel & Singer, 2008). To gain contributions towards understanding the problem, the study explores several entities, such as the process of divergent thinking, decision-making in the design process, and the role of micro trends in human cognition. Finally, it attempts to develop a theory of speculation to emphasize the need for inclusive design methodologies grounded in the Speculative Design philosophy.

A. Mateus et al. (Eds.): WAD 2024, What's Around Design?, pp. 147–155, 2026.
https://doi.org/10.1007/978-3-032-08163-6_11

Considering the initial ideas on Speculative Design by Anthony Dunne and Fiona Raby (2013), it is a possible implementation of Design as a means of speculating how things could be, emphasizing imaginative thinking to explore alternative futures and introduce new perspectives on complex problems. By examining different pathways from present to future, they proposed the concept of 'preferable futures' through Speculative Design. It focuses on the role of designers in collaborating with experts to create diverse future scenarios that stimulate public debate and help communities define their desired futures. To achieve this, Dunne and Raby (2013) proposed the A/B manifesto, focusing on the B-side as an alternative philosophy, and outlined various methodologies behind Speculative Design, such as fictional worlds, cautionary tales, what-if scenarios, thought experiments, counterfactuals, reductio ad absurdum experiments, and prefigurative futures. Finally, Dunne and Raby (2013) mentioned the extended mediums of Speculative Design, including cinema, literature, science, ethics, politics, and art. After a long time since its initial proposition, there has been a significant increase in interest and application of Speculative Design in both industry and academia. This paper aims to initiate a qualitative study of Speculative Design and highlights a fundamental issue within its contemporary applications. Specifically, it focuses on the role of cognitive biases in the process of imagination within Speculative Design and argues for a more inclusive approach to enrich its methodology.

To support this argument, several recent examples can be cited. For instance, Speculative Design practitioners Michaela and Konstantin from Basel-based Circumflex Studio emphasize on the mechanics of speculation which is actually grounding in prior knowledge, emerging tendencies, existing technologies, and human behaviours. They also suggest that by crafting an alternative narrative set in the future, present, or past, the predominant reality is put into question, and inherent biases are revealed (SpeculativeEdu, 2019a). In another interview with the SpeculativeEdu group, Francisco Laranjo, a graphic designer and researcher, highlighted the exploitation of Speculative Design as a standalone design mode or a short-term trend within countries in the global north. He emphasized the importance of integrating Speculative Design's discourse into broader design education to address issues such as decolonization, challenging hegemonic practices, and promoting diversity among faculties (SpeculativeEdu, 2020). Tobias Revell, an artist, designer, and educator, has criticized the divergence of Speculative Design from its critical origins. He highlights its misuse in capitalist business practices and points out prejudices regarding preferable futures and cautionary tales (SpeculativeEdu, 2019b). Beyond the contemporary examples, if we try to analyse the functionality of speculation from diverse fields of knowledge, we can find connections between human cognition and fiction. In her book Only Imagine: Fiction, Interpretation, and Imagination, philosopher Kathleen Stock explores the intricate relationship between fiction, interpretation, and imagination. Delving into the role of intentionality in shaping fictional content, Stock (2017) examines how it influences our perception, interpretation, and appreciation of works of fiction. This specific concept of intention is an integral part of human behaviour if we follow the Augmented Theory of Interpersonal Behaviour by Lilley and colleagues from Design for Behaviour Change (Niedderer et al., 2017). Furthermore, literature from cognitive science explores numerous examples of cognitive biases and their impact on decision-making and perception (Eysenck & Keane, 2020). Finally, within the realm of Design, Donald Norman explores how memories and histories influence beliefs and emotions. Norman (2023) additionally states that the actions of individuals and societies are influenced by past events and circumstances, rather than occurring independently. The origins of people's current beliefs and thoughts can often be traced back over centuries. This explicitly pointed out the connection between imagination as a product of human cognition and past experiences.

Narrowing down these examples, the paper uses qualitative theory triangulation (Robson, 2011) based on a literature review from diverse fields to establish the connection between cognitive biases and Speculative Design. Building on the initial philosophy of Speculative Design and incorporating inclusive actions, consequences, shared learning, and discussions, the paper aims to find a holistic solution to promote diversity. It envisions more inclusive futures by integrating insights from both past and present experiences.

2 Analyzing the Problem Space

Following the objectives of Speculative Design, it becomes clear that it aims to inspire and promote diverse perspectives as alternative solutions to complex problems. If we go through the A/B manifesto (Dunne & Raby, 2013), we can understand that it focuses on the transformation of already existing ideologies around the process of design. Applying a theory triangulation methodology, this paper primarily addresses the exploration of diverse perspectives as the central focus of its argument. It hypothesizes that despite Speculative Design methodologies encouraging new perspectives, contemporary applications may fail to provide a holistic approach due to the influence of cognitive biases. Within this methodology, various entities from diverse fields of knowledge are identified and explored to uncover conceptual similarities with Speculative Design. Subsequently, these entities are analyzed through the lens of cognitive biases to explore their potential impact within the context of Speculative Design.

2.1 Divergent Thinking

Within the field of psychology, divergent thinking is a type of thinking that explores various directions of thought (Guilford, 1966). Sometimes referred to as lateral thinking, divergent thinking is the process of coming up with a range of unique ideas or solutions to a problem. It involves spontaneous, unrestricted thought, requiring the generation of numerous different answers or pathways. Recently, significant advancements have been made in various techniques of divergent thinking, including open ended questions, roleplaying, brainstorming, mind mapping, and supporting unconventional ideas. The conceptual similarities between these techniques and the previously discussed transformation of perspectives through Speculative Design are quite prominent. For example, to spark a question, Speculative Design sometimes follows the tactics of reductio ad absurdum, which attempts to establish a claim by showing that the opposite scenario would lead to absurdity or contradiction (Rescher, n.d.). For his BuyProduct project, Dot Samsen (2009) applies this strategy by following the carbon credit idea through to its logical conclusion, he exposes its absurdity and impossibility[1]. This logical reasoning

[1] Focusing on capitalism, the critique on carbon trading questions the validity of renting trees to offset carbon footprints and demonstrates the absurdity by installing machines on the trees.

influences the process of visualizing and validating opposing scenarios, thereby leading to diverse possibilities. It embodies the essential traits of openness and flexibility inherent in divergent thinking (Kaufman & Sternberg, 2019).

By examining the definition and process of divergent thinking through the lens of cognitive biases, connections can be identified between mood, attitude, and personality in relation to divergent thinking. For instance, findings from various experiments indicate that divergent thinking is influenced by diverse experiences and can be systematically enhanced through the introduction and practice of relevant strategies, or by creating a conducive mood or context for ideation. These findings are consistent with research showing that experience plays a crucial role in improving divergent thinking performance (Behrens et al., 2014). Moreover, literature has demonstrated that episodic specificity induction, involving the detailed recall of past experiences, effectively enhances divergent thinking abilities (Madore et al., 2015). When considering the relationship between divergent thinking and Speculative Design, it becomes evident that idea evaluation is a necessary step or component within the process of idea generation.

2.2 Design Methodologies

Speculative Design primarily focuses on imagining possible futures as a means of exploring alternative perceptions. The process of exploring diverse perspectives is inherent in several already existing design methodologies. For example, if we focus on the Double Diamond design process (Banathy, 1996), particularly the step, 'discovery' within the problem space, we can identify a connection with diverse perspectives. It emphasizes on the critical role of curiosity through various methods such as open-ended discussions, brainstorming to identify relevant stakeholders and sub-stakeholders, and embracing ambiguity. Similarly, in the innovation process, the 'fuzzy front end' or 'pre-design' step, as described by Sanders and Stappers (2008), focuses on various activities aimed at informing and inspiring exploration through open-ended questions. Within the context of Speculative Design, there is again a conceptual similarity with these steps from various design methodologies. To justify the same, we can consider the example of Speculative Design method, counterfactual conditions, a form of mental simulation where one reflects on an event and imagines an alternate ending. Sascha Pohflepp's use of a counterfactual approach in his Golden Institute for Energy. Pohflepp (2009) imagines a scenario where Jimmy Carter won a second term in 1980, leading to the promotion of radical green policies instead of investment in the Star Wars program[2]. The counterfactual mechanism from Speculative Design explores alternative scenarios to uncover potential possibilities that might otherwise be overlooked, which is fundamentally aligned with diverse perspectives found in steps such as 'discovery' and 'fuzzy front end' within design methodologies.

In contemporary times, extensive research examines the relationship between judgment, decision-making in the design process, and cognitive biases (Thomas, 2020; Ellis, 2018). The fundamental critique emphasizes the importance of recognizing and being aware of a designer's personal biases to understand user needs objectively, avoiding

[2] It speculated on the consequences of US President Jimmy Carter advancing renewable energy, exploring the relationship between the state, individuals, and nature.

the imposition of subjective preferences and promoting more inclusive design solutions. Additionally, following a user-centered design approach such as design thinking, we acknowledge the significance of the step, 'empathize', which highlights the importance of mitigating biases through effective observation and communication (Kelley & Kelley, 2013). Finally, when examining the relationship between design methodologies and Speculative Design, it becomes clear that understanding cognitive biases is essential in the process of exploring alternative perceptions.

2.3 Future Scenarios and Microtrends

This section explores how surrounding attributes influence human behavior, potentially shaping the complex process of acquiring new perspectives. Speculative Design stimulates imagination to explore alternative and future scenarios, yet understanding this process involves examining how microtrends (reflect social phenomena) impact human cognition. A microtrend denotes a pervasive tendency within a specific sphere of influence, lasting from a few months to a few years. Evidence from social psychology highlights the innate human desire to belong to a group (Feldman, 1985). Microtrends serve as a means to express belonging, with social media platforms playing a crucial role. The impact of microtrends on human cognition and decision-making can be significant. For example, Mark Penn's book 'Microtrends Squared' illustrates how subtle shifts within specific population segments drive significant societal changes. Penn (2019) identifies fifty new microtrends predicting shifts in the next decade, including the influence of fringe voters, the rise of online influencers, the expansion of the gig economy, and advancements in technology and social movements. Furthermore, the literature demonstrates that emerging trends can influence creative processes, including imaginative thinking (Mason et al., 2015; Manu, 2006). Therefore, the influence of microtrends can be evident in imagining future scenarios within Speculative Design.

To justify the role of cognitive biases within the context of microtrends, we can consider the Bandwagon Effect (Schmitt-Beck, 2015). This psychological phenomenon describes how the adoption of beliefs, ideas, fads, and trends increases as more people adopt them. As individuals observe others believing in something, they are inclined to 'hop on the bandwagon,' often irrespective of underlying evidence. Widely observed in fields like economics, political science, medicine, and psychology, the bandwagon effect serves as a primary cognitive bias influencing the mechanics of microtrends. Moreover, when considering the process of acquiring new perspectives through imagination, it is crucial to be aware of cognitive biases such as the bandwagon effect. As mentioned earlier, the complex role of social factors in shaping intentions in human behavior (Lilley et al., 2017) highlights the need for a structured evaluation in speculative processes.

3 A Theory of Speculation

The fundamental objective of this paper is to develop a theoretical framework around the problem space, focusing on the role of cognitive biases within contemporary speculative practices. Additionally, it aims to initiate a dialogue towards the solution space by introducing a fresh perspective on speculative methodologies. The central argument of this paper is that cognitive biases significantly impact the exploration of diverse perspectives. To deconstruct the problem, the study explores various fields of knowledge to investigate potential opportunity areas, aiming to conceptualize towards a solution. The impact of cognitive biases, such as mood, attitude, and past experiences, on the process of divergent thinking, highlights the importance of incorporating an idea evaluation phase within the process. Furthermore, exploring existing design methodologies illuminates the connection between Speculative Design and its role in the broader design and creative processes. Finally, the impact of microtrends on future scenarios ties into social factors, extending the argument towards social psychology. Following Fig. 1, all these attributes contribute to shaping a new way of thinking, influencing the exploration of diverse perspectives through Speculative Design.

The study proposes a theory of speculation centred on the B-side, as outlined in the A/B Manifesto by Dunne and Raby (2013). As an alternative dimension of the Design field, the B-side fundamentally facilitates discussions and debates through an alternative philosophy, promoting diverse perspectives. For example, attributes of this philosophy such as 'Functional Fictions', 'Parallel Worlds', 'Provocation', and 'Rhetoric' among others aim to stimulate speculative practitioners by inspiring them to explore various paths leading towards diverse contexts and different target entities. 'Functional Fiction' and 'Parallel Worlds' are closely connected to new possibilities that encourage the consideration of varied viewpoints on future scenarios, which are plural and shaped by diverse circumstances. Additionally, 'Provocation' and 'Rhetoric' aim to revive open discussions that expand understanding beyond practical solutions by fostering a space for evaluating ideas.

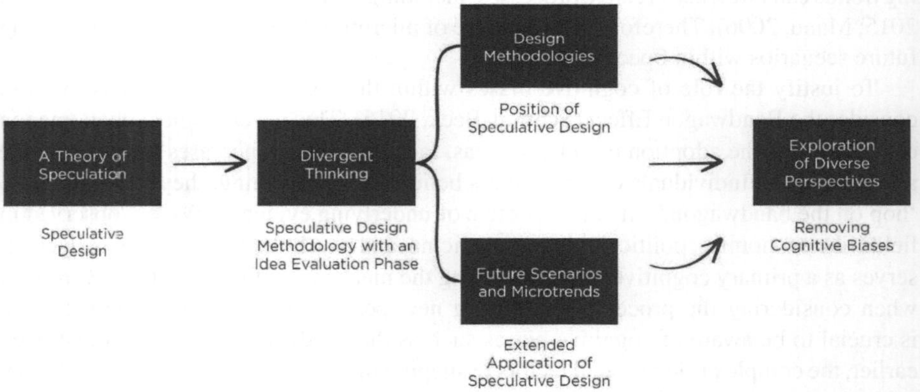

Fig. 1. Visualizing a theory of speculation.

Finally, the study argues for revisiting the initial ideas of Speculative Design to implement this alternative philosophy in collaboration with various experts, including ethicists, political scientists, economists, and others. As a critical medium, it can actively promote exploring and adopting the unexpected, which may offer a more inclusive approach to reflecting on cognitive biases. Discussions may provide a way to evaluate speculative ideas using a participatory approach and the involvement of people from diverse fields. As a result, this approach can promote a range of beneficial outcomes, including fostering inclusive actions, examining the consequences of various factors, encouraging shared learning among participants, and facilitating engaging discussions. This comprehensive approach helps ensure that diverse viewpoints are considered and that the impacts of decisions are thoroughly understood and addressed. This could be a potential approach to addressing the impact of social factors, like microtrends, in the process of exploring diverse perspectives.

However, to fulfil this, one possible way is the strategic implementation of Speculative Design incorporated within the traditional design methodologies. The previous section highlighted the conceptual similarities between Speculative Design and specific steps integrated into existing design methodologies. Speculative Design has the potential to prepare us for complex problems by leveraging foresight through collective social dreaming (Dunne & Raby, 2013), but only if we can visualize its holistic utilization. Overall, while the idea of a theory of speculation through strategic design implementation is essential, exploring its pragmatic applications is beyond the scope of this paper.

Ultimately, in the context of speculative futures, it is of utmost importance to be aware of past and present experiences in order to envision a more inclusive future. Similarly, the recent Protopia Framework proposed by Monika Bielskyte (2021) emphasizes a more holistic approach that is crucial for analyzing future possibilities beyond traditional dystopian and utopian views. The Protopia Framework aims to create a world that is hostile to prejudice and violence by learning from historical injustices, encouraging open and sometimes uncomfortable conversations, and continuously engaging in learning and unlearning despite inevitable setbacks. Furthermore, by incorporating diverse perspectives beyond binaries and cultural borders and celebrating 'Life-Centered Design' (Spoelstra, 2023), we can more comprehensively define the attributes of speculative bio-futures. It is vital to envision inclusive future scenarios by examining various aspects of life, society, and the environment through ethical, legal, and social lenses. In this process, the proposed theory of speculation can create a space to overcome a key limitation of cognitive biases by highlighting the core principles of Speculative Design and its holistic applications.

4 Conclusion

The study aims to investigate how cognitive biases affect the process of speculation to encourage more inclusive future scenarios. To accomplish this, it examines contemporary Speculative Design methodologies and integrates insights from divergent thinking, decision-making in design, and micro trends in human cognition to develop a theory of speculation. This proposed theory builds on the foundational ideas of Speculative Design as introduced by Dunne and Raby (2013) and includes a methodology for effectively addressing the impact of cognitive biases. The paper further mentions future research possibilities, particularly focusing on the need to implement the theory of speculation, yet the methodology for such an application remains a subject for future investigation. Finally, it highlights how the proposed theory can help create more inclusive future scenarios by acknowledging past and present experiences and mitigating the impact of cognitive biases. To achieve a comprehensive understanding of all entities, it emphasizes the role of humans as the most critical agents in this process, stressing the need to examine and address prejudices while maintaining integrity in various contexts.

References

Banathy, B.H.: Designing Social Systems in a Changing World, 1st edn. Springer, New York, NY (1996)

Behrens, J., Ernst, H., Shepherd, D.A.: The decision to exploit an R&D project: divergent thinking across middle and senior managers. J. Prod. Innov. Manag. **31**, 144–158 (2014)

Bielskyte, M.: Protopia Futures [Framework]. medium.com, 18 May 2021. https://medium.com/protopia-futures/protopia-futures-framework-f3c2a5d09a1e

Dunne, A., Raby, F.: Speculative Everything: Design, Fiction, and Social Dreaming. MIT Press, Cambridge (2013)

Ellis, G.: Cognitive Biases in Visualizations. Springer, Cham (2018)

Engel, C., Singer, W.: Better than Conscious?: Decision Making, the Human Mind, and Implications for Institutions. MIT Press, Cambridge (2008)

Eysenck, M.W., Keane, M.T.: Cogn. Psychol. (2020). https://doi.org/10.4324/9781351058513

Feldman, R.S.: Social Psychology: Theories, Research, and Applications. McGraw-Hill Companies (1985)

Guilford, J.: The structure of intellect. Readings Clin. Psychol., 53–81 (1966). https://doi.org/10.1016/b978-1-4832-0087-3.50008-5

Kaufman, J.C., Sternberg, R.J.: The Cambridge Handbook of Creativity. Cambridge University Press, Cambridge (2019)

Kelley, T., Kelley, D.: Creative Confidence: Unleashing the Creative Potential Within us All. Crown Currency, New York City (2013)

Lilley, D., Wilson, G., Bhamra, T., Hanratty, M., Tang, T.: Design interven-tions for sustainable behaviour. Des. Behav. Change, 40–57 (2017). https://doi.org/10.4324/9781315576602-5

Madore, K.P., Addis, D.R., Schacter, D.L.: Creativity and memory: effects of an episodic-specificity induction on divergent thinking. Psychol. Sci. **26**, 1461–1468 (2015). https://doi.org/10.1177/0956797615591863

Manu, A.: The Imagination Challenge: Strategic Foresight and Innovation in the Global Economy. New Riders (2006)

Mason, H., Mattin, D., Luthy, M., Dumitrescu, D.: Trend-Driven Innovation: Beat Accelerating Customer Expectations. John Wiley & Sons (2015)

Niedderer, K., Clune, S., Ludden, G.: Design for Behaviour Change: Theories and Practices of Designing for Change. Routledge (2017)

Norman, D.A.: Design for a Better World: Meaningful, Sustainable, Humanity Centered. MIT Press, Cambridge (2023)

Penn, M.: Microtrends Squared: The New Small Forces Driving Today's Big Disruptions. Simon & Schuster, New York (2019)

Pohflepp, S.: The Golden Institute. cargocollective.com (2009). https://cargocollective.com/saschapohflepp/The-Golden-Institute

Rescher, N. (n.d.). Reductio ad absurdum. Internet Encyclopedia of Philosophy | An encyclopedia of philosophy articles written by professional philosophers. https://iep.utm.edu/reductio/

Robson, C.: Real World Research. Wiley, Hoboken (2011)

Samsen, D.: BuyProduct, 2009 Dot Samsen [Video]. YouTube (2009). https://www.youtube.com/watch?app=desktop&v=TFz3Tz_eSF4

Sanders, E.B., Stappers, P.J.: Co-creation and the new landscapes of design. CoDesign **4**(1), 5–18 (2008). https://doi.org/10.1080/15710880701875068

Luckhoff, C.: Bandwagon effect. In: Raz, M., Pouryahya, P. (eds.) Decision Making in Emergency Medicine, pp. 53–58. Springer, Singapore (2021). https://doi.org/10.1007/978-981-16-0143-9_9

SpeculativeEdu.: interview circumflex studio. speculativeedu.eu., 11 June 2019. https://speculativeedu.eu/interview-circumflex-studio/

SpeculativeEdu.: interview tobias revell. speculativeedu.eu., 15 June 2019. https://speculativeedu.eu/interview-tobias-revell/

SpeculativeEdu.: interview francisco laranjo. speculativeedu.eu., 17 November 2020. https://speculativeedu.eu/interview-francisco-laranjo/

Spoelstra, J.: What is life-centered design? lifecentereddesign.school, 18 July 2023. https://lifecentereddesign.school/blog/what-is-life-centered-design

Stock, K.: Only Imagine: Fiction, Interpretation, and Imagination. Oxford University Press, Oxford (2017)

Thomas, D.D.: Design for cognitive bias (2020)

Pillars of Ontological Biophilic Design

Sergio Davila Urrutia[1,2](✉) and Tyler Fox[1,2]

[1] Science and Arts for Design Division, Universidad Autonoma Metropolitana, Mexico City, Mexico

[2] Human Centered Design and Engineering, University of Washington, Seattle, USA

Abstract. This article explores a future where design is less human-centered, recognizing the limitations of our current focus. We speculate on a bio-centric future, envisioning dreams and goals that extend beyond sustainability. Our conviction is that we should not approach this future good night gently nor alone; instead, we must collaborate with other beings to expand the meaning of our universe. Inspired by science fiction, we explore ideas that teach us about a vibrant world eager to communicate, relate, and participate in our design processes. As researchers, we believe we are at a tipping point with the opportunity to redefine our beliefs, adjust our objectives, and shift our values towards designing a new ontology. We advocate for designers, artists, and creatives to move beyond basic sustainability or biomimicry towards a deeper, more meaningful engagement with other species. The article culminates in three pillars that serve as a framework for imagining and creating future worlds where humans and non-humans coexist in mutually beneficial relationships. This paradigm shift challenges us to reconsider our place in the world, not as dominant exploiters, but as collaborative caretakers. By adopting these principles, we can foster a more regenerative and inclusive approach to design, ultimately leading to a more harmonious existence with the natural world.

Keywords: Biophilia · Ontology · Relationality · Sympoiesis

1 Introduction

In Orson Scott Card's book *Speaker for the Dead (*1986*)* from the *Ender's Game* series, a xenobiologist studies a species called the Pequeninhos (or "little ones") on the planet Lusitania. This species has a profound relationship with the nearby woods. Observations reveal individuals softly hitting trees with sticks in a complex series of movements. This turns out to be a form of language, where the trees resonate and communicate back to the Pequeninhos. Spoiler alert: it is later discovered that the Pequeninhos and the trees are actually the same species, just at different stages of life. This narrative beautifully illustrates the idea of viewing trees as equals, finding ways to communicate with them in their own language, and preserving sacred spaces for them. Orson Scott Card advocates against "otherness," which is the fear and opposition towards those who are different from us. This fear dissipates when we begin to see others as equals, breaking down the concept of otherness (Andréu, 2013).

A. Mateus et al. (Eds.): WAD 2024, What's Around Design?, pp. 156–168, 2026.
https://doi.org/10.1007/978-3-032-08163-6_12

In this article, we call for a relational practice of design, including and perhaps most importantly, within Speculative Design. If speculative design posits possible futures with the goal of shaping preferable futures, we seek more and varied relational designs between humans and nonhumans. For example, "The Quiet Enchanting" by Superflux (2022) asks questions about human relations to the world and envisions a world in which a reciprocal relation between humans and nonhumans is achieved.

This research will offer a schematic to help designers, especially ourselves, move our design in the desired direction. A view of design that explores how we relate to other species. Such design must particularly explore how we negotiate with them to share resources, shape our environment, add value to each other's existence, and explore the creation of shared meaning. Relational philosophers such as Gilbert Simondon and Alfred North Whitehead help elucidate how being emerges from relations, not despite them (Simondon, 2020; Whitehead, 1929). However, such perspectives add to the conversation; they are not foundational. There is also much to learn from indigenous tribes and their sages, who teach stewardship of animals, plants, and even the spirits of inanimate objects. As we describe below, we understand stewardship to be of critical importance in our relations to nonhumans. It is worth noting that such behaviors are not unidirectional nor limited to human-to-nonhuman interactions. Symbiotic relations, hormone signaling, and allelopathic interactions can be argued as different forms of stewardship.

We see opportunities to expand our understanding of a relational ontology from multiple perspectives, a way to build new ontologies through intercultural understandings that privilege relations between entities, nonhuman and human alike. Embracing alternative definitions, approaches, ontologies, and epistemologies can help humans foster inspiring relationships with the environment. While shamans may lack scientific knowledge, they possess a relationship with nature that is not based on accumulation, capital, or economy. These alternative perspectives could offer novel solutions to the environmental crisis and climate change or simply more meaningful ways of living. Reconsidering our relationship with the environment and negotiating with other species could lead to more definitive solutions rather than merely striving for sustainable consumption. The Anthropocene is the root cause of the environmental crisis, and ontological design offers a path to create new alternatives to the market priorities. This approach transcends mere economic considerations and the accumulation of resources.

Perhaps the solution lies in rethinking our relationship with the environment and other species. There is already a term that reflects a shift in this relationship: Biophilia. This concept could mark the beginning of a new way to define human presence on the planet. Guided by principles such as stewardship, collaboration, and fierce connectivity, humans might integrate and transcend the era of exploitation and extraction, fostering a more harmonious coexistence with the Earth.

2 Background

This concept of *Biophilia* was popularized by the biologist Edward O. Wilson. In his book "Biophilia" (1984), Wilson describes biophilia as "the urge to affiliate with other forms of life." He argues that this tendency is a fundamental part of human nature, shaped by evolutionary processes, and that it has important implications for our health, well-being, and environmental stewardship. The biophilia hypothesis suggests that humans have an innate tendency to seek connections with nature and other forms of life. Wilson posits that this affinity is deeply rooted in our evolutionary history, as our ancestors' survival depended on their interaction with and understanding of the natural environment.

In landscape architecture, the concept of biophilia is embraced and expanded upon by authors like Charles Waldheim and Ian McHarg. Charles Waldheim integrates biophilia into his work by advocating for the seamless incorporation of natural systems into urban design (Waldheim, 2016). He emphasizes creating urban environments that foster human-nature connections, enhancing ecological performance and human well-being. Ian McHarg, in his seminal work "Design with Nature" (1969), lays the foundation for biophilic design by promoting the integration of natural processes and ecological principles into planning and design. McHarg's approach underscores the importance of respecting and working with natural landscapes, ensuring that human developments harmonize with the environment to foster sustainable and life-affirming spaces. Both authors highlight the critical role of biophilia in creating landscapes that nurture the human spirit while promoting ecological health.

Psychological benefits of this connection are significant. Interacting with natural environments can reduce stress, improve mood, and enhance overall well-being.

Studies have shown that exposure to nature can lower blood pressure, reduce anxiety, and improve cognitive function, highlighting the importance of our bond with nature for mental and physical health (Williams, 2017).

Furthermore, the biophilia hypothesis emphasizes the importance of environmental conservation. A deeper connection with nature fosters a sense of responsibility for protecting natural ecosystems. By nurturing our biophilic tendencies, we are more likely to engage in behaviors that promote environmental stewardship. (Williams, 2017).

In summary, Edward O. Wilson's biophilia hypothesis underscores the intrinsic human affinity for nature. Recognizing and embracing our biophilic tendencies can lead to healthier, more fulfilling endeavors that promote well-being and sustainability, ultimately fostering a more harmonious relationship with the natural world.

This is the transition from the Anthropocene to the Ecocene, where humans recognize other species as sentient and collaborate with them to expand meaning and comprehension of our world (Küpers, 2020). This shift represents a new exploration of the relationship between the perceiver and the perceived, the body and the world, and subjectivity and objectivity—concepts encapsulated in Merleau-Ponty's idea of the chiasm, and Simondon's (Merleau-Ponty's pupil) concept of the individual-milieu dyad, wherein individuations (becomings) of individuals reverberate through their associated milieux. Merleau-Ponty argues that there is a fundamental intertwining between the perceiver and the perceived world, suggesting that we are not separate from the world we perceive but are deeply intertwined with it. This intertwining is the chiasm, where the boundaries between subject and object become blurred. His thesis for overcoming dualism could be initiated by blurring the lines of individualization and hierarchization among species. A vast universe might be discovered when humans experience the world through the eyes and thoughts of other species (Merleau-Ponty, 1968).

Simondon argues that philosophy's attention should focus on individuations not individuals. That is to say, we should pay attention the how we become, rather than who or what becomes. Individuals are linked to their associate milieu. In this relational ontology, the world is metastable, energies and potential are activated across individuals and milieux alike, leaving behind excess and other untapped potential that will individuate in the future. Individuation does not stop. Simondon argues that this process can be seen across all modes of existence, living, nonliving, technical, human, and nonhuman. Technology in particular is important for Simondon, as it is a mediator between humans and nature (Simondon, 2020).

Together with the chiasm, we can see that our relations with nonhumans, the surrounding associated milieu, shape our becoming, which in turn shapes the milieu. Nothing individuates in isolation.

Thus, the main thesis of this article is that by deeply, emotionally, intellectually, and collaboratively relating with other species, the experience of existence will become more meaningful. There is an entire universe to be discovered, and the distinctions between species are less important than the relations between them. The authentic experience of the world is one of interconnectedness and mutual constitution—it is a poem written by many authors.

3 Ontological Design

Arturo Escobar has contributed to the above-mentioned poem. Escobar's ontological design calls for multiple ways of being and knowing in design practices.

(Escobar, 2018). This "Pluriversal" approach challenges the dominance of Western, modernist design paradigms and aligns closely with indigenous worldviews mentioned above, emphasizing stewardship of nature.

Key to Escobar's thinking is the concept of relational ontology, which sees the world as fundamentally interconnected, where beings don't pre-exist in their relations but come into being through them. This perspective resonates with Simondon's concepts, as well as with biophilic thinking that sees humans as intrinsically connected to nature. However, it pushes us to go further than either, moving beyond simply incorporating nature into human-centered designs and instead reimagining our very relationship with the natural world as a fundamental act of design. Escobar, Osterweil, and Sharma point to such a lineage of relational philosophy in the work of Spinoza, Whitehead, Deleuze & Guattari, and the American Pragmatists, among others, stating "There is much to be learned from these exponents of a wiser, alternative West, who still help us to unlearn the mainstream West" (Escobar et al., 2024, p. 141).

This ontological approach to biophilic design pushes us to consider not just how we can use nature in our designs, but how our designs shape our relationship with nature and our understanding of our place within it. It challenges us to create designs that are not just sustainable, but regenerative, interdependent, fostering the flourishing of human life through the flourishing of non-human life.

4 Biophilic Creative Ecosystem Diagram

In order to illustrate the solutions of the Biophilic Creative Ecosystem, we decided to cross-reference the concept of meaning creation with the idea of emotional relationships. This approach allows us to reflect on cases and experiences that can be situated within this framework. In the diagram (Fig. 1), the X-axis represents relationality, ranging from *Otherness*, where an individual is perceived as separate and disconnected from its counterpart, to increasing relationality. The Y-axis represents *Meaning Creation.* This intersection of axes creates four quadrants: Biomimicry in the bottom left, Biotech in the top left, Sustainability in the bottom right, and Biophilia in the top right. Here is the approach for each of these concepts:

Biomimicry: Biomimicry is limited by an extractive understanding of biological forms, a perspective that adopts forms and functions with less attention to context. Biomimicry is related to a hylomorphic epistemology wherein form and matter are what constitutes a being—living or nonliving, natural or created. Simondon critiques hylomorphism, arguing instead that philosophy (and, by extension, the rest of us) should focus not on the individual (the resulting being seen in form and matter), but in individuation: the relational processes between interacting forces, entities, and context by which things come to be. In short, Simondon argues for a relational ontology that eschews hylomorphism and similar epistemologies (Simondon, 2020) A practical example of biomimicry is Velcro, the hook-and-loop fabric inspired by burrs in nature. Practical, clever, and useful to be sure, but not an item that directly enhances our relation to nonhuman entities, at least not living ones (Velcro.com, 2020).

Biotech: Biotech can be understood equally as extractive, yet the materiality of intervention is not mimicry. Living material is manipulated, engineered even, to take on new properties. "Plascetamol" is a speculative design product from the winning 2024 Bio Design Challenge project "Consume Our Consumption." The team isolated and amplified the capacity of the bacteria *Zophobas morio* to consume plastic. They then imagined the, presumably, over-the-counter medication Pascetamol that humans might take to help remove the microplastics in our gut (UT Austin 2024. (n.d.). Biodesign Challenge. Retrieved July 27, 2024, from https://www.biodesignchallenge.org/ut-austin-2024. This fantastic project exemplifies the engineering mindset found within biotech: a process of amplification of nonhuman properties to be mined, bottled, and sold. Such processes, we argue, no matter how fascinating and innovative, do little to enhance meaningful relations between humans and nonhumans. Or, perhaps more precisely, the meaningful foci of the creators offer a knowing wink that this only addresses our overproduction and overconsumption of plastic.

Sustainability: Sustainability encompasses practices like recycling, circularity, and resource management, which are crucial for addressing the environmental impacts of industrialized modern life. However, these approaches primarily focus on maintaining current systems rather than fostering a deeper, regenerative relationship with nature. They emphasize the preservation of existing conditions without addressing the broader interdependence inherent in ecosystems or promoting the creation of new, meaningful connections. True sustainability should extend beyond mere maintenance to actively enhance and regenerate our interactions with the natural world.

Biophilia: In contrast to conventional sustainability practices, which often focus on maintaining the status quo and minimizing harm, the concept of Biophilia emphasizes forging deeper, more meaningful relationships with other species and the natural world. While sustainability tends to center on preserving existing systems and processes, often within the framework of industrialized life, Biophilia encourages a transformative

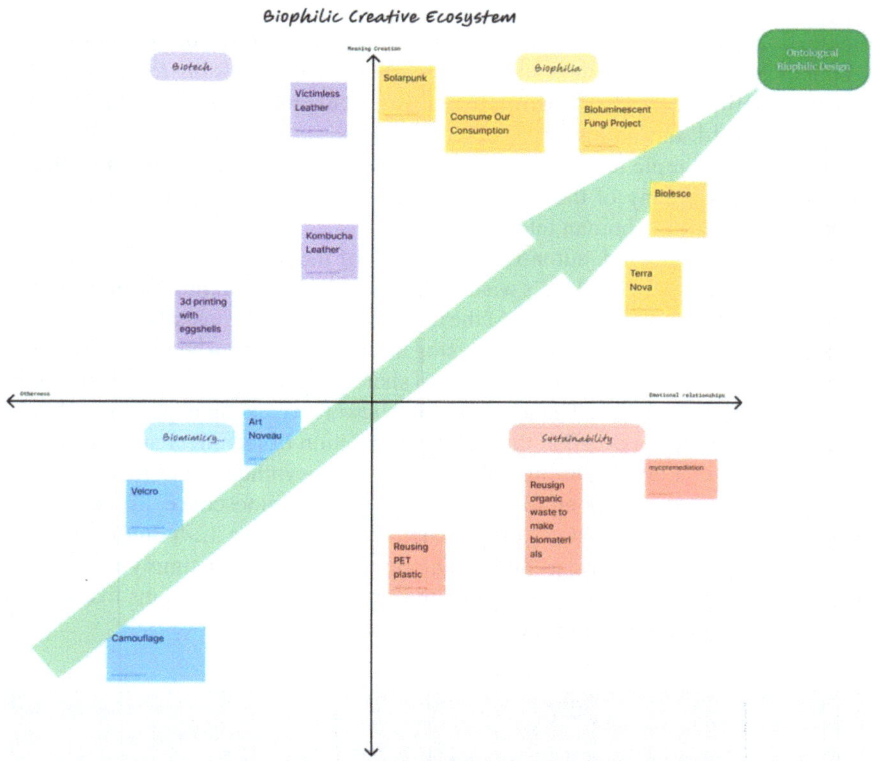

Fig. 1. Biophilic Creative Ecosystem Diagram. Author 1 & Author 2. 2024.

approach that fosters interdependence and regeneration. Our approach briefly defines Biophilia as "meaningful relationships with other species," it extends beyond mere coexistence or conservation. The biophilic approach seeks to cultivate a profound connection with the natural world, recognizing and valuing the intricate web of life that sustains us.

This perspective advocates for a shift from maintaining static systems to actively engaging in practices that enhance the vitality and health of ecosystems. By embracing Biophilia, we acknowledge the dynamic interdependence between species and strive to regenerate natural processes rather than just mitigate their impact. This regenerative approach not only aims to repair ecological damage but also to build resilient, flourishing environments where human and non-human life can thrive together. In essence, Biophilia represents a commitment to nurturing the rich, reciprocal relationships that underpin the health of our planet, fostering a deeper appreciation for the interconnectedness of all living beings.

4.1 Biolesce (Author 2, 2015)

Biolesce is a series of iterative installations and sculptures featuring the expressive potential of bioluminescent algae, made by Author 2. *Pyrocystis fusiformis* is a single-celled, non-motile, ocean-dwelling dinoflagellate that expresses a flash of blue light when physically agitated. In the artworks comprising Biolesce, algae are physically agitated through relatively simple technological arrangements of motors, servos, and pumps for a human audience. Many of the installations are interactive, using a biosensor (like a heartbeat sensor) for the human interactants.

This way, the embodied, autonomic processes of the human interactant (heartbeat) trigger physical pulses in the algae's water, which results in a bioluminescent response—also an embodied, autonomic process of the algae. A visitor to the installation can see blue light exhibited through living plant life pulsing in time to their own heartbeat. The goal of this series of sculptures is to create moments of shared experience across human and non-human experiential registers. The artwork asks viewers to *feel* an experience alongside the algae, a feeling made possible through the mediation of technical apparatuses.

P. fusiformis is diurnal. During the day, it is photosynthetic, and then after a few hours in the dark, it can respond to movement with bioluminescence. To make this art experience possible, it necessitates proper care. In addition to a 12-h on/off light cycle, over time the algae need nutrients and clean seawater. This is not simply metaphorical, given the current state of the oceans around the world, stewardship seems like a missing aspect of human relations to the earth (Figs. 2 and 3).

Fig. 2. Biolesce 0.5 2024, on display at the Pensacola Museum of Art, 2023, photo credit Richard Rodriguez, Historic Trust.

Fig. 3. Biolesce 0.5 2024, on display during its 'day phase' at the Pensacola Museum of Art, 2023 photo credit Richard Rodriguez, Historic Trust.

4.2 Bioluminescence in Fungi Project (Author 1, 2024)

The *Bioluminescence in Fungi Project* aims to foster interdisciplinary collaboration between product designers, biologists, and mycologists to explore the potential of bioluminescent fungi as a sustainable and aesthetically pleasing source of illumination in urban environments. Moreover, the collaboration between fungi and humans aims to be based on principles of fair trade and life affirming solutions.

The team for this project is based in Mexico where the growing number of documented species of bioluminescent fungi is growing every year. So far, the list goes in 16 species making Mexico the second largest country in documented bioluminescent fungi just after China.

The primary objective of the project is to establish a comprehensive understanding of bioluminescent fungi, select suitable species for cultivation, determine optimal growth conditions, and explore innovative applications for these fungi in human settlements. Our overarching goal is to enhance the city environment, promote biophilia, and reduce light pollution through the integration of bioluminescent fungi into public spaces.

Finally, the laboratory dedicated to growing mushrooms will become a.

"Makerspace." Here, some solutions for shelter, colonization and acclimatization will be designed. The bioluminescent mushrooms will live in luminaries for public space (Fig. 4), in furniture, niches in parks and gardens, and even have some shelter in people's homes.

Fig. 4. Bioluminescence in bicycle pathways (2023). *Bioluminescence in Fungi Project.* [AI-generated image]. Credits to Mario Castro and Sergio Dávila.

5 Ontological Biophilic Design

Our diagram also includes an arrow indicating the desired direction: Ontological Biophilic Design. This approach involves designing products, services, systems, and strategies that foster meaningful experiences in collaboration with other species, thereby redefining human behavior towards a more interconnected and respectful relationship with the natural world.

Ontological Biophilic Design delves into how we can engage with other species within the creative process. The proposed diagram (Fig. 1) serves as a compass to show what kind of design would be expanding our understanding and deepening our insights. This is far from a design that just replicates nature and keeps it at bay. The aim is to shift the focus from human-centered design to one that acknowledges and explores the vast kingdoms of nature, inviting creativity that arises from a collaborative relationship with other species.

In a manner reminiscent of the film *The Arrival* (Villeneuve, 2016), where a character reflects that the most profound aspect of encountering extraterrestrials was not their technology or physiology, but the new perspectives gained through their shared experience. Ontological Biophilic Design invites us to similarly expand our worldview. It encourages us to consider not only the tangible benefits of collaborating with other species but also to embrace their unique ontologies, definitions, and perspectives. By listening to the trees, collaborating with fungi and algae, and learning from the language of crickets, we may uncover profound insights that transform our understanding of the universe and our place within it.

There are certain convictions in the Ontological Biophilic Design. We will call them *pillars* as they give the basic structure for these meaningful relationships and shared interspecies experiences. They would be the following 3 concepts:

5.1 Sympoiesis

In "Staying with the Trouble" (2016), Donna Haraway critiques the concept of teleology in nature, arguing against the idea that life has a predetermined purpose or endpoint. Teleological thinking often implies a linear progression toward a specific endpoint or purpose. She argues that this perspective oversimplifies the dynamic and non-linear nature of evolutionary processes. Instead, she promotes the concept of "sympoiesis," emphasizing the interconnected and co-evolving nature of living systems. Sympoiesis is defined as "making-with," it underscores the collaborative and co-evolving nature of life, where organisms and environments shape each other in complex and ongoing processes.

Haraway introduces the Chthulucene as a new epoch that highlights multispecies storytelling and the interdependence of all life forms. She advocates for making kin to foster ethical and sustainable relationships and encourages us to stay with the trouble by engaging with the complexities of contemporary life. Haraway's work challenges us to rethink our understanding of nature, recognize the interconnectedness of all life forms, build alliances that transcend traditional biological and social boundaries, and embrace a more integrative and reciprocal approach to living in a damaged world.

She urges us to confront the difficulties and uncertainties of living in a damaged world and to find ways to coexist and thrive within it. This involves embracing uncertainty, resisting deterministic thinking, and fostering creativity and resilience.

5.2 Stewardship

In the book "Designs for the Pluriverse" Arturo Escobar discusses the cosmologies and worldviews of various indigenous groups, including the Koguis and Arahuacos from the Amazonian region (2018). These tribes refer to themselves as the "older brothers," positioning themselves as spiritual guardians of the Earth. They believe it is their responsibility to perform rituals and prayers to ensure the balance and well-being not only for themselves, but for all beings of the world. This includes praying for the sun to rise each day, which they view as a sacred duty essential for the continuation of life for all beings. This worldview emphasizes a deep connection to nature and a profound sense of duty to maintain harmony in the world, highlighting the indigenous understanding of interdependence and stewardship.

The privileged position of humankind in the world should be reconsidered. The responsible direction would be to take our place as caretakers of nature. Stewardship requires a shift in human behavior towards the environment and other beings. This new way of being would humble us as a species and change our priorities from capital accumulation and resource exploitation to assistance, midwifery, and stewardship.

5.3 Relationality

This relational understanding of the world aligns with the recent work of Arturo Escobar, Michal Osterweil, and Kriti Sharma (2024). To return to biophilic design, we see form and function as being privileged over relations. We do not argue against "integration with natural systems" or "sustainable resource use" but lament the lack of focus on human-nonhuman relations. How might designers adopt a relational understanding of designing the mutual individuation of humans and nonhumans alike?

Before moving forward, we wish to applaud the work of McHarg and Waldheim on their biophilic design principles for the built environment. We see the call for biophilia as a motivating force in design as a deeper understanding of the natural world than other biological paradigms, such as biomimicry. Biomimicry is limited by an extractive understanding of biological forms, a perspective that adopts forms and functions with less attention to context. Biomimicry is related to a hylomorphic epistemology wherein form and matter are what constitutes a being—living or non-living, natural or created. Simondon critiques hylomorphism, arguing instead that philosophy (and, by extension, the rest of us) should focus not on the individual (the resulting being seen in form and matter) but on individuation: the relational processes between interacting forces, entities, and context by which things come to be. In short, Simondon argues for a relational ontology that eschews hylomorphism and similar epistemologies.

This relational understanding of the world aligns with the recent work of Arturo Escobar, Michal Osterweil, and Kriti Sharma. To return to biophilic design, we see form and function as being privileged over relations. We do not argue against "integration with natural systems" or "sustainable resource use," but we also lament the lack of focus on human-nonhuman relations. How might designers writ large adopt a relational understanding of designing the mutual individuation of humans and nonhumans alike?

6 Conclusions

The concept of Ontological Biophilic Design offers an underlying structure for transforming our relationship with the natural world, one that closely aligns with the goals of speculative design in envisioning preferable futures. By embracing the pillars of sympoiesis, stewardship, and relationality, we move beyond mere sustainability or biomimicry towards a deeper, more meaningful engagement with other species. These pillars serve as a framework for imagining and creating future worlds where humans and non-humans coexist in mutually beneficial relationships. This paradigm shift challenges us to reconsider our place in the world, not as dominant exploiters, but as collaborative caretakers. Projects like Biolesce and the Bioluminescence in Fungi Project exemplify this speculative, biophilic approach, illustrating how we might create shared experiences and collaborative relationships with other life forms.

As we face ongoing environmental challenges, adopting an ontological biophilic approach in both design and worldview offers new pathways for addressing these issues. It invites us to listen to the language of trees, collaborate with fungi and algae, and learn from the myriad forms of life around us. In doing so, we open possibilities for innovative solutions and new ways of being that not only address current problems but also enrich our understanding of the universe and our place within it. Through this lens, speculative design becomes a powerful tool for exploring and shaping a future where the boundaries between species blur, and where human and non-human flourishing are inextricably linked. In doing so, we may not only find innovative solutions to our current problems but also discover new ways of being that enrich our understanding of the universe and our place within it.

References

Andreu, L.I., Rudnicki, R.W.: Beyond space colonialism: otherness in orson scott card's speaker for the dead. In: 2013–2014 Contest "So you think you can write?" School of Literature and Language. Louisiana Tech University (2013). https://www2.latech.edu/~bmagee/writing_contest/20132014/400--research_paper.pdf. Accessed 26 Jul 2024

Ávila, M.: Designing for interdependence: a poetics of relating. Bloomsbury Visual Arts (2022)

Beck, D.: Spiral Dynamics in Action: Humanity's Master Code. John Wiley & Sons, New York (2018)

Bierend, D.: In Search of Mycotopia: Citizen Science, Fungi Fanatics, and the Untapped Potential of Mushrooms. Chelsea Green Publishing (2023)

Escobar, A.: Designs for the Pluriverse: Radical Interdependence, Autonomy, and the Making of Worlds. Duke University Press (2018)

Escobar, A., Osterweil, M., Sharma, K.: Relationality an Emergent Politics of Life Beyond the Human. Bloomsbury Publishing USA (2024)

Gagliano, M.: Thus Spoke the Plant: A Remarkable Journey of Groundbreaking Scientific Discoveries and Personal Encounters with Plants. North Atlantic Books (2018)

Godfrey-Smith, P.: Other Minds: The Octopus, the Sea, and the Deep Origins of Consciousness. Farrar Straus and Giroux, New York (2017)

Haraway, D.J.: Staying with the Trouble: Making Kin in the Chthulucene. Duke University Press (2016)

Hạnh, Thich N., Levitt, P.: The Heart of Understanding: Commentaries on the Prajñaparamita Heart Sutra. Parallax Press (2016)

Kohn, E.: How Forests Think: Toward an Anthropology Beyond the Human. University of California Press (2018)

Kraus, R.: Happy City. Little Simon (1987)

Küpers, W.: From the Anthropocene to an 'Ecocene': Ecophenomenological perspectives on embodied, anthrodecentric transformations towards enlivening practices of organising sustainably. Sustainability **12**(9), 3633 (2020). https://doi.org/10.3390/su12093633

Makanadar, A.: Neuro-adaptive architecture: buildings and city design that respond to human emotions, cognitive states. Res. Globalization **8**, 100222 (2024). https://doi.org/10.1016/j.resglo.2024.100222

McCoy, P., Rogers, R.D.: The Mycocultural Revolution: Transforming Our World with Mushrooms, Lichens, and Other Fungi. Microcosm Publishing (2022)

McHarg, I. L.: Design with Nature. Natural History Press (1969)

Merleau-Ponty, M.: The Visible and the Invisible. Northwestern University Press (1968). Lefort, C. (ed.) Lingis, A. (Trans.) (Original work published 1964)

Card, O.S.: Speaker for the Dead. Tor Books (1986)

Rupprecht, C., Cleland, D., Tamura, N., Chaudhuri, R., Ulibarri, S.: Multispecies Cities: Solarpunk Urban Futures. World Weaver Press (2021)

Sheldrake, M.: Entangled Life. Jonathan Cape (UBD) (2023)

Simondon, G.: Individuation in Light of Notions of Form and Information (Adkins, T. (Trans.)). Univ Of Minnesota Press (2020)

. Simard, S.: Finding the Mother Tree: Discovering the Wisdom of the Forest. Vintage Books, A Division of Penguin Random House LLC (2022)

Thackara, J.: Cómo prosperar en la economía sostenible: Diseñar Hoy El mundo del futuro. Experimenta (2016)

. Villeneuve, D. (Director): Arrival [Film]. Paramount Pictures (2016).

Velcro.com: A Mind-Blowing Biomimicry Example | VELCRO® Brand Blog. USA, 27 July 2020. https://www.velcro.com/news-and-blog/2020/07/a-mind-blowingbiomimicry-examples/

Waldheim, C.: Landscape as Urbanism: A General Theory. Princeton University Press (2016)

Whitehead, A.N.: Process and Reality, An Essay in Cosmology; Gifford Lectures Delivered in the University of Edinburgh During the Session 1927–28. The Macmillan Co., University Press (1929)

Wilber, K.: The Future of Spirituality: Why It Must be Integral. Sounds True (2013)

Williams, F.: The Nature Fix: Why Nature Makes Us Happier, healthier, and More Creative. W.W. Norton & Company (2017)

Wilson, E.O.: Biophilia. Harvard University Press (1984)

UT Austin 2024. (n.d.): Biodesign Challenge. https://www.biodesignchallenge.org/ut-austin-2024. Accessed 27 Jul 2024

Nature-Based Solutions Inspirations

Prologue: Designing Our Shared Future: Landscape Design and the Promise of Nature-Based Solutions

José Miguel Lameiras, NbS Expert FCUP.

We stand today at a defining crossroads, faced with the pressing realities of climate change, urban expansion, and ecological decline. As designers, we bear not only the creative privilege but also the profound responsibility to shape environments that sustain and uplift communities, ecosystems, and future generations. Our task is not merely aesthetic—it is fundamentally ethical and deeply connected to the well-being of all living beings.

Landscape design is uniquely positioned as a discipline bridging art, ecology, and social consciousness. It holds the extraordinary power to envision and implement transformations from the broad scale of urban planning to the intimate detail of a neighborhood garden. At every scale, our designs must embrace sustainability, resilience, and the adaptive capacity of nature itself.

Nature-based solutions offer us a guiding principle: to design with, rather than against, nature. By harnessing natural processes, restoring ecological functions, and integrating biodiversity into the fabric of our cities and towns, we enhance both environmental resilience and human well-being. Trees become more than shade—they become tools for urban cooling, stormwater management, air purification, and community healing. Green spaces are no longer merely ornamental—they are integral infrastructure, supporting biodiversity, climate adaptation, and social cohesion.

Yet, sustainability is more than ecological. True sustainability embraces social equity, recognizing that thoughtful design must uplift communities, providing equitable access to green spaces and fostering meaningful engagement. The parks, streets, plazas, and gardens we create are the platforms upon which daily lives unfold, influencing the social dynamics and quality of life in profound and lasting ways. Our responsibility extends beyond physical spaces; it encompasses creating inclusive environments that celebrate diversity and foster a sense of belonging.

In confronting climate change, designers must think holistically, designing landscapes that anticipate and respond to climatic stresses, from flooding to heatwaves. By integrating nature-based solutions such as green roofs, permeable pavements, rain gardens, and resilient native plantings, we equip our cities to adapt and thrive despite an uncertain future.

As we move forward, let us be driven by a vision of cities and landscapes not just surviving, but flourishing—where natural ecosystems are revitalized, communities strengthened, and sustainability embedded deeply into the fabric of our designs. Let us commit ourselves to accountability, ensuring every project reflects our ethical duty to future generations.

Together, we can—and we must—lead by example, pioneering design solutions that blend ecological intelligence, creativity, and compassion. This is our calling and our legacy: to design landscapes that are regenerative, socially just, and resilient.

Let us design responsibly, boldly, and compassionately.

Introduction to Part 4: Nature-Based Solutions in Design and Urban Regeneration

Américo Mateus, ISMAT, Portugal; Gonçalo C. Moniz, Coimbra University

In the face of escalating environmental challenges, urban areas are called to embrace innovative strategies that not only mitigate ecological and social crises but also foster resilience and well-being. Nature-Based Solutions (NBS) stand at the forefront of this transformation, offering an integrative approach to urban regeneration that harmonizes natural systems with human development. More than mere green interventions, NBS represent a paradigm shift—one that reconnects communities with their landscapes through co-creation, co-design, and collaborative governance.

The What's Around Design Conference track on Nature-Based Solutions brings together exemplary practices from across Europe, showcasing projects in Italy, Portugal, Turkey, and Spain. These initiatives illustrate the diverse spectrum of NBS, from participatory approaches that empower communities to highly technological solutions that enhance ecological performance and urban resilience. Yet, despite their varied methodologies, all share a fundamental principle: the centrality of citizens and local stakeholders in shaping their environments.

Co-creation and co-design emerge as critical processes in the successful implementation of NBS. Urban and territorial challenges are complex, deeply embedded in the social, economic, and ecological fabrics of cities. Citizens, local organizations, and stakeholders possess invaluable knowledge about the lived realities, the rhythms, and the intricacies of their territories. They are not passive recipients of interventions; they are co-authors of sustainable urban futures. When engaged meaningfully, they provide insights that ensure NBS are not only contextually relevant but also socially embraced, fostering long-term stewardship and maintenance of these solutions.

The selected works in this track illuminate the potential of NBS to address multifaceted urban challenges. They span the revitalization of degraded neighborhoods through community-driven greening projects, the integration of blue-green infrastructure in flood-prone areas, and the application of digital tools to enhance environmental monitoring and engagement. These projects embody the interplay between nature, technology, and human agency, demonstrating that sustainable urban regeneration is not solely a matter of policy or design expertise but a collective endeavor rooted in inclusivity and adaptability.

Furthermore, the role of design in NBS implementation cannot be overstated. Design, in its broadest sense, is a mediator between human aspirations and ecological imperatives. It shapes how spaces are perceived, utilized, and valued. The works presented here showcase the critical role of design in facilitating participatory processes, translating scientific and technical knowledge into accessible strategies, and crafting interventions that are both functional and poetic. Through design, NBS become tangible expressions of resilience—interwoven with local identities, histories, and aspirations.

This collection of case studies serves as an inspiration and a call to action. As cities grapple with climate change, biodiversity loss, and social fragmentation, NBS offer a way forward—an approach that embraces nature as an ally rather than an obstacle, and communities as partners rather than beneficiaries. The insights shared in this track reaffirm that regenerative urban futures are not only possible but already in motion, shaped by the synergy of ecological intelligence and human creativity.

We invite readers to explore these transformative projects, to engage with the methodologies presented, and to envision the boundless possibilities of Nature-Based Solutions in shaping more sustainable, inclusive, and resilient urban landscapes.

From Green Spaces to Healthier Cities: Community-Driven Nature-Based Solutions

Xavier Ferrari Tumay[1(✉)] and Federica Maria Lorusso[2]

[1] University of Genova, Genova, Italy
xavier.ferraritumay@unige.it
[2] University of Campania Luigi Vanvitelli, Caserta, Italy

Abstract. This article explores the role of Nature-based Solutions (NbS) in enhancing urban health and well-being, with a focus on the synergy between community-driven innovation. By examining case studies in various urban contexts in Italy (both top-down and bottom-up approaches), we highlight the importance of involving local communities in the development of NbS to address environmental challenges while simultaneously promoting social equity. This article explores the role of Nature-based Solutions (NbS) in enhancing urban health and well-being, with a focus on the synergy between community-driven innovation and urban policy. The findings demonstrate how holistic approaches can lead to sustainable and health-promoting environments that reflect the needs and aspirations of residents.

Keywords: Biodiversity · Nature-Based Solutions · Participation · Design-Driven Strategy · Urban Environment

1 Introduction

Stefano Mancuso (2020), through the concept of "Fitopolis", invites us to consider the city as a living organism, in which Nature-Based Solutions (NbS) play a vital role in maintaining the balance between nature and the built environment. This holistic approach to urban design can facilitate the creation of healthier and more livable urban spaces.

Modern cities are facing an unprecedented series of environmental challenges, including climate change, pollution, and biodiversity loss. These issues demand innovative solutions that go beyond traditional urban planning approaches. Franco La Cecla (2015), in his book "Against Urban Planning", sharply criticizes modern urbanism, describing it as an art that has lost touch with the real needs of people and local communities. According to La Cecla, contemporary urban planning is often driven by technocratic and bureaucratic logics, which prioritize grandiose and impersonal projects while neglecting the social and cultural fabric of cities. He emphasizes how this approach has led to the creation of alienating urban spaces, unable to meet the daily needs of residents or effectively address environmental challenges.

A clear example of these issues is represented by overtourism and mass tourism, which have a significant impact on the economic and environmental sustainability of

A. Mateus et al. (Eds.): WAD 2024, What's Around Design?, pp. 173–186, 2026.
https://doi.org/10.1007/978-3-032-08163-6_13

cities. Recent studies have highlighted how mass tourism can overwhelm urban infrastructure, increase the cost of living for residents, and degrade both the natural and built environment. According to Milano et al. (2019), overtourism not only compromises the quality of life of local residents but also contributes to the loss of cultural identity in cities and the transformation of historic neighborhoods into commercial areas devoid of authenticity. This phenomenon underscores the inability of traditional urban planning to sustainably manage economic and environmental pressures, leading to the creation of urban spaces that, while attracting investments and tourists, ultimately alienate residents and compromise the quality of the urban environment.

To address these challenges, La Cecla proposes a "listening urbanism," where solutions are co-designed with local communities, respecting the complexity and diversity of urban contexts. In this perspective, Nature-Based Solutions (NbS) emerge as fundamental tools for rethinking cities, not only in terms of environmental sustainability but also in terms of social equity and collective well-being. NbS, therefore, do not merely represent a technical alternative but a new paradigm of urban planning that centers the relationship between humans and the environment.

The concept of Green Urbanism has emerged as a critical response to conventional urbanization practices, which often prioritize economic development at the expense of environmental sustainability and social well-being. As discussed by McCann (2017), Green Urbanism proposes a paradigm of urban planning that holistically integrates the principles of ecological sustainability, social equity, and economic resilience. This approach recognizes that urban sustainability cannot be achieved through fragmented or sectoral solutions but requires a structural transformation of cities, including the promotion of sustainable mobility, the expansion and protection of green spaces, and the creation of inclusive and accessible public spaces.

Another central pillar of Green Urbanism is the mitigation of urban temperatures, both in summer and winter, through the strategic planning and implementation of green spaces. Urban green spaces, such as parks, gardens, and green roofs, have a significant impact on reducing the urban heat island effect, a phenomenon where urban areas become significantly warmer than surrounding rural areas. As discussed by Bowler et al. (2010), during the summer, urban vegetation can lower air temperatures through shading and the process of evapotranspiration, improving thermal comfort for residents. In winter, green areas can act as barriers against cold winds, reducing the need for building heating and, consequently, greenhouse gas emissions. However, McCann (2017) emphasizes that these benefits are not only environmental but also social and economic: the integration of urban green spaces is seen not only as a strategy to mitigate the impacts of climate change, such as the urban heat island effect, but also as a means to promote social equity and community cohesion. In this context, Nature-Based Solutions (NbS) play a crucial role in Green Urbanism, as they provide solutions that not only address environmental challenges but also enhance urban quality of life. These solutions also serve as catalysts for social interaction, strengthening the sense of community and belonging among residents.

McCann (2017) further emphasizes that the effective implementation of Green Urbanism requires a radical shift in urban policies and governance practices. NbS must be integrated into a participatory planning framework that actively involves local communities in the decision-making process. This approach not only enhances the legitimacy of environmental policies but also ensures that the adopted solutions genuinely address the needs and aspirations of residents.

The research by Kabisch et al. (2016) underscores the crucial importance of Nature-Based Solutions (NbS) in addressing climate change challenges in urban areas, but also highlights several gaps and difficulties that limit the effectiveness of these solutions. One of the primary challenges concerns the measurement of the long-term impacts of NbS. Although it is evident that these solutions can provide significant benefits, there is a scarcity of long-term data that allows for a comprehensive and accurate evaluation of their impact over time. This lack of data hinders the ability to reliably predict the future benefits of NbS and to adjust implementation strategies accordingly.

Another significant gap is the lack of standardized indicators for measuring the effectiveness of NbS. While various indicators, such as the reduction of urban temperatures or carbon storage capacity, are used to monitor NbS, their application varies widely across projects, making it difficult to compare results. The lack of standardization limits the ability to generalize successes achieved in one context and apply them in others, thereby reducing the effectiveness of NbS as a global tool for climate mitigation and adaptation. Moreover, NbS require an approach that integrates multiple disciplines, from natural sciences to social and economic sciences. However, there is often a lack of adequate interdisciplinary knowledge, resulting in fragmented planning and implementation of NbS. This shortage of integrated expertise can lead to solutions that fail to fully leverage the potential of NbS to address urban challenges in a holistic manner.

Another crucial challenge is community engagement. Although the involvement of people living in areas affected by NbS is essential for their success, there are still significant gaps in how to effectively engage the diverse stakeholders, especially in complex and densely populated urban contexts. Furthermore, adequate tools to facilitate continuous and meaningful community involvement are not always available, which can result in projects that do not fully address local needs or fail to garner the necessary support.

In this context, community engagement through Citizen Science initiatives is essential for the success of NbS. Active citizen participation not only increases the legitimacy of the proposed solutions but also enables the collection of valuable data and the co-design of interventions that truly meet local needs. Raymond et al. (2017) propose an impact evaluation framework to support the planning and assessment of NbS projects, which is based on the inclusion of local communities and the measurement of long-term environmental and social benefits.

Political and institutional barriers present another significant obstacle. Often, NbS fail to be implemented due to conflicts of interest, lack of political will, or difficulties in coordinating across different agencies and sectors. These barriers can prevent the integration of NbS into existing urban policies, relegating them to marginal solutions rather than central components of a sustainable urban strategy.

The concept of Citizen Science emerges as a promising strategy to overcome some of these gaps. By actively involving citizens in data collection, monitoring, and the co-design of NbS, Citizen Science can help bridge knowledge gaps, improve the quality and quantity of available data, and ensure that the proposed solutions are closely aligned with the needs and expectations of local communities. Moreover, the participatory approach of Citizen Science can contribute to overcoming political and institutional barriers by increasing the legitimacy of NbS and fostering greater public and political support for their implementation.

A concrete example of the integration of Nature-Based Solutions (NbS) into urban planning is the "3–30-300" model proposed by Cecil Konijnendijk (2023). This model emphasizes the importance of ensuring that every citizen can benefit from a daily connection with nature through three key principles: each individual should be able to see at least three trees from their home, live in an urban area with 30% tree canopy cover, and have access to a large public green space within a 300-m radius. The application of this model not only aims to improve urban quality of life and mental well-being but also contributes to mitigating the negative environmental impacts associated with urbanization. A constant connection with nature can reduce stress, improve mental and physical health, and foster a sense of community among residents.

Similarly, Pauleit et al. (2017) describe four distinct "shades of green" in NbS, representing different levels of nature integration within the urban context. These shades range from minimal integration, where nature is used in a limited manner, to maximum integration, where nature becomes a predominant element of the urban landscape. This classification demonstrates the flexibility of NbS in adapting to various urban contexts, allowing cities to adopt solutions that range from gradual interventions to radical transformations, based on specific local needs and priorities. In this way, NbS can be strategically implemented, taking into account the socio-ecological characteristics of different urban areas and contributing to more resilient and sustainable urban planning.

2 NbS to Enhance Urban Public Space: Italian Case Studies

In this contribution, we have analysed several national case studies concerning the implementation of NbS, focusing on interventions that have converted and enhanced abandoned or underutilised urban public spaces into new environments for social aggregation and interaction. Drawing from various repositories of NbS case studies, such as OPPLA, the EU's online repository for NbS, we have compared some case studies that help counter environmental challenges with the active participation of the community, in order to demonstrate how such interventions can be successfully implemented both through top-down approaches, such as institutional initiatives or government-led projects, and bottom-up community-led actions.

2.1 Bottom up

Le Serre dei Giardini, Kilowatt, Bologna

Le Serre dei Giardini (The Greenhouses of the Gardens), located in 'Giardini Margherita', the main public park in the city of Bologna, is a collaborative space inaugurated in 2015 by Kilowatt cooperative, a local organisation that operates in the fields

Fig. 1. Aquaponic table to allow people to get in touch with the system, understand how it works and spark their interest in taking care of it. le serre dei giardini, Bologna.

of social innovation, urban regeneration, education and food, which took over and redeveloped the 650 square metres of abandoned former municipal greenhouses in Giardini Margherita to give them back to the city (Leone et al., 2020). To rehabilitate the public space granted by the Municipality of Bologna for 15 years, Kilowatt invested approximately 2 million euros, also supported by some funding from the Emilia Romagna Region dedicated to the development of local infrastructure. The primary objective was to transform these long-abandoned areas into a meeting place for cultural activities and experiments capable of connecting citizens, research centres and businesses on topics related to nature, environmental sustainability, culture, innovation and art.

The initiative represents a notable urban regeneration experience distinguished by its focus on environmental sustainability and active community participation. The project, in fact, is a hybrid space hosting a cultural centre which, through numerous projects and initiatives, including the 'community gARTen' for the management of the urban garden, the coworking centre, the educational services for children, and the promotion of summer cultural events, aims to address one of the main challenges of this generation: climate change and the depletion of natural resources.

Kilowatt contributes to combating environmental challenges through various sustainable practices, such as the implementation of nature-based solutions that embrace numerous aspects of space management and activities. First and foremost, Kilowatt is committed to water resource conservation. In 2019, it inaugurated an aquaponic cultivation system within its spaces (Fig. 1) in collaboration with the startup Aquaponic Design (Impianti di Coltivazione in Acquaponica, 2024). This soil-less cultivation method allows for the simultaneous cultivation of vegetables and freshwater fish farming. Aquaponic cultivation system is highly environmentally sustainable as it does not involve the use of fertilizers and pesticides, saves up to 90% of water resources used for cultivation, significantly reduces fossil fuel consumption and CO_2 emissions, and offers higher cultivation density per square meter. Furthermore, the urban garden created within the greenhouses actively involves the local community, not only providing fresh and organic produce, but also acting as an educational and social space where community members can learn and collaborate, strengthening social bonds and promoting a culture of sustainability.

Among various initiatives, Kilowatt for the management of the greenhouse bar, pursues the exclusive use of plant-based and organic raw materials through collaboration with small local producers, so that its actions support the local economy and have no further negative impact on the environment. This choice is also reflected in the management and maintenance of the spaces. Since 2020, Kilowatt has adopted the use of Ecolabel-certified cleaning products made from raw materials of vegetable origin, which are biodegradable and not toxic to users and the environment.

This attention to detail extends to the complete elimination of plastic in packaging and the reduction of single-use compostable products, which are entirely replaced by reusable materials such as glass, by investing in efficient washing systems. Moreover, the installation of a public fountain with chilled and sparkling water has also reduced plastic bottle sales, providing an ecological and convenient solution for visitors to the Le Serre dei Giardini.

Finally, Kilowatt launched an awareness-raising campaign to promote the use of bicycles as a sustainable means of transportation to reach Le Serre dei Giardini. Specifically, it installed 63 bicycle racks, blocking the entrance to cars and scooters. This initiative encourages sustainable mobility, reducing the ecological footprint of visitors and promoting a healthier and more conscious lifestyle.

In short, through these practices, Le Serre dei Giardini not only help limit the environmental impact, but engages children, users and visitors with practical and playful activities such as cultural and musical events, co-design activities, workshops and talks and invites them to spend time inside the sustainable bar, concretely teaching them to take care of nature and environmental sustainability even through small daily actions. Therefore, this project represents a virtuous model of how public space management can be integrated with ecological and community initiatives, offering a tangible example of sustainable urban regeneration.

Giardini Condivisi, C.E.Ri.S.M. & PyKa Onlu, Palermo

Giardini Condivisi (Shared Gardens) project, initiated in 2017 in the city of Palermo, southern Italy, is a community initiative implemented by the music association C.E.Ri.S.M (European Center for Research and Musical Experimentation) and the cultural ludo-didactic association PyKa Onlus, through the co-financing of the Presidency of the Council of Ministers - Department of Youth and National Civil Service - under the "Youth for the social" measure.

The initiative was launched by a loose network of non-profit associations and cooperatives in Palermo working in collaboration with the Department of Agricultural, Food and Forestry Sciences of the University of Palermo, to propose regulations to the Municipality of Palermo to govern urban gardens and shared gardens in municipally-owned areas, with a view to safeguarding, caring for and enhancing the public green heritage, while encouraging good practices in environmental sustainability.

Fig. 2. Temporary installation at Giardini Condivisi, Palermo.

Giardini Condivisi project aims to achieve two main goals. First, it seeks to involve residents in reclaiming and revitalizing urban areas that are often neglected or abandoned, such as neighborhood gardens, through local events and initiatives launched on the Facebook group. By creating and maintaining shared gardens managed collectively, these spaces become meeting places that encourage social cohesion and relations between different generations and cultures.

Second, the project aims to nurture local talent aged 16 to 30, selected through a public call for applications, by establishing the TeatroTascabile theater-music company. Through theater and music workshops, participants receive training as singers, dancers, actors, set designers, costume designers, sound technicians, and makeup artists, ultimately producing performances for their community. These goals are pursued through an integrated approach that combines art and nature to promote social interaction and inclusion.

The project, conceived for the extensive green spaces in via Paganini 4/6 that surround the TeatroTascabile company headquarters, soon evolved into the first sensorial garden in the city of Palermo, this space was designed to be not only an incubator for therapeutic activities such as art-therapy, chromo-therapy and gardening, but also a place for experimentation, training and inclusion for all citizens, thanks to the maintenance of the green spaces, the creation and installation of interactive works (Fig. 2) and the launch of the theatre activities carried out by the TeatroTascabile company. Hence, the shared garden is a recreational, therapeutic, and educational space accessible to all, including children, the elderly, and people with physical and mental disabilities. It serves as a place to relax, reflect, read, meditate, contemplate nature and the city, and socialise, as well as an outdoor stage for the theatrical performances organised by the local company.

The sensory garden was designed by a heterogeneous team of professionals, including landscape planners, artists, set designers and architects. It temporarily houses art installations and electronic devices that allow both the perception of music through senses other than hearing and the decoding of nature's vibrations into light and sound. In addition, the fruition of sensory pathways and installations aims to stimulate synaesthesia between touch, sound, fragrance and taste offering garden users the opportunity to empathise with the disabilities of others, understanding the difficulties and hypersensitivities of differently-abled individuals. *Giardini Condivisi* project represents a clear example of how the integration of culture, art, theatre, technology, and nature-based solutions can transform degraded urban spaces into environments that promote mental and physical well-being, social inclusion, and educational growth of the local community.

2.2 Top Down

Gavoglio Urban Park, UNaLab, Genoa

Fig. 3. Aerial photo of the requalified urban park of Gavoglio, in the Lagaccio district of Genoa.

In 2016, the city of Genoa launched a pilot urban regeneration project employing nature-based solutions. The aim was to convert a 16,200 m^2 area, featuring a 60-m elevation change, of the old Gavoglio military barracks - closed since the 1990s - located in the Lagaccio district, into a smart, multifunctional, inclusive, and sustainable public urban park.

The project was financed through the "Pact for the City of Genoa - Interventions for the economic development, social and territorial cohesion of the City of Genoa", with a contribution of €6,400,000 from the City of Genoa and co-financing of €1,710,250 from the EU's UNaLab project, under the Horizon 2020 Program (Campailla, Zarino, 2021, *Nature-based Solutions Are Growing in the Gavoglio Park | UNaLab*, n.d.).

After the closure of the barracks, the area remained unused for years, raising the hopes of many citizens who wished to see it transformed into a public space and a place for civic engagement. For this reason, since 2015, before the site's remediation, citizens and stakeholders were actively engaged in a participatory process of co-creation to explore needs and perspectives on the desirable functions of green spaces. A shared feasibility project was developed through workshops, thematic meetings, and community-driven initiatives, focusing on the restoration of historical heritage and the creation of new services. At the same time, design ideas were developed taking into account the area's morphology, the physical limitations of the sites, and available funds. This process highlighted the need for accessible green public spaces and the mitigation of hydro-geological risk in the Lagaccio neighbourhood, one of Genoa's most densely populated districts characterised by chaotic urbanisation and a high rate of immigration. Additionally, like several areas in the city of Genoa, the Lagaccio district area suffers from high flood risk, as it is crossed by the Lagaccio watercourse which channels the runoff from other significant watercourses and is largely tombed along its path.

Consequently, the project core focused on regenerating the entire military site through NbS compatible with the local ecosystem and constructed by using recycled materials aimed at increasing the soil's drainage capacity and the infiltration of rainwater to counteract the frequent flooding that occurs in the city of Genoa. The Gavoglio Park pilot requalification project implemented 12 types of NbS, which, following adequate monitoring, could be replicated in other areas of the city with similar problems and characteristics. The adopted solutions include water-permeable pavements, sand-draining play areas, rain gardens, infiltration basins, bioswales, tree groups, xerophilous orchards and meadows, reforestation, green walls, live wicker and palisade structures, an underground rain retention basin, and greened metal gabions filled with crushed debris from the demolition of barracks buildings.

In 2022, the project returned to the community an urban area abandoned for over 40 years, creating new green spaces for socialization and environmental sustainability (Fig. 3). Thanks to this intervention, which would not have been possible without substantial funding and institutional efforts, Gavoglio Park is currently fully integrated into the urban fabric of the Lagaccio neighbourhood, contributing to the improvement of the quality of life, the reduction of hydrogeological risk, the preservation of native species, and the valorisation of abandoned and demolished areas. The participatory process underlying the project ensured that Gavoglio Park met the real needs of the community, while simultaneously promoting sustainability and social cohesion.

Oasi in Ospedale, WWF, Palidoro, Bari, Palermo, Padua

The *Oasi in Ospedale* (Oases in the Hospital) project is an initiative promoted by WWF, aimed at creating green spaces within pediatric hospitals across Italy to improve the quality of young patients' hospital stays and enhance urban biodiversity, while simultaneously fostering awareness of the inseparable connection between human health and the health of nature (*Oasi Negli Ospedali Dei Bambini, Grazie al WWF* 2023).

Fig. 4. One of the "oasis in the hospital" at the Bambin Gesù pediatric hospital in Palidoro, Rome

The initiative was first launched in 2022 at the Bambin Gesù Pediatric Hospital in Palidoro (Fig. 4), near Rome, with the Urban Nature "Nature Does Care" edition and expanded through a fundraising campaign activated by WWF that same year, to enable the creation of new green Oasis in the Giovanni XIII Pediatric Hospital in Bari, the Cristina Benfratelli Hospital in Palermo, and the Pediatrics Complex of the University Hospital of Padua.

The project is based on the concept that contact with natural environments and caring for them can promote physical and psychological regeneration, thereby improving children's mental health and cognitive development.

Recent studies emphasise that green areas in hospitals are indispensable both for accelerating patients' recovery and alleviating the mental fatigue of healthcare personnel (Chang & Chien, 2017). The American Society of Landscape Architects advocates that green spaces should be an integral part of the design of new hospitals. It is now widely recognized that spending time in nature, surrounded by landscapes filled with trees, flowers, small animals and insects, ponds, and fountains, can have healing benefits that enhance the psychological and physical well-being of patients, helping them to reduce both physical and mental pain, lower stress, anxiety, and anger levels, while also positively impacting physiological parameters such as blood pressure, muscle tension, heart rate, and brain activity.

Oases in the hospital are spaces ranging in size from 100 to 250 m^2 that feature native natural elements such as trees, flower borders, hedges, and aromatic plants, as well as NbS such as small pools of water, nest houses to provide places-shelter for insects, and feeders to attract small fauna, such as pollinators like butterflies and birds. Additionally, the oases offer seating areas for moments of tranquillity and socialisation between young patients and their families, as well as olfactory boxes and sensory pathways to provide necessary sensory, motor, psychomotor, and coordination stimuli during rehabilitation sessions.

The spaces dedicated to Oases in Hospital are designed to meet both educational, therapeutic and rehabilitative goals. They serve as outdoor laboratories where, through educational, playful, and creative activities, young patients can observe various living organisms, learn about and take care of nature, and participate in rehabilitation programs - particularly those requiring long-term care - directly in contact with nature.

By actively caring for nature and ecosystems, children can regain a range of cognitive and motor knowledge and skills that may have been lost due to illness, as well as learn respect for nature and biodiversity, strengthen self-esteem and positive self-perception, and finally socialize and enjoy positive and psychologically beneficial experiences through participation in group activities.

proGIreg Project: Planning and Implementing Nature-Based Solutions (NBS)

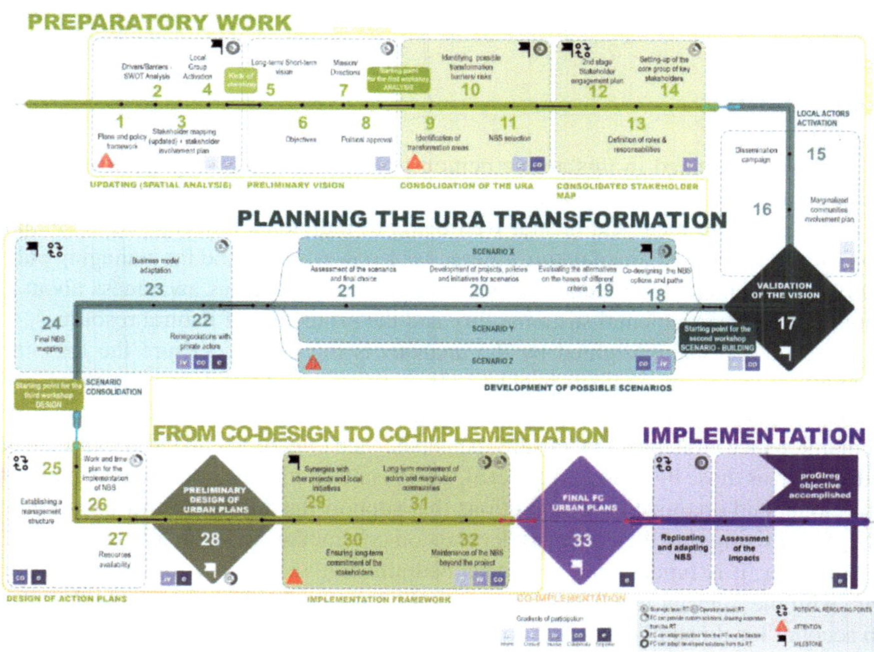

Fig. 5. The Roadmap for Follower Cities

The proGIreg project, funded by the European Union's Horizon 2020 program, focuses on post-industrial urban regeneration through productive green infrastructure. It involves four "Front Runner Cities" – Dortmund, Turin, Zagreb, and Ningbo – which host urban living labs for the development and implementation of nature-based solutions. Other

"Follower Cities," such as Cascais, Cluj-Napoca, Piraeus, and Zenica, follow the project to replicate the solutions in their own areas.

The planning of NbS began with a spatial analysis of the involved cities, following six key phases, including data collection and interpretation. The urban areas selected for the living labs face common issues like degradation, social disconnection, and abandoned green spaces, often linked to former industrial zones. The co-design process was conducted through workshops in the Front Runner Cities, actively involving local communities. The process highlighted the importance of tailoring NbS to the specific context of each area, balancing citizen engagement with the technical complexities of the solutions. Six key co-design guidelines were established, including transparency, inclusiveness, and experimentation. The project developed a roadmap for the Follower Cities (Fig. 5), outlining the steps for implementing NbS based on lessons learned from the Front Runner Cities. This roadmap focuses on how cities can integrate NbS into local urban planning, with a replication toolkit to address challenges and barriers.

The project followed a common methodology for implementing NbS, including continuous monitoring and adjustments to ensure the success of the solutions. The Front Runner Cities incorporated risk monitoring tools and timelines to facilitate the implementation of the solutions.

3 Conclusions

The analysis conducted in this article demonstrates how community involvement can effectively address climate challenges, leading to the creation of healthy environments that enhance the physical and mental well-being of individual citizens and the broader community. Collective care of nature emerges as a powerful method for managing public green spaces, which values active citizen participation and raises awareness about the importance of environmental sustainability and the protection of natural resources.

A crucial aspect highlighted by the analysis is the need to ensure the long-term continuity of the positive effects generated by Nature-Based Solutions. For these interventions to maintain their effectiveness over time, it is essential to develop monitoring and maintenance strategies that ensure the persistence of ecological and social benefits. Studies such as those by Frantzeskaki et al. (2019) underscore the importance of creating institutional and organizational conditions that support the durability and scalability of NbS, emphasizing that their success depends on continuous and adaptive governance.

Moreover, it is fundamental that urban administrative policies are in constant dialogue with the relevant territory. Urban policies must be flexible and adaptable, taking into account local peculiarities and community dynamics. Only through effective coordination between local administrations, citizens, and other stakeholders is it possible to implement NbS that genuinely meet the needs of the territory and promote integrated sustainability. As discussed by Wolfram et al. (2021), a multiscalar governance approach that integrates local policies with regional and national strategies is crucial to overcoming institutional challenges and ensuring that NbS contribute to long-term urban resilience.

Finally, participatory activities have not only led to the development of pedagogical and social green spaces but have also significantly improved the quality of life in urban areas. These initiatives have successfully converted underutilised urban public spaces into vibrant environments of social interaction, contributing both to environmental sustainability and the well-being of local communities. The case studies examined in this article illustrate how both top-down and bottom-up approaches can be effective in the implementation of NbS in urban contexts. Whether driven by institutional initiatives or community-led actions, these projects demonstrate the potential of NbS to transform urban spaces into areas that reflect the needs and aspirations of their inhabitants.

Bibliography

Bowler, D.E., Buyung-Ali, L., Knight, T.M., Pullin, A.S.: Urban greening to cool towns and cities: a systematic review of the empirical evidence. Landscape Urban Plann. **97**(3), 147–155 (2010)

Campailla, S., Zarino, S.: UNaLab – Nature-Based Solutions nel nuovo parco urbano a Genova. (2021). Reticula, 28. ISPRA (2021)

Chang, K.G., Chien, H.: The influences of landscape features on visitation of hospital green spaces-a choice experiment approach. Int. J. Environ. Res. Public Health **14**(7), 724 (2017). https://doi.org/10.3390/ijerph14070724

Frantzeskaki, N., Borgström, S., Gorissen, L., Egermann, M., Ehnert, F.: Nature-based solutions for urban climate resilience: linking research and practice in a co-creation framework. Environ Sci Policy **101**, 83–91 (2019). https://doi.org/10.1093/biosci/biz042

Impianti di Coltivazione in Acquaponica. Le Serre dei Giardini - Kilowatt - Impianti di Coltivazione in Acquaponica. Impianti Di Coltivazione in Acquaponica (2 May 2024). https://www.aquaponicdesign.it/soluzioni/brisighella-contrada-consulting/

Kabisch, N., et al.: Nature-based solutions to climate change mitigation and adaptation in urban areas: perspectives on indicators, knowledge gaps, barriers, and opportunities for action. Ecol. Soc. **21**(2) (2016). http://www.jstor.org/stable/26270403

Konijnendijk, C.C.: Evidence-based guidelines for greener, healthier, more resilient neighbourhoods: introducing the 3–30–300 rule. J. Forestry Res. **34**(3), 821–830 (2023). https://doi.org/10.1007/s11676-022-01523-z

La Cecla, F.: Contro l'urbanistica. Einaudi (2015)

Leone, L., Scapolan, A.C., Montanari, F., Mannucci, P.V.: The case of Le Serre dei Giardini Margherita, Bologna. Collaborative Spaces at Work: Innovation, Creativity and Relations (2020)

Mancuso, S.: Fitopolis: La città vivente. Giunti Editore (2020)

McCann, E.: Mobilities, politics, and the future: critical geographies of green urbanism. Environ. Planning A: Econ. Space **49**(8), 1816–1823 (2017). https://doi.org/10.1177/0308518X17714850

Milano, C., Novelli, M., Cheer, J.M.: Overtourism and tourismphobia: a journey through four decades of tourism development, planning and local concerns. Tourism Planning Dev. **16**(4), 353–357 (2019). https://doi.org/10.1080/21568316.2019.1599604

Nature-based solutions are growing in the Gavoglio park|UNaLab. (n.d.). https://unalab.eu/en/news/nature-based-solutions-are-growing-gavoglio-park

Oasi negli ospedali dei bambini, grazie al WWF. (2023, 3 ottobre). WWF Italia. https://www.wwf.it/pandanews/ambiente/oasi-negli-ospedali-dei-bambini-grazie-al-wwf/

Pauleit, S., Zölch, T., Hansen, R., Randrup, T.B., Konijnendijk Van Den Bosch, C.: Nature-based solutions and climate change – four shades of green. In: Kabisch, N., Korn, H., Stadler, J., Bonn, A., (A c. Di), Nature-Based Solutions to Climate Change Adaptation in Urban Areas, pp. 29–49. Springer International Publishing (2017). https://doi.org/10.1007/978-3-319-56091-5_3

Progetto Oasi in Ospedale|WWF Italia. (2024, 11 luglio). WWF Italia. https://www.wwf.it/cosa-facciamo/progetti/oasi-in-ospedale/

Raymond, C.M., et al.: A framework for assessing and implementing the co-benefits of nature-based solutions in urban areas. Environ Sci Policy **77**, 15–24 (2017). https://doi.org/10.1016/j.envsci.2017.07.008

Wolfram, M., Borgström, S., Farrelly, M.: Cities and nature-based solutions: governance innovation in urban climate change resilience. Urban Planning **6**(1), 30–40 (2020). https://doi.org/10.1016/j.jclepro.2020.125569

R.ES.P.I.R.A - Co-Design and Implementation of Climate-Regulating Urban Furniture

Chiara Farinea(✉), Tim Bruder, and Andrea Conserva

Institute for Advanced Architecture of Catalonia (IAAC), Barcelona, Spain
{chiara.farinea,tim.bruder,andrea.conserva}@iaac.net

Abstract. Leading cities and metropolitan regions around the globe are increasingly restricting access to motorized traffic, thereby transforming urban landscapes and creating opportunities for the development of spaces that prioritize citizen well-being, social interaction, and physical activity. This shift is accompanied by environmental challenges, including more frequent extreme heat events due to climate change, and air pollution, both of which pose health risks and reduce overall living comfort in urban areas. This paper presents the development of a set of modular urban furniture designed to foster inclusive public spaces while simultaneously enhancing environmental conditions through the incorporation of climate shelters, urban greenery and advanced technologies for fabrication and air quality improvement. *R.ES.P.I.R.A* identified a catalogue of urban furniture responding to pressing needs through a collaborative analysis with local residents, including usability studies and the identification of potential sites. Throughout the design phase, work-in-progress states were communicated and evaluated with the residents to ensure optimal and aligned results. The project includes the fabrication of multiple elements of the catalog and their implementation at two different sites. The prototypes were produced using digital fabrication technology such as 3D printing and CNC milling, eco-friendly materials like wood and cellulose fiber composites, and novel coating techniques for air purification and decarbonization. Following installation, they were tested and evaluated by participants involved in the co-design process and are currently co-mantained. The project serves as a practical demonstration of a model process for creating sustainable urban environments via democratic design and community involvement.

Keywords: Nature Based Solutions · Urban Furniture · Urban Oasis · Digital Fabrication · Participatory Design

1 Introduction

1.1 Context

In response to global warming, the collective aim of developing sustainable cities has become increasingly urgent. Urban heat island effects and declining biodiversity present significant challenges for metropolitan areas around the globe, leading to the launch of numerous urban greening projects.

A. Mateus et al. (Eds.): WAD 2024, What's Around Design?, pp. 187–200, 2026.
https://doi.org/10.1007/978-3-032-08163-6_14

Nature-based solutions (NBS) refer to utilizing natural ecosystems to resolve societal challenges related to climate change, while also promoting human well-being and thus have proven to play a critical role in the forming of regenerative and sustainable cities (IUCN 2016). Research has underscored the symbiotic relationship between ecological design and citizen engagement, highlighting that the active involvement of residents in urban design and care-taking not only enhances environmental resilience but also improves overall livability (Chanse 2009). Urban furniture presents opportunities for a playful, bottom-up approach to creating inclusive and ecological urban spaces. It serves as a small-scale implementation of Nature-Based Solutions, effectively amplifying and complementing larger landscape interventions. In the context of biophilic cities, where residents actively engage with and develop emotional connections to their natural surroundings (Beatley 2013), urban furniture can serve as a bridge between human activity and ecological responsibility.

Urban furniture refers to the elements placed in urban spaces that significantly influence their functionality, social dynamics, and aesthetics. These elements include benches, tables, bins, street lights, pergolas, and playgrounds, among others. Urban furniture should be inclusive, offering accessibility and functionalities to a diverse range of users. Given the variety of urban spaces, from large plazas to narrow streets, adaptability in arranging urban furniture is crucial. Urban furniture has the potential to support various activities, including outdoor working, eating, sitting, talking, and playing, while also providing shading and enhancing security. With the new challenges that urban spaces are facing, incorporating features that improve air quality and offer protection from extreme weather can further enhance the palette of functionalities and relevance of urban furniture.

1.2 State of the Art

Innovative projects like the Particle Cruiser, which integrates air-filtering plants and photocatalytic coatings to neutralize contaminating particles in the air, exemplify the potential of mobile urban furniture to adapt to changing environmental and user conditions (Hayek 2024).

Terracool urban oasis, demonstrate the emergence of new urban furniture typologies, including modular cooling systems that regulate environmental conditions in public spaces (Temel & Fahy 2024). Similar to Philippe Rahms idea of *Jade Eco Park*, where a catalog of natural and artificial heat, humidity and pollution reducing climatic devices was designed (Rahms 2016).

As an inspiration for community based design counts the *Living Pavilion,* an event space including horticultural elements, performances embracing local artists and place making methodologies (Beer 2019).

2 Methodology

The primary research objectives of this study are to explore the integration of innovative technologies with urban and natural environments, and to investigate novel functionalities, design approaches, and implementation processes that contribute to the creation of sustainable, inclusive, and adaptive urban solutions. The project brings together a consortium of diverse partners, with the Institute for Advanced Architecture of Catalonia (IAAC) as the lead design partner, and the Noumena Group as the project coordinator and digital fabrication facility. Furthermore, R.E.S.P.I.R.A involved three community groups in co-design and co-implementation processes: the Association of Neighbours and Residents of Poblenou, the Platform of Roquetes Entities, and the Fundació Comtal. This research aims to hybridize community involvement, technology use, and ecological design to advance environmental design practices. Participatory design not only promotes inclusiveness and a sense of community but also fosters broad engagement and ensures that the user's input is carefully considered. Projects designed with community input are more likely to be maintained and cared for by the community, enhancing both their sustainability and longevity. The methodology outlined in this paper involves a four-step participatory process, and is complemented by the digital fabrication and implementation of the designed urban furniture. Digital technologies and fabrication methods, as simulation tools, parametric design software and additive manufacturing, provide the opportunity to create and produce non-standardized design systems that are specifically tailored to local needs (Farinea 2021). In a co-design process, personalized and customized design can lead to increased user satisfaction and engagement (Ganji, Shah, and Coutroubis 2018). Furthermore, designers can create digital prototypes and utilize 3D printers or CNC machines to bring them to life as physical models (Bhaskaran and Jayaram R.S. 2022). This facilitates rapid iteration and thus feedback from the participants, which is essential for refining designs within a co-design framework.

3 Participatory Process

The participatory process involved three different entities, located in three different neighborhoods in Barcelona and each of them with different urban needs and challenges. The main objectives were to define the function, aesthetic direction, and potential application sites for the urban furniture catalog. Involving local communities in the design process also aimed to introduce the participants to novel fabrication technologies and material concepts and create a sense of ownership and identity with the project. To ensure community input was integrated throughout the entire process and thus meet the objectives, a four-phase participatory methodology was established: Co-analysis, co-design, co-implementation, and co-evaluation.

4 Co-Analysis

The co-analysis phase focused on identifying and documenting potential sites for the implementation of urban furniture and evaluating their suitability. This involved assessing environmental conditions, the urban context of each site, existing vegetation, patterns of daily use, and pedestrian traffic. These observations helped determine the specific needs of each space and the potential user groups. The workshop followed the methodology of "focus group discussion (FGD)", where participants provided local insights by responding to guided questions. Each neighborhood association conducted an analysis of the sites within their area. The analyzed sites varied in size and usability, accommodating events of different scales. Furthermore, participants identified main target groupsConcerns were raised about potential accessibility issues due to the placement of urban furniture, as well as whether the furniture might hinder rather than support the frequent events held in these spaces. This groundwork laid the foundation for the co-design phase, which resulted in the development of a catalog featuring adaptable furniture design solutions, tailored to meet the specific needs of each site.

4.1 Co-Design

The co-design workshops focused on defining the objectives, functions and morphology of the decarbonising urban elements, as well as identifying different user categories (target groups) to enhance the inclusiveness of the space. During the initial session of the co-design phase, participants were introduced to the Crazy Eights exercise. This method is commonly used in design sprints and other brainstorming workshops and involves participants quickly sketching eight different ideas in eight minutes. The goal was for participants to move beyond their initial ideas, which are often the least innovative, and to explore a wide range of diverse solutions to the given challenge.

The created ideas served as inputs for the designers, who created a set of solutions that were then evaluated by citizens in the next workshop. Participants evaluated both their own work and the designers' proposals in subsequent workshops using a dot voting methodology. Each participant was given a total of 10 points, which they could either allocate entirely to a single solution or distribute among several options. The ideas that receive the most votes are considered the most reflective of the group's preferences. Among all participants, the biomorphic designs proved to be the most popular. Furthermore, a modular design concept not only addressed the adaptability to all tested spaces but was also the most favored by the participants.

After several iterations of feedback, the typology and morphology of the furniture were well-advanced, leading to a discussion on the placement of the different modules. For this purpose, a site plan was printed, along with small-scale prints of the design elements. Participants and designers collaboratively experimented with different positioning on the plan, and through discussion, arrived at an optimal layout (Fig. 1).

To deepen their technical understanding of digital fabrication tools and to guide their design instincts, the participants of the co-design workshops were invited to tour the laboratories and production halls of fabrication partner Noumena in Barcelona and observe the manufacturing process firsthand.

4.2 Co-Implementation

The results of the co-design process were then optimized for fabrication and produced via 3D printing and CNC milling. After production, the urban furniture modules were implemented at two different sites. The first site, a plaza in the city center, was selected for its wide space and high foot traffic. Despite being surrounded by tall buildings, the plaza receives direct sunlight throughout the whole day. For the opening event, a spacious layout was chosen (Fig. 2). In the following week, various configurations were tested, ranging from the furniture in a continuous line (Fig. 3), to nearly enclosed

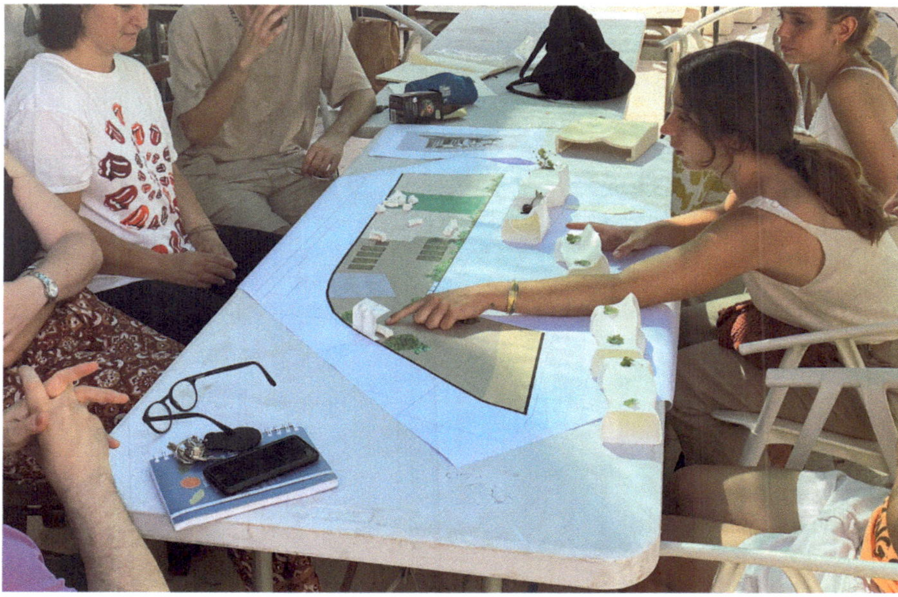

Fig. 1. Photograph During The Co-Design Workshop Session, Evaluating Morphology And Positioning By Small Scale Representational Model of the Work in Progress Design

circular formations. After one week, the furniture was transported to its final location. This second site presented a very different setting: an enclosed area with limited spatial freedom, and surrounded by ongoing art and planting projects and gardens. Here the furniture was forming space for relaxation and social interaction for and after activities organized by the association (Fig. 4).

4.3 Co-Evaluation

The project included a comprehensive evaluation of the furniture, conducted during the opening events and again one month after implementation. This timeframe allowed sufficient opportunity for people to interact with the furniture in various settings and events, experiment with different arrangements, and demonstrate its effectiveness in enhancing environmental conditions. Participants were given an evaluation questionnaire in which they were asked to rate their satisfaction across various assessment criteria, including the value of their input in design decisions, the transparency of the decision-making process, the design and construction quality of the final outcome, its functionality, and the overall effectiveness of the intervention.

Participants particularly appreciated the inclusive design approach that involved all community members. They valued the opportunity to collaborate with their neighbors, united by the shared goal of improving a community space. The furniture served as an attraction for new people that showed interest in the project and community work.

During the implementation events, several external co-evaluation workshops were held to evaluate the design outcome with the community of potential promoters and

Fig. 2. The Spacious Layout Chosen For The First Public Event on A Central Plaza

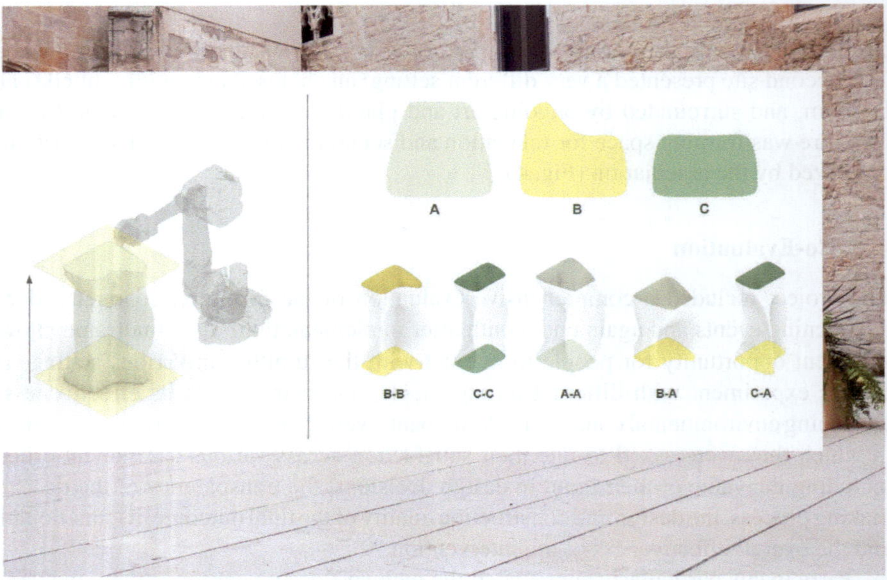

Fig. 3. Linear Layout of Benches and Pergola, Showing The Different Modules as A Conscientious Monolithic Object

Fig. 4. Placement of the Furniture in its Final Location. The Furniture Support the Organization to Host Social and Educational Events

experts in the field. These sessions proved to be highly valuable by offering diverse feedback on how to design and refine the project further. They also facilitated the creation of networks among various stakeholders operating within the same urban territory, focusing on public space and urban furniture from different disciplinary and sectoral perspectives.

4.4 Participatory Process Conclusions

The four-step approach ensured ongoing community involvement throughout the project. Focus group discussions helped identify local needs early on, while creative exercises in the co-design phase encouraged a wide range of ideas and genuine input. The implementation phase allowed the designs to adapt to different neighborhood contexts, and the evaluation confirmed that participants were satisfied with both the process and the outcomes. Overall, the methods ensured the final designs were relevant, well received, and that the furniture continues to be widely used to this day.

5 Catalog Overview

The catalog aims to provide flexible, reconfigurable furniture that responds to diverse social and spatial needs identified through the participatory process. The design should aim to address various requirements, such as accommodating different group sizes, encouraging social interaction while also providing options for privacy, supporting activities ranging from short rests to extended stays, and responding to spatial constraints in diverse urban contexts. The final catalog features a system of ten distinct urban furniture elements to address this: a large table, two urban lounge chairs, three benches designed for various seating arrangements, a small table with a stool, a pergola module, a planter, and a connective piece for forming large, circular bench configurations (Fig. 5). The catalog prioritizes functional variety through design gestures, accommodating a broad range of social interactions and privacy needs. The benches, in particular, are designed to foster various social dynamics. Benches positioned opposite or adjacent to each other and facing the same direction promote communication, while those facing different directions provide a sense of anonymity. High-comfort benches with backs encourage extended sitting periods or even quick naps, while standard-backed benches support moderate sitting durations and can also serve as space dividers. Open benches without backs are intended for shorter rests, accessible from both sides. Additionally, double seaters are designed for couples or friends, while single seaters are ideal for individuals working or reading alone. This variety allows urban planners to create diverse and engaging seating arrangements that enhance the urban environment. Additionally, the table supports activities such as eating, playing games, or reading.

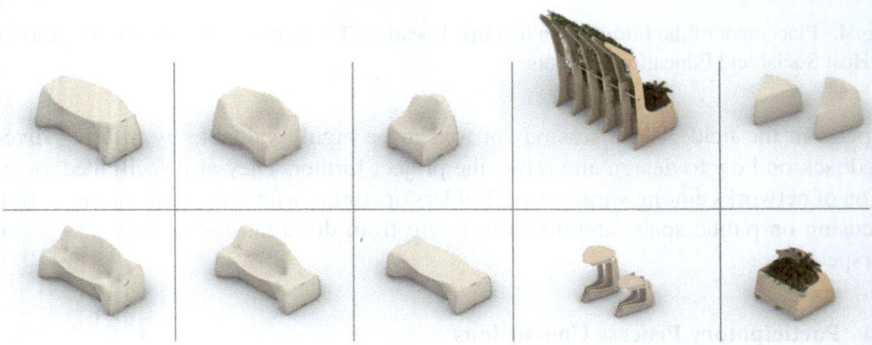

Fig. 5. The 3D Models of All Ten Elements of the Urban Furniture Catalog Developed For the Project

The six seating modules, despite their diverse morphological designs, share three types of side profiles (Fig. 6). This allows certain bench modules to connect with others, forming extended bench configurations for larger urban spaces. This modularity enables the use of single modules in narrow urban spaces and streets, as well as larger group placements in wider plazas (Fig. 7).

The seating modules as well as the pergola modules are movable by two people.

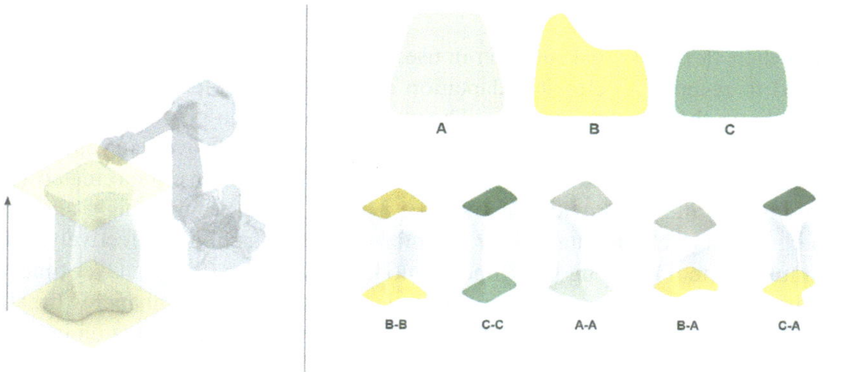

Fig. 6. Printing Direction and Identical Side Profiles of the Bench and Table Elements, Allowing the Alignment of Several Modules.

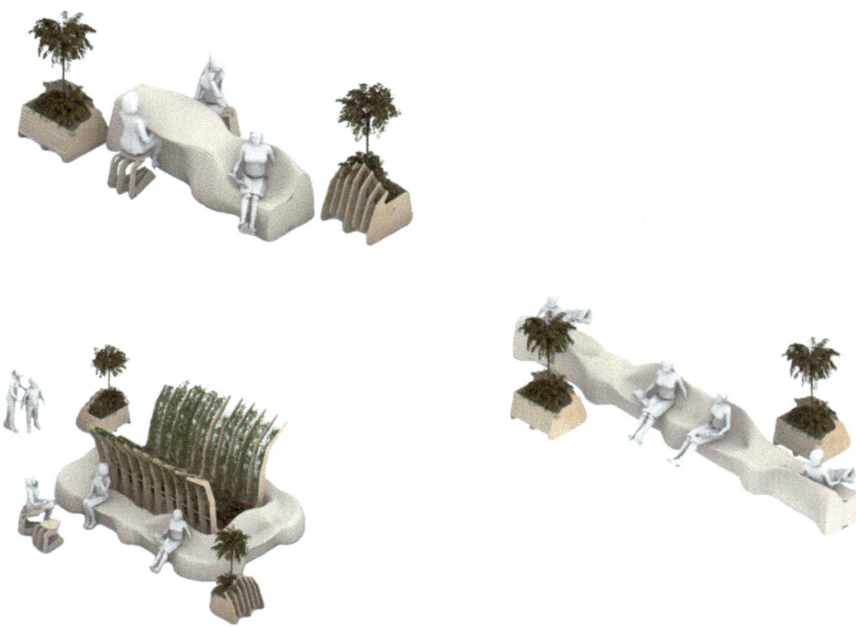

Fig. 7. Display of a Selection of Different Configurations, From Tiny Space to Open Plaza, Depending on the Quantity and Arrangement of Catalog Elements

6 Climate Adaptation

Besides their flexibility and function in use, both the bench and pergola modules were designed to respond to climate adaptation needs on its implementation space, more specifically to the heat island effect. The result of the design process with the citizens were pergola modules that offer adaptive shading, an adjustable overhang and large plant pots attached. Initially, rattan lines stretched between the six main pillars created a pattern of varying opacities to provide shading.The pattern, as well as the height and angle of the pergola, can be computationally optimized to block sunlight effectively, utilizing the simulation tool Ladybug for Rhino3D modeling software (Fig. 8). Over time, plants from the connected planter grow over the structure and are further enhancing the shading effect. The four different modules developed in the case study can function individually or be connected to form a coherent structure. The riffle pattern integrated into the benches emphasizes connectivity between the bench modules through creating a cohesive visual experience (Fig. 9). Inspired by the calming rhythm of Zen garden motifs, these waves contribute to the overall aesthetic of the elements while increasing the surface area and thus maximize air-purifying qualities.

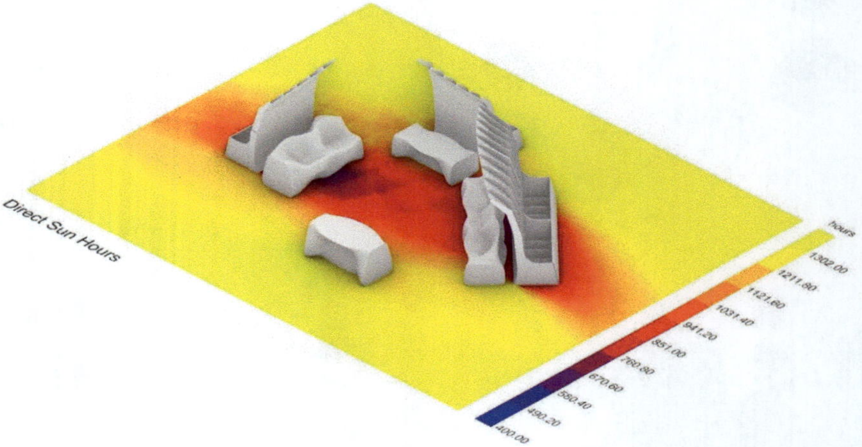

Fig. 8. Simulation of Direct Sun Hours Touching the Plaza With One Possible Implemented Arrangement

Fig. 9. Close-Up Photograph of the Linear Arrangement, Showing the Riffle Structure and the Deposition Lines of the 3d Printing Technology

7 Digital Fabrication of Prototypes

7.1 Additive Manufacturing and CNC

The digital fabrication techniques used in this project were additive manufacturing (commonly known as 3D printing) for the benches and CNC milling of wood for the pergola. To facilitate the fabrication of the seating modules, the benches were flipped 90 degrees and printed from bottom to top. Geometric modifications ensured that the slope angles remained below 45° to avoid printing issues. The benches are made from a cellulose fiber-filled plastic composite and are printed in between 12 and 23 h per module (Fig. 10). CNC milling enabled the economical manufacturing of the customized shapes for fabricating the wooden pergola modules and boards on the sides of the benches.

All elements were coated in multiple layers of decarbonizing paint PURE.TECH. This paint is infused with minerals that actively purify the air by capturing and neutralizing greenhouse gases. In addition to its significant environmental benefits, the coating provides robust protection for the furniture against weathering and potential vandalism.

Fig. 10. Photograph During the 3D Printing Process, Giving an Idea of Material Aesthetics and Printing Quality

7.2 Assembly

The core part of the benches was fabricated in one piece via consecutive 3D printing. Wooden boards to close off the open sides of the benches were attached afterwards. Due to the size, the pergola used a more sophisticated assembling approach. One module consists of 16 pieces of plywood, assembled using a waffle structure and mechanical screws, which allows for rapid on-site assembly and disassembly (Fig. 11). The CNC files are uploaded along with a step by step manual on the project website, allowing replication and modification of the structure. Each element can be transported with a forklift by entering underneath the elevated surface of the pergola module, even after the attached pots are filled with soil and on their full weight.

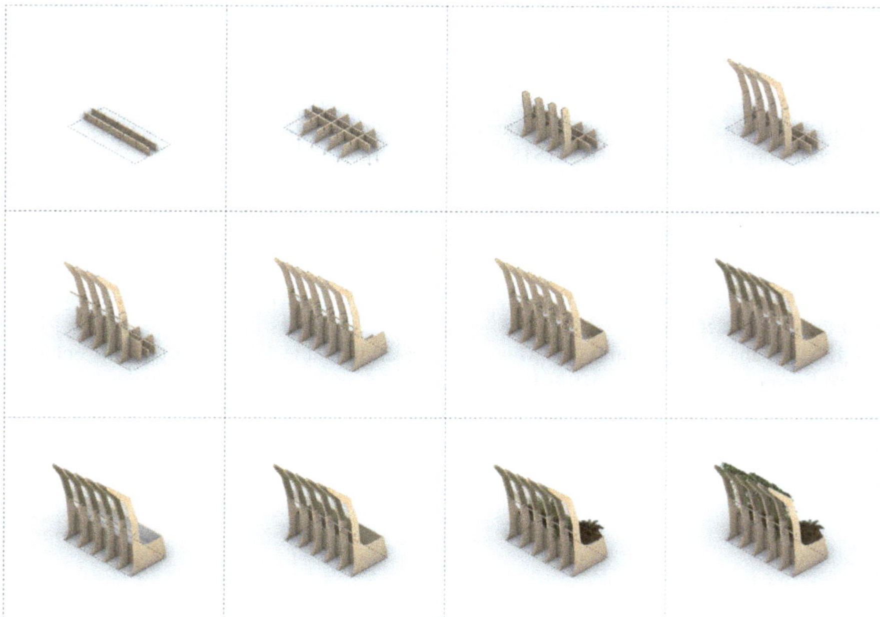

Fig. 11. The Step By Step Manual for Assembly and Potential Disassembly of the Pergola Module

8 Conclusion

The workflow presented in this paper serves as a successful model for the ecological democratization of urban infrastructure. The urban furniture typology, with its small scale and close proximity to daily human interaction, serves as a great example of how public spaces can be transformed through community engagement. This approach not only integrates people's creativity but also enhances their understanding of innovative fabrication technologies. By involving citizens in the design and implementation process, the project fosters a deeper connection between individuals and their environment, demonstrating how urban infrastructure can be both sustainable and responsive to the needs of its users.

Acknowledgement. R.ES.P.I.R.A is a project developed by IAAC, Noumena, La Associació de Veïns i Veïnes del Poblenou, La Plataforma d'Entitats de Roquetes, Espacio Benet Corner, with support from the BitHabitat Foundation.

Bibliography

Chanse, V.: Design for Ecological Democracy by Randolph Hester 2006, MIT Press. Cambridge, Massachusetts. Landscape Journal 28, no. 2 (January 1, 2009): 235–236 (2009)

Beatley, T., Newman, P.: Biophilic cities are sustainable. Sustainability. **5**(8), 3328–3345 (2013). https://doi.org/10.3390/su5083328

Hayek, J., Daneluzzo, M.: Incorporating Collective Intelligence In Urban Design And Planning. Responsive cities conference, Barcelona (2023)

Dilara, T.D, Fahy, L.: Terracool Urban Oasis, Fabricate, pp. 64–71 (2024)

Rahm, P.: Thermal sensations: the case of the jade eco park in Taichung (Taiwan). In: The Urban Microclimate as Artifact, by Sascha Roesler and Madlen Kobi, pp. 102–19. De Gruyter (2018). https://doi.org/10.1515/9783035615159-102

Beer, T., et al.: The Living Pavilion Research Report. The University of Melbourne, Victoria, Australia (2019)

Bhaskaran, B., Jayaram, R.S.: Additive Manufacturing: A Review, pp. 2349–3976 (2022)

Farinea, C.: Enhancing the integration of nature-based solutions in cities through digital technologies. TECHNE – J. Technol. Archit. Environ. 165–169 (2021). https://oaj.fupress.net/index.php/techne/article/view/10703

Ganji, E.N., Shah, S., Coutroubis, A.: Mass customization and technology functions: enhancing market responsiveness. In: 2018 IEEE International Conference on Engineering, Technology and Innovation (ICE/ITMC) (2018). https://doi.org/10.1109/ice.2018.8436313

Production of Microbial Bricks for Ecological Architectural Restoration in Rural Areas Through Community Engagement

Gozde Damla Turhan-Haskara(✉) and Hande Yıldız Çekindir

Department of Interior Architecture and Environmental Design, Izmir University of Economics, Izmir, Turkey
gozde.turhan@ieu.edu.tr

Abstract. This study investigates the ecological architectural restoration in rural areas through nature-based solutions and community-driven innovation. It focuses on a specific workshop entitled "Microbially induced carbonate precipitation (MICP) for microbial brick production", held as part of a comprehensive Caravanserai Summer School 2024 program that comprises seminars and workshops on ecological architectural restoration and various nature-based materials to physically reinforce cultural heritage and to support dissemination of ecological construction knowledge through design practice. This particular workshop among others addressed circular biobased materials as alternatives to conventional construction materials. It was aimed to teach participants to develop eco-friendly bricks, focusing on sustainable material formulation with zero waste and 100% biodegradability. The MICP workshop provided participants to develop cementitious bricks without the use of cement, while contributing to the promotion of biobased materials and production methodologies. The results have shown that the participants achieved microbial activity for brick production at an experimental scale through MICP to further experiment and improve the material properties. The findings indicate potential for these biobased materials to serve as ecological alternatives to conventional construction materials, with future projections focusing on broader application and refinement of these techniques in rural architectural projects.

Keywords: Nature-Based Solutions (NBS) · Biobased · Ecological Architectural Restoration · Community-Driven Innovation · MICP

1 Introduction

The increasing environmental challenges posed by conventional construction practices underscore the need for sustainable alternatives that not only address ecological concerns but also foster community engagement. This paper investigates nature-based solutions (NbS) within the realm of ecological architectural restoration, with a specific focus on rural areas. NbS, which involves the protection, management, and restoration of ecosystems to tackle societal challenges, offer a promising pathway for achieving these

A. Mateus et al. (Eds.): WAD 2024, What's Around Design?, pp. 201–211, 2026.
https://doi.org/10.1007/978-3-032-08163-6_15

goals. In rural settings, where cultural heritage and traditional architectural practices are deeply rooted, the application of NbS can play a critical role in enhancing both the built environment and the surrounding natural landscape.

A central component of this study is the workshop on microbially induced carbonate precipitation (MICP) for microbial brick production, conducted as part of a broader summer school program. This workshop exemplifies the potential of biobased materials in promoting sustainable construction practices. By following the natural processes of microbes to create cementitious materials, the workshop provided participants with hands-on experience in developing eco-friendly bricks that are entirely biodegradable and free from traditional cement. The initiative not only emphasized the technical aspects of material production but also highlighted the importance of participatory design and co-creation in addressing the environmental and social challenges faced by rural communities.

The findings from this workshop indicate that the use of the MICP process holds significant promise for ecological restoration in rural architecture. The participants successfully experienced producing microbial bricks, demonstrating the potential of these biobased materials as alternatives to conventional construction methods. The broader implications of this work suggest that, with further refinement and application, these materials could play a vital role in rural architectural projects, contributing to the preservation of cultural heritage while promoting sustainability and resilience. This study sets the stage for future research and projects aimed at expanding the use of nature-based solutions in ecological architectural restoration across diverse rural contexts.

2 Theoretical Background

2.1 Nature-Based Solutions (NbS) in Construction

Nature-based solutions (NbS) are actions that protect, manage, and restore ecosystems to address societal challenges while benefiting people and nature (Vidal et al. 2023). NbS plays a crucial role in climate change mitigation and adaptation, offering benefits for biodiversity and human well-being (Reiter et al. 2021). The concept serves as an umbrella term encompassing related ideas such as ecosystem services and ecological engineering (Cassin 2021). NbS has gained traction in scientific literature since 2017, particularly in European urban contexts, with flood mitigation and biodiversity conservation being primary focus areas (Hanson et al. 2020). While NbS is closely linked to green infrastructure and ecosystem services, its integration in research remains limited, with most studies focusing solely on environmental benefits rather than holistically assessing social, economic, and environmental impacts (Hanson et al. 2020). As countries revise their Nationally Determined Contributions, there is an opportunity to increase global climate ambition by strengthening the role of NbS (Seddon et al. 2019).

Considering the negative impact of construction activities in urban environments in particular, NbS can reverse the trend of natural landscape degradation and impermeable surfaces (Xing et al. 2017). The construction industry has significant negative impacts on the environment, including pollution, resource consumption, and ecosystem disruption (Mogbo et al. 2023). Construction activities contribute to air and noise pollution, waste generation, and deforestation (Kaja & Goyal 2023). They also lead to habitat destruction, ecosystem fragmentation, and hydrological system alterations (Mogbo et al. 2023). Resource consumption is identified as a major environmental concern in the construction industry (Marema Keno Sesaba 2024). To mitigate these impacts, various strategies have been proposed, including Environmental Management Systems, Environmental Impact Assessments, and sustainable construction practices (Kaja & Goyal 2023). Implementing effective environmental resource management models can help analyze and reduce environmental impacts at both micro and macroeconomic levels (I. Vasilca et al. 2021). Stakeholders in the construction sector are urged to adopt and enforce regulations, increase awareness, and implement mitigation strategies to minimize the industry's environmental footprint (Kaja & Goyal 2023; Marema Keno Sesaba 2024). Successful NbS implementation often involves stakeholder participation and integration with broader social-ecological systems (Baptist et al. 2021).

2.2 Ecological Architectural Restoration in Rural Areas

Ecological architectural restoration in rural areas is crucial for preserving cultural heritage, promoting sustainability, and enhancing the environment. Montero-Parejo et al. (2018) emphasize the importance of maintaining architectural quality while preserving rural characteristics in building design to create sustainable environments. Skataric et al. (2021) further highlight the significance of architectural quality and the preservation of rural characteristics in sustainable building design, particularly in agrarian areas.

Chen & Kitagawa (2023) discuss how architects involved in rural projects contribute design thinking that integrates construction operations with local elements, benefiting the sustainability of rural construction. Niu (2023) explores the creation of rural ecological architecture from a landscape aesthetics perspective, emphasizing the importance of eco-architecture for rural development. Additionally, Lei et al. (2022) focus on optimizing industrial structure adjustments in rural areas, including architectural and landscape considerations, to enhance sustainability.

Parlato et al. (2022) stress the sustainable promotion of traditional rural buildings as heritage attractions, emphasizing a holistic approach that considers social, environmental, and economic factors. Sahebzadeh et al. (2017) delve into the sustainability features of vernacular architectures, highlighting elements like fabric, materials, and courtyards that contribute to sustainable design practices in rural areas.

Ecological architectural restoration in rural areas requires a comprehensive approach that integrates architectural quality, cultural preservation, and sustainability principles. By combining local elements, design thinking, and eco-friendly practices, rural architectural restoration projects can contribute to the preservation of heritage, the enhancement of rural landscapes, and the promotion of sustainable development.

2.3 Biobased Materials in Construction

Biobased materials are increasingly recognized for their sustainability and potential to reduce the environmental impact of construction. These materials, derived from renewable biological resources such as plants, animals, and microorganisms, offer eco-friendly alternatives to conventional building materials. For instance, wood and bamboo have long been used in construction, but recent innovations have expanded the list of materials used such as bacterial cellulose, mycelium, algae (Turhan 2022). These materials are not only biodegradable but also have a lower carbon footprint compared to traditional materials such as concrete and steel (Hempcrete Direct 2021; Zhao et al. 2020).

The integration of biobased materials into construction practices is driven by their ability to minimize environmental degradation. Unlike conventional construction, which often involves significant resource extraction and high energy consumption, biobased materials can be locally sourced and require less energy to produce. This contributes to reducing greenhouse gas emissions and other pollutants associated with traditional construction methods. Moreover, biobased materials provide excellent insulation, further reducing the energy required for heating and cooling buildings (Brennan 2022; Nguyen et al. 2018).

Adopting biobased materials aligns with the principles of the circular economy and sustainable development (Turhan 2023). The use of these materials promotes resource efficiency by encouraging the reuse and recycling of natural resources. As a result, construction projects can reduce waste and lessen their environmental impact. With ongoing research and advancements in biobased construction methods, these materials hold great promise for creating sustainable, healthy, and resilient buildings, particularly in the context of ecological restoration and rural development (Rowell 2012).

2.4 NbS Facilitated by Community Engagement

Nature-based solutions facilitated by community engagement have gained significant attention in recent years. By engaging diverse populations, nature-based solutions not only benefit the environment but also enhance social and community cohesion (Howarth et al. 2020). Indigenous knowledge and local communities are central to the co-creation of eco-creative nature-based solutions, emphasizing the importance of traditional knowledge in driving impactful outcomes (Porri et al. 2022). Community engagement frameworks are essential for the success of natural and nature-based projects, providing structured processes, tools, and measures to support effective stakeholder collaboration at the community level (Thorne et al. 2022).

Additionally, community-based conservation projects, such as comanagement and community-based natural resource management, have shown success in promoting conservation efforts while benefiting local communities (Brooks et al. 2012). In the conservation and management of historic areas, the roles of actors involved in the restoration process and the benefits achieved should be critically examined. Despite the positive outcomes of conservation policies, participation, education and innovation need to be encouraged. More active participation of the community, especially citizens, students and artisans, in restoration processes should be encouraged. Collaborative restoration processes can play an important role in heritage conservation by increasing community participation. These processes should ensure active participation in both decision-making and intervention to secure the public interest.

The Caravanserai Summer School 2024 - collaborative and sustainable restoration project, an organization of the Depo Pergamon Initiative, held as one of the projects at rural areas over the past three years, focuses on collaborative and sustainable restoration, specifically aiming to carry out architectural renovation using ecological methods (Fig. 1). This initiative emphasizes community involvement and environmentally friendly practices in preserving rural heritage. As villages face abandonment due to urban migration or are altered by the use of standardized, ready-made materials for more convenient construction, this project seeks to rediscover and adapt local and traditional building techniques. The goal is to enhance living conditions while preserving the villages' original architectural character. This restoration project in a rural area in Türkiye, İzmir, Bergama, Ferizler Village, serves as a model for ecological renovation, utilizing natural and local materials like stones, timber, earth, straw, manure, water, and sunlight. It also emphasizes waste reduction by repurposing old building elements for new functions.

Through student participation in workshops, alternative techniques are being taught, experimented with, and displayed, helping to revive ancient craftsmanship by focusing on existing structures. In addition to the students, local residents were also involved in the process by providing materials and transferring traditional knowledge.

This particular workshop, microbially induced carbonate precipitation (MICP) for microbial brick production, started with a presentation on biobased emerging materials as an alternative to the petroleum-based construction materials. It continued with introducing the procedure for formulating and developing microbial bricks. During the site visit, the possible locations that require maintenance, as well as the potential application of the microbial bricks were discussed. There were 8 participants who were from the 2nd and 3rd years of architecture and interior architecture departments of diverse universities in Türkiye. The workshop was conducted in a hot and dry climate where the temperature of the day was measured as 35 °C degrees. It was a one day long workshop, including the maintenance and curing processes. In this particular case, the community played an effective role, as co-creators of NbS.

Fig. 1. Applications from the summer school - collaborative and sustainable restoration project

3 Methodology

This study employed an experimental approach within the context of a workshop on microbially induced carbonate precipitation (MICP) for microbial brick production. This methodology (Fig. 2) aimed to provide both a practical demonstration of biobased material production and an evaluation of its potential for ecological restoration in rural settings. Throughout the workshop, participants were encouraged to engage in discussions about the potential applications of microbial bricks in rural architecture, as well as the perceived viability of MICP-based construction materials. The environmental impact of the experimental procedures was minimized by using non-toxic materials and ensuring proper disposal of any waste.

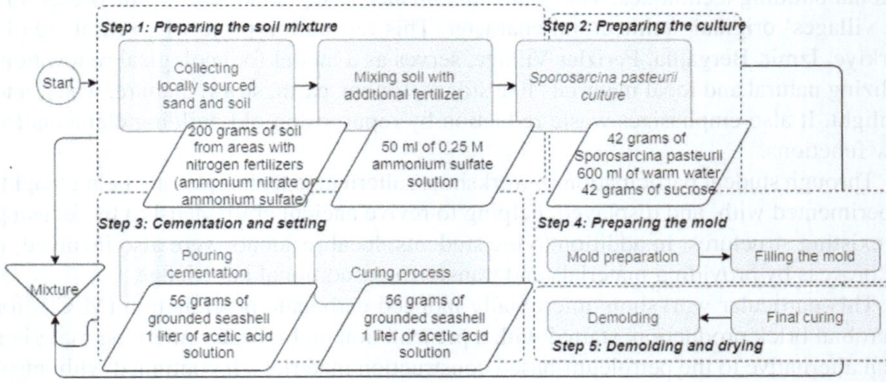

Fig. 2. MICP methodology

The methodology involved several key stages:

Step 1. Preparing the soil mixture: The first step involved the collection of 200 g of soil from areas where nitrogen fertilizers like ammonium nitrate or ammonium sulfate are regularly applied, such as agricultural fields or gardens. The collected soil was thoroughly checked and sifted to ensure it was free from large debris that might interfere with the brick-making process. The collected soil was placed into a mixing container. Then, 50 ml of 0.25 M ammonium sulfate solution was added and mixed thoroughly to enrich the soil with additional nitrogen, thereby promoting bacterial activity crucial for MICP (Fig. 3).

Fig. 3. Preparing the soil mixture

Step 2. Preparing the culture: 42 g of Sporosarcina pasteurii culture was mixed within 600 ml of warm water. The culture was fed with 42 g of sucrose. The mix was stirred for 10 min and then added to the soil (Fig. 4).

Fig. 4. Preparing the culture

Step 3. Cementation and setting:The cementation solution was prepared by mixing equal volumes of ammonium sulfate and calcium chloride (derived from vinegar-seashell powder) solutions. The prepared solution was poured over the soil mixture in the mold until it was fully saturated. The soil was kept moist but not waterlogged to ensure even calcium carbonate precipitation throughout the mixture. The saturated mixture was allowed to sit undisturbed for several hours to facilitate calcium carbonate formation. Additional cementation solution was added every 2–3 h, with the process repeated 4–6 times over the day to maximize the cementation process (Fig. 5).

Fig. 5. Cementation and setting

Step 4. Preparing the mold: A wooden mold was prepared, capable of holding the soil mixture and withstanding the pressure during the setting process. The mold was cleaned and dried before use to prevent contamination. The soil mixture was then packed into the mold, ensuring it was tightly and evenly packed. The soil was pressed down and shaken to remove any air gaps that could weaken the structural integrity of the brick (Fig. 6).

Fig. 6. Mold preparation

Step 5. Demolding and drying: After the final application of the cementation solution, the mold was left undisturbed for 24 h to allow for complete setting. The brick was then carefully removed from the mold and placed in a well-ventilated area to air dry for 5 days (Fig. 7).

Fig. 7. Demolding and drying examples

4 Results and Discussion

The results from the experimental workshop provided valuable insights into the feasibility and challenges of producing microbial bricks using microbially induced carbonate precipitation (MICP) in rural settings. The participants successfully cultivated microbial activity within the bricks, leading to the formation of cementitious structures without the use of traditional cement. This outcome suggests that biobased materials can serve as a viable alternative to conventional construction methods, particularly in ecological restoration projects. However, some samples exhibited cracking due to low hydration, highlighting the need for careful moisture management and further experimentation to optimize the material's properties. The self-ability to repair these cracks using MICP fillers in further studies could also highlight the material's potential in real-world scenarios, which could further enhance its durability and applicability. The findings from this workshop also underscore the importance of context-specific considerations when applying MICP technology in rural areas. The hot and dry climate of the workshop location posed challenges for maintaining the necessary moisture levels during the curing process, suggesting that climate-adaptive strategies may be required for successful implementation in diverse environments. Additionally, the hands-on approach of the workshop fostered a deeper understanding of the material's behavior and its potential use in rural architectural restoration such as restoration of masonry walls, as well as reinforced the importance of community engagement and participatory design in the development and application of ecological construction techniques.

5 Conclusion

The workshop on microbial brick production using MICP highlights the potential of biobased materials as ecological alternatives to conventional construction methods, particularly in the context of rural architectural restoration. Through co-creation and participatory design, the initiative not only advanced technical knowledge but also fostered social equity and environmental stewardship by actively involving participants in the development of eco-friendly materials. The success of this workshop indicates that with further refinement and climate-adaptive strategies, these materials could play a crucial role in preserving cultural heritage while promoting ecological sustainability in rural areas.

In conclusion, ecological architectural restoration in rural areas requires a holistic approach that integrates ecological, cultural, and social dimensions. By leveraging local knowledge, honoring traditional practices, and fostering community engagement, such initiatives can contribute to the long-term sustainability and resilience of rural landscapes and communities. Nature-based solutions facilitated by community-driven innovation offer a promising pathway to addressing the environmental challenges of our time while promoting inclusive and sustainable development.

Acknowledgements. We would like to thank the Depo Pergamon Initiative and especially its co-founder Jose Manuel Garcia Torres, as well as their team, for their invitation to the Caravanserai Summer School 2024 and organizational support for the workshop. The summer school program is organized by Depo Pergamon, Meydan Architecture + Design, Strüktür and Poçolana Works and academically supported by Izmir University of Economics and Yaşar University, with the institutional support of the Embassy of Spain in Türkiye.

References

Baptist, M., et al.: Salt marsh construction as a nature-based solution in an estuarine social-ecological system. Nature-Based Solutions **1**, 100005 (2021)

Brennan, J.: The potential of straw bale construction for sustainable housing. J. Green Building **17**(3), 45–60 (2022)

Brooks, J., Waylen, K.A., Mulder, M.B.: How national context, project design, and local community characteristics influence success in community-based conservation projects. Natl. Acad. Sci. **109**(52), 21265–21270 (2012)

Cassin, J.: History and development of nature-based solutions: concepts and practice. Nat. Based Solutions Water Secur. (2021)

Chen, Y.F., Kitagawa, K.: Locally based architectural construction strategies in rural china: textual analysis of architects' design thinking. Sustainability **15**(13), 10573 (2023)

Hanson, H.I., Wickenberg, B., Olsson, J.A.: Working on the boundaries—How do science use and interpret the nature-based solution concept? Land Use Policy **90**, 104302 (2020)

Hempcrete Direct.: Hempcrete as a sustainable building material (2021). https://www.hempcretedirect.com

Howarth, M., Griffiths, A., Silva, A.D., Green, R.: Social prescribing: a 'natural' community-based solution. Br. J. Comm. Nurs. **25**(6), 294–298 (2020)

Kaja, N., Goyal, S.: Impact of construction activities on environment. Int. J. Eng. Technol. Manage. Res. **10**(1), 17–24 (2023)

Lei, Y., Li, X., Tsai, S.B.: Processing and optimizing industrial structure adjustment of intangible cultural heritage by big data technology in the internet Era. Scientific Programming, 1–11 (2022)

Montero-Parejo, M.J., Jeong, J.S., Blanco, J.H., García-Moruno, L.: Rural landscape architecture: traditional versus modern façade designs in Western Spain. Landscape Architecture - The Sense of Places, Models and Applications (2018)

Nguyen, T., Lum, S., Lee, J.: Energy efficiency in buildings through biobased materials. Sustain. Cities Soc. **41**, 458–470 (2018)

Niu, M., Wang, L.: Analysis of the creation of rural ecological architecture from the perspective of landscape aesthetics: the case of diaojiao building in Xijiang Qianhu Miaozhai. In: E3S Web of Conferences, vol. 439, p. 01009 (2023)

Parlato, M.C.M., Valenti, F., Porto, S.M.: Sustainable promotion of traditional rural buildings as built heritage attractions: a heritage interpretation methodology applied in South Italy. Sustainability **14**(23), 16206 (2022)

Porri, F., McConnachie, B., Walt, K.V.D., Pattrick, P.: Eco-creative nature-based solutions to transform urban coastlines, local coastal communities and enhance biodiversity through the lens of scientific and indigenous knowledge. Cambridge Prisms: Coastal Futures, vol. 1 (2022)

Reiter, S.M., Cheng, L.M., Pouponneau, A., Taylor, S., Wedding, L.M.: A framework for operationalizing climate-just ocean commitments under the Paris agreement. Front. Climate **3**, 724065 (2021)

Rowell, R.M.: Handbook of Wood Chemistry and Wood Composites. CRC Press (2012)

Sahebzadeh, S., Heidari, A., Kamelnia, H., Baghbani, A.: Sustainability features of Iran's vernacular architecture: a comparative study between the architecture of hot-arid and hot-arid-windy regions. Sustainability **9**(5), 749 (2017)

Seddon, N., Sengupta, S., García-Espinosa, M., Hauler, I., Herr, D., Rizvi, A.R.: Nature-based Solutions in Nationally Determined Contributions: Synthesis and Recommendations for Enhancing Climate Ambition and Action by 2020. IUCN and University of Oxford, Gland, Switzerland and Oxford, UK (2019)

Skataric, G., Spalević, V., Popović, S.G., Perosevic, N., Novićević, R.: The vernacular and rural houses of agrarian areas in the zeta region, montenegro. Agriculture **11**(8), 717 (2021)

Thorne, S., Kovács, D., Gailani, J.Z., Suedel, B.C.: Informing the community engagement framework for natural and nature-based projects : an annotated review of leading stakeholder and community engagement practices. Engineer Res. Dev. Center (2022)

Turhan, G.D.: Mechanically-informed computational form-finding for biobased materials in the case of bacterial cellulose. Dissertation. Ph.D. in Design Studies, Graduate School, Izmir University of Economics, Izmir (2022)

Turhan, G.D.: Life Cycle Assessment for the Unconventional Construction Materials in Collaboration with a Large Language Model. 41st eCAADe 2023, 2, 39–48. Graz, Austria, 20–22 September 2023 (2023)

Vidal, A., Martinez, G., Drion, B., Gladstone, J., Andrade, A., Vasseur, L.: Nature-based Solutions for Corporate Climate Targets. Views Regarding the Corporate Use of Nature-Based Solutions to Meet Net-zero Goals. IUCN, Gland, Switzerland (2023)

Xing, Y., Jones, P.J., Donnison, I.S.: Characterisation of nature-based solutions for the built environment. Sustainability **9**(1), 149 (2017)

Zhao, X., Qian, X., Sun, D.: The emerging role of microbial bricks in sustainable construction. Environ. Sci. Technol. **54**(7), 4126–4134 (2020)

KINE[SIS]TEM'17 - Bridging Biomimetic Design and Environmental Analysis in Architecture

Maria João de Oliveira(✉)

ISTAR-Iscte - Information Sciences, Technologies and Architecture Research Centre (ISTA), Lisbon, Portugal
mjoao.oliveira@iscte-iul.pt

Abstract. Nature has always served as a model for sustainable and efficient human innovation. The Kine[SIS]tem'17 International Conference and Summer School exemplified this by integrating biomimetic principles into architectural design. The event, consisting of a two-day international conference and a nine-day design studio, focused on passive, low-cost shading systems inspired by natural forms and processes. Keynote presentations by renowned experts—Manuel Kretzer, Alex Haw, Bob Sheil, and Alberto T. Estévez—offered interdisciplinary insights on topics such as process-oriented design, human biological responses, and adaptable architectural solutions. The design studio challenged participants to create nature-inspired shading systems for a specific site on the University campus. After analyzing environmental conditions, participants developed three distinct systems: *Mechanical Leaves*, *The Lotus Project*, and *Bioshading*. *Mechanical Leaves* employed electronically controlled units inspired by palm leaves to allow real-time shading adjustments but faced challenges with weight. *The Lotus Project*, drawing inspiration from the Codariocalyx motorius plant and Lotus flower, balanced sun protection and wind regulation with aesthetics, though it struggled to fully block environmental factors. *Bioshading*, featuring a pyramidal structure inspired by animal pupils, responded dynamically to changes in light intensity but required complex fabrication processes. Each shading system demonstrated its own strengths and limitations, reflecting the dynamic balance between aesthetic appeal, environmental responsiveness, and practical application. Overall, Kine[SIS]tem'17 highlighted the potential of biomimetic design in addressing contemporary architectural challenges, emphasizing the importance of contextual adaptation, material considerations, and balancing innovation with functionality.

Keywords: Biomimetics · Shading Systems · Nature-Based · Performance-Based Design · Digital Fabrication

1 Introduction

Nature, as the epitome of sustainability and efficiency, has continuously inspired human innovation for centuries. By leveraging minimal resources and offering a vast inventory of forms, structures, and systems, nature has driven creativity across various fields—formal, structural, and material—leading to the development of new materials, tools,

A. Mateus et al. (Eds.): WAD 2024, What's Around Design?, pp. 212–226, 2026.
https://doi.org/10.1007/978-3-032-08163-6_16

mechanisms, systems, and methodologies (Oxman 2010). The Kine[SIS]tem'17 International Conference and Summer School marked a pioneering event in Portugal, dedicated to exploring how natural concepts, systems, adaptations, and geometries can be integrated into architectural functions and aesthetics. Kine[SIS]tem'17 ([SIS]'17), held over eleven days from July 19th to 30th, 2017, was structured in two phases. The first phase, titled "From Nature to Architectural Matter" (Oliveira, Rato, & Leitão 2017); (Oliveira, Rato, & Leitão, KINE[SIS]TEM'17 From Nature to Architectural Matter International Conference 2018), was a two-day international conference featuring keynote addresses from four eminent figures: Manuel Kretzer (Professor of Material and Technology, Fachbereich Design Dessau), Alex Haw (Architect and CEO, AtmosStudio), Bob Sheil (Director, The Bartlett School of Architecture), and Alberto T. Estévez (Founder and Director, ESARQ – School of Architecture, UIC Barcelona). This conference brought together academics, students, and professionals from fields as diverse as biology, architecture, computer science, and sociology, facilitating interdisciplinary discussions on cutting-edge research and projects related to the application of natural principles in architecture.The second phase of [SIS]'17 was the Passive Shading Systems Design Studio, a nine-day international summer school focused on the development of passive, low-cost shading systems. This phase consisted of two stages: context creation and analysis, followed by the design studio. The preparatory stage involved defining the intervention site, collecting data on typology, morphology, and occupancy, conducting environmental analyses, and studying solar radiation patterns (Author 2018). During the design studio, participants immersed themselves in the creative process, developing and fabricating three nature-inspired passive shading systems. Biomimicry in architecture, which draws inspiration from the intricate systems and materials found in nature, was a central theme of the Passive Shading Systems Design Studio. This studio, hosted during the [SIS]'17 International Conference (IC), provided an experimental platform for students to apply these nature-inspired principles to real-world architectural challenges. The event unfolded over four thematic sessions that guided the conceptual framework and design approaches of the studio.

1. [SIS]'17 - From Nature to Architectural Matter IC

The International Conference (IC) featured four thematic sessions, each consisting of seven paper presentations followed by a keynote lecture and a thirty-minute discussion. These thematic sessions—Time Matter and Matters, Human Nature, Biologically-Inspired Design, and Architecture as an Interdisciplinary Field—were crucial in shaping the concepts explored in the design studio (Oliveira & Osório 2017). The presentations encouraged participants to transcend conventional problem-solving approaches and delve into new dimensions of architectural practice that respond to natural elements and processes. In the Time Matter and Matters session, Manuel Kretzer, through a dystopian lens, emphasized a process-oriented design approach, suggesting that innovative architectural ideas could emerge from seemingly bleak scenarios (Kretzer 2017). This perspective provided the foundation for the design studio's speculative yet practical approach to addressing environmental challenges through shading systems. The

Mechanical Leaves project mirrored Kretzer's ideas by dynamically responding to sunlight, aiming to anticipate environmental conditions. Bob Sheil's lecture on the significance of time in design resonated with this theme, inspiring projects like The Lotus Project to incorporate time-sensitive environmental responses that reflect the natural cycles of sunlight and shade (Sheil 2017). Tsiamis and Haselman challenged the static notion of space, arguing that architecture must evolve to reflect the dynamic nature of human experience (Tsiamis & Haselman 2017). Similarly, Dickinson's paper, "Design as an Adaptive Time-Scape," highlighted the urgency of incorporating adaptability and sustainability into design processes to address global challenges (Dickinson 2017). Alex Haw's keynote on Human Nature offered a more human-centered critique of contemporary design, advocating for a greater focus on the biological and emotional needs of users (Haw 2017). This prompted design studio participants to integrate human experiences into their shading systems, exploring how environmental comfort and social engagement could be enhanced. For instance, the Bioshading project transformed an underused terrace into a vibrant space for social interaction, providing a comfortable, shaded environment that exemplified a human-centered application of biomimicry. Blanchi and Mortamais introduced a psychotropic surface that reacts to human emotional input through pneumatic actuators (Blanchi & Mortamais 2017). Meanwhile, Méndez and Niño presented seismic-resilient structures inspired by the human skeleton (Méndez & Niño 2017).

The conference also showcased advancements in material science and bio-inspired design. Vera Parlac discussed the development of artificial materials that mimic natural conditions (Parlac 2017). Daguin, Chaltiel, and Barney presented innovations in 3D printing (Daguin, Lin, & Fayiga 2017; Bravo & Chaltiel 2017; Barney, Lesina-Debiasi, & Pazik 2017), while Kozlov explored resilient bionic structures (Kozlov 2017). The evolving nature of architectural practice was emphasized, highlighting the importance of collaboration across disciplines. Blonder, Osório, and Davidová presented research on natural systems applied through parametric design (Blonder 2017; Osório & Oliveira 2017; Davidová 2017), while Giodice's work on origami-based structures underscored the importance of digital fabrication for structural innovation (Giodice & Romeo 2019). The integration of digital tools and interdisciplinary approaches was further exemplified by projects like Cardoso's agricultural design in urban settings and Zavoleas' study on nanopatterns in structural biology (Polites et al. 2017; Zavoleas 2017).

2. Passive Shading Systems Design studio

Concept and Execution: The goal of the Passive Shading Systems Design Studio was to develop shading solutions for an outdoor terrace exposed to extreme sunlight during peak hours. Participants were encouraged to apply biomimicry principles to create adaptable, responsive, and sustainable systems that would minimize the need for artificial cooling solutions. Using advanced environmental analysis tools like Ladybug and leveraging recycled materials provided by Coretech, students explored how shading systems could be optimized both aesthetically and functionally. A defining feature of the studio was its emphasis on nature-inspired geometries and behaviors. Participants experimented with forms found in plant structures and biological organisms that naturally regulate exposure to sunlight, airflow, and temperature. The integration of digital fabrication techniques, such as CNC milling and 3D printing, enabled the production of highly customized shading systems that responded to specific environmental stimuli.

Preparation, Analysis and Creation of the Context: The preparation, analysis, and creation of context for the Passive Shading Systems Design Studio were pivotal stages in developing nature-inspired shading systems. Conducted on the university campus, this stage involved a rigorous assessment of both the physical site and environmental conditions to ensure an informed design process. The design studio tutors meticulously evaluated the terrace's architectural and environmental characteristics, laying the groundwork for subsequent design activities. This thorough analysis enabled participants to engage with a well-defined context, enhancing the relevance and effectiveness of their designs.

Site Selection and Preliminary Considerations: The site for the shading system was selected based on architectural criteria and specific solar conditions on campus. The chosen location was a second-level terrace, covering an area of 100 square meters and situated between two seven-story façades clad in white lioz marble. This choice was motivated by the terrace's underutilization, primarily due to discomfort caused by excessive heat and solar radiation. Users typically limited their use of the terrace to its edges, seeking relief from intense sunlight. To enhance user comfort and increase terrace occupancy, a canopy skeleton was designed as the foundational structure for the shading systems to be developed by participants. The initial task involved defining the precise location for this canopy. An environmental analysis was conducted using Ladybug, an open-source plugin for Grasshopper, facilitating a detailed assessment through a wind rose diagram and solar path analysis (Fig. 1). This analysis was crucial for mapping the terrace's environmental conditions and understanding the impact of solar exposure on shading requirements.

Environmental and User Impact Analysis: The context of the terrace revealed significant factors affecting irradiance, reflection, and refraction. The white lioz marble façades exacerbated light reflection and refraction, intensifying heat and causing considerable visual discomfort. This effect constrained potential spatial uses and underscored the need for effective shading solutions. Following the environmental analysis, user occupancy and pedestrian movement patterns were studied. Mapping and graphical analysis of these patterns provided insights into how the terrace was used, informing the design of the shading system to better accommodate user needs. One crucial dimension that the original analysis lacked was the social impact of the shading systems. Prior to intervention, the terrace was underused, primarily due to discomfort from direct sunlight. The introduction of these shading systems aimed not only to address environmental factors but also to transform the space into a social hub. As the shading structures improved the terrace's comfort, they encouraged greater interaction and social engagement, fostering a new dynamic for the space. For example, the Mechanical Leaves project empowered users to interact with the shading system directly, adjusting the levels of sunlight they wished to experience. This interactive aspect transformed users' relationship with the environment, making them active participants in controlling their space. Similarly, the Bioshading project was explicitly designed to invite social use by creating comfortable and attractive shaded areas where people could gather and relax. These projects underscore the idea that architectural interventions can play a crucial role in shaping

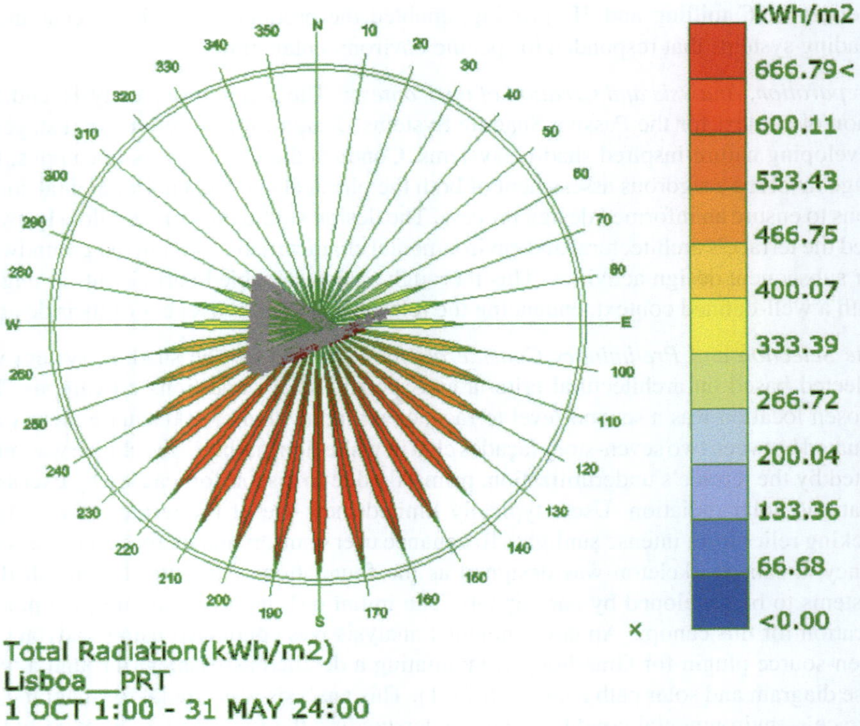

Fig. 1. Radiation graphic extracted from ladybug analysis

both environmental and social dynamics, demonstrating that nature-inspired designs can address both human and environmental needs simultaneously.

Design Principles and Material Constraints: Self-Sufficiency: The canopy framework needed to be entirely self-assembled, without reliance on additional materials such as screws or adhesives. Sustainability: Materials for the canopy were to be recycled or reusable to minimize environmental impact. Resource Efficiency: The design aimed to use minimal resources to reduce production and maintenance costs. Sponsored by Coretech, which provided thermal and acoustic panels made from 100% recyclable automotive by-products, the design was constrained by the amount of available material. The canopy framework utilized sixteen boards, each measuring 2200 × 1200 × 16 mm.

Canopy Framework Design and Assembly: In accordance with these principles, the canopy was conceived as a self-supporting structure composed of four triangular frames and a bench (Fig. 2). These components were preassembled as isolated modules, transported to the terrace, and assembled using a locking and sealing system designed for ease of construction. The canopy was strategically positioned based on the wind and solar analysis, with one frame oriented toward the West, two frames facing South/Southwest, and a horizontal frame linking the others to function as a rooftop. This configuration aimed to optimize shading effectiveness and user comfort while addressing the challenges posed by the terrace's environmental conditions.

Fig. 2: [SIS]'17 Shading System canopy framework.

3. Design Studio

The Passive Shading Systems design studio challenged participants to analyze, design, re(design) based performance, fabricate using CNC, Laser and 3D printing elements, and assemble passive shading systems, leveraging the site-specific conditions of the campus terrace. The design challenge was informed by a pre-designed canopy framework, specific location constraints, and a focus on three core principles: self-sufficiency, the use of recycled or reusable materials, and cost-effectiveness. Participants worked within these parameters to create innovative shading solutions that could be integrated into the pre-assembled canopy framework. Participants were divided into three groups, each tasked with addressing the shading system design from a unique perspective. The process began with a critical analysis of four main issues: structural constraints of the canopy framework, material availability (which ranged between valchromat, OSB, Corethec, pine wood and betula plywood), nature-inspired design, and the relevance of these elements to and into the shading system. This structured approach facilitated a comprehensive exploration of each group's design strategy. The developed projects:

Mechanical Leaves (Team: Diogo Bulhões and Louay Youssef)

The Mechanical Leaves project addressed the rooftop module of the canopy, an area with the least direct impact on thermal comfort according to the environmental analysis. The primary goal was to develop a shading system that could dynamically adjust to varying sunlight conditions, providing users with the option to control their exposure to sunlight or achieve a complete blackout. Inspired by the movement of palm tree leaves, which shift to optimize sunlight exposure, the Mechanical Leaves system incorporated an electronic control mechanism powered by a photovoltaic solar panel. This allowed for adjustable shading based on real-time sunlight intensity. The system featured two symmetrical fan units mounted on opposite edges of the canopy's rooftop frame (Fig. 3). Each fan consisted of five blades, with the first blade on each fan being static. The fans were designed to replicate the overlapping movement of palm tree leaves, facilitating variable sunlight exposure. The mechanical system faced several challenges, including the limited availability of linear motors and the need for precise material selection to avoid deformation. An eight-millimeter thick red fiberwood board was chosen for the fan blades due to its durability and low friction properties. The system was driven by a single linear motor connected to a continuous lashing wire rope and activated by a pulleys circuit. This design allowed for bilateral symmetry in the fan movement, mimicking the natural overlapping process of palm leaves. The fans were parametrically modeled to anticipate physical constraints and potential collisions with the existing structure. A 1:1 prototype was fabricated to test the mechanical and material properties. The final design featured a Lemniscata-inspired mechanism, where the fan systems' symmetrical movement was achieved through a motor-driven lashing wire rope system. This setup allowed users to adjust the shading from fully open to fully closed positions, providing flexible control over sunlight exposure.

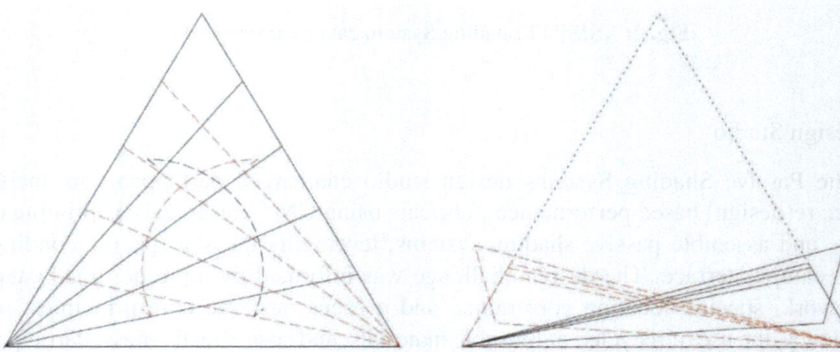

Fig. 3. Mechanical Leaves blades system.

The Lotus Project (Team: Luísa Almeida and Adriana Coutinho)

The Lotus Project was developed for the west-facing canopy module, which required protection from both sun and wind. The design was informed by the environmental analysis, which indicated that while the south edge of the module needed larger openings for sunlight, the north corner required substantial protection. Drawing inspiration from the Codariocalyx motorius plant and the Lotus flower, the system featured autonomous, adjustable shading elements. The Codariocalyx motorius plant's ability to rotate its leaves to capture sunlight and the Lotus flower's daily cycle of opening and closing petals were used as design precedents (Fig. 4). The shading system included a parametric design that allowed for control over the size and distribution of openings based on radiation and wind analysis. The facade was constructed using an eight-millimeter perforated fiberwood panel reinforced with a thirty-millimeter expanded cork agglomerate insulation board. This double-skin approach provided enhanced thermal comfort and concealed the rotation blocking system. The system's interactive elements consisted of bidimensional Y-shaped components that could be locked or unlocked by users via a cylindrical turn button. This feature allowed users to manually control the spin elements and adjust the shading according to their preferences. The Lotus Project's facade system achieved a balance between passive dynamic shading and user interaction, optimizing light and wind exposure while maintaining user comfort.

Fig. 4. Blocking system, Y elements, controlled by a cylindrical left-right turn bottom.

Bioshading – A Performative Mockup (Team: Carmo Cardoso, Pedro Sequeira, and Tania Papasotiriou)

The Bioshading project focused on two triangular frame modules facing South/Southwest, which experienced extreme levels of solar radiation. The design aimed to manage this intense sunlight while allowing controlled light exposure. Inspired by the adjustable nature of animal pupils, the Bioshading system featured a tridimensional structure composed of two pyramids of different sizes, connected at two common vertices to create a rotational axis. This design enabled the system to transition between a closed pyramidal shape and an open diamond frame, mimicking the adjustable opening of animal pupils. The shading system included a hierarchical grid of voids, designed to regulate light intensity and shade distribution based on environmental conditions. The structure was composed of double MDF layers with foldable laser-cut sheets. The voids followed a linear proportional size and were aligned in horizontal rows by threaded rods and a wood pulley network. The system's shading effectiveness was achieved through a progressive hierarchy of voids, which allowed for controlled solar ray passage between 20% and 50% of the total area. This design improved thermal comfort by 2° to 3° Celsius, translating to a 20% to 25% increase in comfort according to Ladybug simulations (Fig. 5).

Fig. 5. Bioshading unit's construction and assemblage

4. Combining Mechanical Leaves, The Lotus Project, and Bioshading

The Passive Shading Systems design studio Design Studio was a testament to the power of biomimetic design in developing passive shading systems that address complex environmental challenges. Each project—Mechanical Leaves, The Lotus Project, and Bioshading—demonstrated unique solutions grounded in nature-inspired principles, while collectively contributing to a comprehensive shading strategy for the university campus terrace.

Mechanical Leaves was designed to address the shading needs of the canopy's horizontal surface, which experienced less direct solar impact compared to other areas. The project's design was characterized by its use of interactive, electronically controlled shading elements, inspired by the dynamic movement of palm tree leaves. This system allowed users to control the shading levels, leveraging a photovoltaic panel for power. Despite its innovation, the Mechanical Leaves project faced challenges related to the weight of its mechanical components and the complexity of integrating the system into the existing canopy structure. In contrast, The Lotus Project focused on the west-facing panel of the canopy, where it was crucial to balance protection from intense afternoon sun and wind. This system drew inspiration from the Codariocalyx motorius plant and the Lotus flower, employing a dynamic arrangement of spinning elements to regulate light and wind exposure. The project's strength lay in its ability to offer controlled shading while enhancing the aesthetic experience within the canopy. However, its design could not completely block solar radiation or wind, which, while a limitation, also contributed to creating interesting light patterns. Bioshading addressed extreme solar radiation on the south/southwest-facing modules through a tridimensional system inspired by the functional movements of animal pupils. This design utilized pyramidal units that adjusted based on light intensity, but faced practical challenges related to material weight and resistance to wind forces. Despite these challenges, Bioshading effectively met its goals by offering a passive shading solution that adapted to varying light conditions.

The integration of these projects into a cohesive shading system showcased the strengths and limitations of each approach (Fig. 6). Mechanical Leaves provided an innovative, interactive solution but struggled with weight issues. The Lotus Project achieved a delicate balance between shading and aesthetic design but could not fully block environmental factors. Bioshading presented a functional adaptation to extreme solar conditions but required complex fabrication techniques. Together, these projects highlighted the diverse applications of biomimetic design principles in creating effective shading systems.

Fig. 6. [SIS]'17 Shading System canopy, completed by the three passive shading systems proposed solutions

2 Conclusions

The Passive Shading Systems Design Studio exemplified the successful application of biomimetic design in architectural shading systems. While the design studio provided a rich opportunity for exploration, several areas could be improved for future iterations.

The terrace presented a unique challenge due to its reflective marble façades, which caused significant solar gain. This setting was ideal for the design studio, but future iterations could benefit from exploring different site contexts. By exposing participants to a broader range of environmental conditions and challenges, they would be better equipped to tackle diverse design problems.

Contextual responsiveness was a critical factor in each project's approach. Participants tailored their design solutions to specific environmental conditions and constraints, demonstrating the importance of adapting to local solar geometry and structural limitations. For instance, Mechanical Leaves, which focused on interactive shading, leveraged the freedom provided by solar geometry. In contrast, The Lotus Project and Bioshading adapted to more constrained conditions, reflecting the need for context-aware designs.

The design of the shading systems was partially limited by the pre-assembled canopy structure provided. Although this framework offered an immediate base for developing the shading devices, greater flexibility in the structural design could foster more creative and integrated solutions. By allowing participants to explore different structural configurations, the studio could enhance innovation.

Fabrication challenges varied across the projects. Mechanical Leaves encountered issues with weight and mechanical complexity, while The Lotus Project optimized material use but could not achieve complete blockage of sunlight. Bioshading, on the other hand, faced difficulties due to its use of multiple materials, which added complexity to its construction. While the incorporation of recycled materials was a commendable step toward sustainability, it somewhat constrained the design possibilities. Future studios could encourage participants to source their own materials or provide a wider range of options to stimulate more diverse and innovative shading designs.

Each project had distinct strengths and weaknesses that contributed to the overall learning experience. Mechanical Leaves excelled in mimicking natural dynamics but struggled with weight. The Lotus Project created engaging light patterns but was unable to fully block sunlight. Bioshading adapted well to changing light conditions but faced challenges with material performance. These insights emphasize the need for a balanced approach that integrates functional performance with practical considerations.

The [SIS]'17 International Conference (IC) and the Passive Shading Systems Design Studio successfully demonstrated how biomimicry can inspire innovative architectural solutions that address both environmental and human needs. The connection between the keynote presentations and the design studio projects was evident in the way participants translated abstract concepts into tangible designs that responded to real-world challenges.

By addressing the social and environmental dimensions of space, the shading systems not only mitigated solar exposure but also transformed the terrace into a more engaging and dynamic social environment. Thus, the design studio made meaningful contributions to the fields of architecture and design by merging sustainability with human-centered design.

By critiquing the studio's proposal and offering recommendations for future iterations, this paper contributes to the ongoing discourse on how nature-inspired design can inform the architectural profession and pedagogy. Ultimately, the studio serves as a model for future educational initiatives that seek to balance environmental performance with social engagement, pushing the boundaries of what biomimetic design can achieve in architectural practice.

3 Insights

The Passive Shading Systems Design Studio made significant contributions to the scientific domain of architecture by investigating how biomimicry can address environmental sustainability challenges. By integrating environmental analysis software with nature-inspired designs, the studio enhanced the understanding of how architectural solutions can achieve both energy efficiency and aesthetic appeal. The data collected from the shading systems' performance in real-world conditions demonstrated their potential to reduce heat gain and energy consumption, thereby contributing to ongoing discussions in climate-responsive architecture. The iterative design process involved experimentation with recycled materials and advanced fabrication technologies, highlighting how digital tools can improve the precision and efficiency of environmental design solutions. This combination of natural principles and digital methodologies paves the way for future research in bio-inspired design.

The studio underscored the importance of aligning shading systems with local environmental conditions, including solar geometry and structural constraints. For instance, Mechanical Leaves was designed for a horizontal surface, allowing flexibility in managing solar exposure but imposing significant weight-related challenges. Conversely, The Lotus Project adapted its vertical shading units to the triangular geometry of the canopy when applied to a west-facing panel, optimizing shading effectiveness despite constraints on design flexibility. Bioshading utilized a rotational mechanism based on a horizontal axis, accommodating various latitudes and effectively employing the geometry of the triangular canopy frame, though it faced challenges with wind resistance. Fabrication and material constraints played a crucial role in the success of each project. Mechanical Leaves struggled with the heavy weight of its mechanical components, while The Lotus Project optimized material use by varying only the size of its spinning units. Bioshading encountered difficulties due to the complexity of using multiple lightweight, foldable materials for its three-dimensional units. These insights reveal that while each system had its weaknesses—such as the weight of Mechanical Leaves and the wind issues faced by Bioshading—they also exhibited unique strengths. Mechanical Leaves excelled in biomimetic design with its autonomous operation, The Lotus Project created visually captivating light patterns, and Bioshading effectively achieved its biomimetic design goals.

The experimental methodology employed in the [SIS]'17 Design Studio, which encompassed problem analysis, iterative design, and prototyping, was critical for developing successful shading solutions. This approach guided participants through a structured process of analyzing problems, drawing on nature-inspired solutions, and refining designs through prototyping. The Biomimetic System Design Model (BSDM) (Oliveira, Rato, & Leitão 2021) (Oliveira M. J., Towards a Bio-Shading System Concept Design Methodology, Doctoral Dissertation 2020) integrates these lessons into a circular design framework that includes three main domains: Architecture, Nature, and Artifact. The Architecture domain focuses on defining and responding to architectural challenges; the Nature domain involves abstracting natural models into design concepts; and the Artifact domain emphasizes the physical construction and emulation of these concepts. From a pedagogical perspective, the design studio served as a model for integrating theoretical knowledge with practical application. Students engaged with real-world challenges, such as excessive solar radiation, and developed solutions that were both functional and innovative. The hands-on nature of the studio, combined with advanced tools like Ladybug and CNC fabrication, provided a holistic learning experience that bridged the gap between theory and practice.

Future iterations of the studio could further enhance its interdisciplinary approach by involving experts from environmental science, engineering, and sociology. This collaboration would encourage students to explore the multifaceted nature of design challenges, deepening their understanding of how architecture intersects with environmental, social, and technical domains.

References

Barney, R., Lesina-Debiasi, L., Pazik, K.: Flips + drips: multi-material deposition augmented through tactile 1:1 volumetric modeling. In: Kine[SIS]tem From Nature to Architectural Matter International Conference, pp. 65–75. Lisbon: DINÂMIA'CET-IUL (2017). http://hdl.handle.net/10071/13603. Accessed 15 Apr 2019

Blanchi, Y., Mortamais, E.: A psychotropic surface based on soft shape changing material emotional input and pneumatically driven actuator. In: Kine[SIS]tem From Nature to Architectural Matter International Conference, pp. 130–136. Lisbon: DINÂMIA'CET-IUL (2017). http://hdl.handle.net/10071/13603. Accessed 15 Apr 2019

Blonder, A.: Fabric materiality FRP shifting towards structures of bio-inspired properties. In: Kine[SIS]tem From Nature to Architectural Matter International Conference, pp. 46–55. Lisbon: DINÂMIA'CET-IUL (2017). http://hdl.handle.net/10071/13603. Accessed 15 Apr 2019

Bravo, M.T., Chaltiel, S.: Paste matter 3D printing in monolithic shells fabrication methods. In: Kine[SIS]tem From Nature to Architectural Matter International Conference, pp. 10–18. Lisbon: DINÂMIA'CET-IUL (2017). http://hdl.handle.net/10071/13603. Accessed 15 Apr 2019

Daguin, M., Lin, S., Fayiga, O.: Percolation printing: defining volume through time and viscosity to create non-layered volumetric prints. In: Kine[SIS]tem From Nature to Architectural Matter International Conference, pp. 56–54. Lisbon: DINÂMIA'CET-IUL (2017). http://hdl.handle.net/10071/13603. Accessed 15 Apr 2019

Davidová, M.: Co-habited, co-lived, co-designed and co-created eco-systemic responsiveness in systemic approach to architectural performance: a case study on interaction of performative solid wood envelope ray and algae. In: Kine[SIS]tem From Nature to Architectural Matter International Conference, pp. 36–45. Lisbon: DINÂMIA'CET-IUL (2017). http://hdl.handle.net/10071/13603. Accessed 15 Apr 2019

Dickinson, S.: Design as an adaptive time-scape. In: Kine[SIS]tem From Nature to Architectural Matter International Conference, pp. 314–322. Lisbon: DINÂMIA'CET-IUL (2017). https://repositorio.iscte-iul.pt/handle/10071/13603. Accessed 13 Mar 2019

Giodice, M., Romeo, F.: Origami textures for adaptive plate and shell structures. In: Kine[SIS]tem From Nature to Architectural Matter International Conference, pp. 205–215. Lisbon: DINÂMIA'CET-IUL (2019). http://hdl.handle.net/10071/13603. Accessed 15 Apr 2019

Haw, A.: Human nature. In: Kine[SIS]tem From Nature to Architectural Matter International Conference, pp. 76–82. Lisbon: DINÂMIA'CET-IUL (2017). https://repositorio.iscte-iul.pt/handle/10071/13603. Accessed 13 Mar 2019

Kozlov, D.: Architectural bionics: from living nature to architecture. In: Kine[SIS]tem From Nature to Architectural Matter International Conference, pp. 102–110. Lisbon: DINÂMIA'CET-IUL (2017). http://hdl.handle.net/10071/13603. Accessed 15 Apr 2019

Kretzer, M.: The optimist's guide to the anthropocene. In: Kine[SIS]tem From Nature to Architectural Matter International Conference, pp. 2–9. Lisbon: DINÂMIA'CET-IUL (2017). https://repositorio.iscte-iul.pt/handle/10071/13603. Accessed 13 Mar 2019

Méndez, W., Niño, E.: Human-inspired seismic-controlling autonomous structures. In: Kine[SIS]tem From Nature to Architectural Matter International Conference, pp. 172–181. Lisbon: DINÂMIA'CET-IUL (2017). https://repositorio.iscte-iul.pt/handle/10071/13603. Accessed 13 Mar 2019

Oliveira, M.D., Rato, V., Leitão, C.: Bioshading system design method (BSDM). In: Em, D.S.-K. Ting, J.A.S., (Edits.), Climate Change Science Causes, Effects, and Solutions for Global Warming, pp. 195–222. Netherlands: Elsevier (2021). https://doi.org/10.1016/C2020-0-00720-6

Oliveira, M.J.: Towards a Bio-Shading System Concept Design Methodology, Doctoral Dissertation. ISCTE-Instituto Universitário de Lisboa, Lisbon (2020)

Oliveira, M.J., Rato, V.M., Leitão, C.: KINE[SIS]TEM'17 from nature to architectural matter international conference. Nexus Netw. J. **20**(2), 489–495 (2018). https://doi.org/10.1007/s00004-018-0369-9

Oliveira, M.J., Rato, V., Leitão, C.: From natur(e) to architectural matter: responsive. In: Estévez, A.T., (Ed.), 3rd International Conference on Biodigital Architecture. 3rd, pp. 303–312. Barcelona: Bubok Publishing S.L (2017). https://repositorio.iscte-iul.pt/bitstream/10071/16024/1/B17_046%20From%20Natur%28e%29%20to%20Architectural%20Matter.pdf. Accessed 15 Mar 2019

Oliveira, M., Osório, F.: (Edits.). KINE[SIS]TEM From Nature to Architectural Matter. Lisbon: DINÂMIA'CET-IUL, ISCTE-IUL - Instituto Universitário de Lisboa. Obtido em 31 de 07 de 2024, de (2017). https://repositorio.iscte-iul.pt/handle/10071/13603

Osório, F., Oliveira, S.: Origami folded surfaces: kinetic systems behind the folding. In: de Oliveira, M.J., Osório, F.C., (Ed.), Kine[SIS]tem From Nature to Architectural Matter International Conference, pp. 28–35. Lisbon: DINÂMIA'CET-IUL (2017). http://hdl.handle.net/10071/13603. Accessed 15 Apr 2019

Osório, F., Oliveira, S.: Origami folded surfaces: kinetic systems behind the folding. In: Kine[SIS]tem From Nature to Architectural Matter International Conference, pp. 28–35. Lisbon: DINÂMIA'CET-IUL (2017). Accessed 15 Apr 2019. http://hdl.handle.net/10071/13603

Oxman, N.: Material-based Design Computation. PhD Thesis. Massachusetts Institute of Technology, MIT Cambridge (2010)

Parlac, V.: Material dynamics: shape memory materials in dynamic skins. In: Kine[SIS]tem From Nature to Architectural Matter International Conference, pp. 19–27. Lisbon: DINÂMIA'CET-IUL. http://hdl.handle.net/10071/13603. Accessed 15 Apr 2019

Polites, M., Belle, M., Maiorino, M., Paramo, R., DiFalco, B., Cardoso, C.: Heliotropic nature's. In: Kine[SIS]tem From Nature to Architectural Matter International Conference, pp. 294–300. Lisbon: DINÂMIA'CET-IUL (2017). https://repositorio.iscte-iul.pt/handle/10071/13603. Accessed 13 Mar 2019

Sheil, B.E.: Digital doubles, colliding. In: Mid-Air. In Kine[SIS]tem From Nature to Architectural Matter International Conference, pp. 150–162. Lisbon: DINÂMIA'CET-IUL (2017). https://repositorio.iscte-iul.pt/handle/10071/13603. Accessed 13 Mar 2019

Tsiamis, M., Haselman, R.: Adaptability in the built environment through the use of transformable architecture - an exploration into the architectural why and engineering how. In: Kine[SIS]tem From Nature to Architectural Matter International Conference, pp. 216–234. Lisbon: DINÂMIA'CET-IUL (2017). https://repositorio.iscte-iul.pt/handle/10071/13603Accessed 13 Mar 2019

Zavoleas, Y.: Re-morphing the amorphous: creating new urban substance. In: Kine[SIS]tem From Nature to Architectural Matter International Conference, pp. 265–274. Lisbon: DINÂMIA'CET-IUL (2017). https://repositorio.iscte-iul.pt/handle/10071/13603. Accessed 13 Mar 2019

Craft 3.0; Positioning Craft and Design Practice(s) Within 'Reduction' and Nature-Based Design

Robert Phillips(✉) and Fernando Galdon(✉)

School of Design, The Royal College of Art, Kensington, UK
{robert.phillips,fernando.galdon}@rca.ac.uk

Abstract. Extractive industries and excessive production are furthering ecological collapse. In the age of the planetary climate crisis, impactful new approaches to ecological practices are imperative. Designers and creatives are not only accountable for innovative solutions, but their work is also intertwined with environmental collapse. By evading ecological implications, designers are complicit in (potential) created problems, requiring a deeper comprehension of their decisions, actions, and their 'craft'. Our next paradigm (around design) must be nature-based 'design for planet'. As the world considers this impact, we examine the designer's role (as a practice of their craft) considering the earth and non-human species. It is within this expanded notion that we see 'nature-based design' and Ecological Citizenship as a timely planetary-scale, multi-species contemporary craft.

Authors characterise *Craft* $^{3.0}$ as reducing ecological impacts, positioning it as a discourse of: reduction, considerate to its surroundings, contexts, beyond human species, supporting skills to contextualise within environments for positive benefit. *Craft* $^{3.0}$ seeks to differentiate tactics to prioritise ecosystem(s), intent on mitigating against negative consequences. The *Craft* $^{3.0}$ position is mapped through; interviews and insights from contemporary literature, creating a contemporary approach to craft and tacit skills. *Craft* $^{3.0}$ establishes trans disciplinary skills nurturing knowledge of materials, their cultivation, application and growth/regeneration. In turn this impacts: usability, ecological issue(s) and 'craft practices', working contextually within environments for ecological remediation.

Keywords: Sustainability · Climate · Nature Based · Steward

1 Research Objective

Craft $^{3.0}$ is a practice intent on achieving ongoing reductions in [environmental] impact(s), and takes account of contextual ecological consideration(s). It seeks to unite trans-disciplinary skills nurturing knowledge of materials, their cultivation, application, growth, and ecological issue(s).

A. Mateus et al. (Eds.): WAD 2024, What's Around Design?, pp. 227–239, 2026.
https://doi.org/10.1007/978-3-032-08163-6_17

2 Introduction

Human civilization is presently overshooting critical planetary boundaries, facing a multidimensional crisis of ecological breakdown, dangerous climate change, ocean acidification, deforestation, and biodiversity collapse [1]. We have become detached from the environment, materials we consume and the 'craft' we practise due to; urbanisation, the neoliberal economy, our working roles, and distance from materiality, repair, or waste we produce. Our e-waste, fast fashion, and waste exportation through colonialist retro-extractivist approaches (amongst others) contribute to this breakdown [2]. Safeguarding the planet provides biodiversity, sustaining all life [3]. Despite an increased awareness of the need to safeguard the environment, our disconnection from nature, and seasonal material dependencies, are becoming corrosive to our contemporary lives. Authors: position *Craft* $^{3.0}$, the natural world's value within it, summarising peer interview analysis, results, and transferable principles. A position/discourse paper framing: skill sets for ecological consideration and working practices within environment(s).

2.1 Designers' Role

As designers, our outputs are intrinsically interlinked with the materials of the world. We need a deep contextual understanding of designing within different human/non-human settings; being critically informed, beyond our own perspective, and in consideration of caring for places and notions of 'othering'. Material complexities encompass: application, overuse, and subsequently preservation. Traditionally, our comprehension of the natural world has leveraged: material selection, procurement, and ever-deepening extractive agendas. Creative contemporaries are positively prioritising responses to the burgeoning climate emergency. Practitioners are learning carbon-footprint reduction methods, building progressive platforms to provide ethical sourcing of materials, e.g., their extraction, provenance, by whom, and its conditions.

Authors advocate for employing 'designed-in' means to facilitate environmental remediation and see this transition as a skill, a tacit knowledge… as a 'craft'. We are in times of transition to more ecologically centred approaches. Transition Design believes that "while many have called for design-led societal change, few have articulated how to undertake and catalyse such change, nor have they identified the areas of knowledge and investigation required to do so" [4]. Transition Design should be based upon a deep understanding of the social history of technologies. In this context, we position a design perspective of 'reduction' and coin *Craft* $^{3.0}$.

2.2 Background

Morris defined craft as "skill or ability in something... Proficiency; expertness... Indicates work, art, or practice of, for example, woodcraft, stagecraft" [5]. The *Crafts Council* (UK craft charity), "believe craft skills and knowledge enrich and uplift us as individuals" [6]. There is no mention of 'pragmatism or just problem solving' but the elevation of craft to "magic in handling materials with thought and care, an ideal all of us can understand, appreciate, and pursue" [7]. Historically, craftsmanship is skilled work, "a desire to do a job well for its own sake" [8]. In contemporary terms, craft has been positioned as a process which "explore[s] and challenge[s] technology, to question and develop cultural and social practices, and to interrogate philosophical and human values" [9]. Crafts practices are pervasive in; unexpected places [10], counteracting obsolescence [11] and providing "social, cultural, political and ecological resistance" [12].

We identify three evolutions, *Craft* $^{1.0}$, *Craft* $^{2.0}$, and *Craft* $^{3.0}$. We characterise a practice based on skills, cultural heritage, and use of local materials as *Craft* $^{1.0}$. An evolution of this approach, from the 1980s, within the creative industries, pushing boundaries of contemporary technologies using global supplies include the works of: *Tom Dixon*, *Iris Van Herpen*, and others. In this evolution, craft was placed in gallery contexts, or in one-off articulations, focusing on high conceptual and technical scope achieving specific aesthetics. This interpretation of craft embraced the capital project and its intrinsic valuation-speculation dichotomy characterised as *Craft* $^{2.0}$.

We see *Craft* $^{3.0}$ as a skill set for the climate emergency reliant on circularity, after-life(s), and impact reduction. Building-off prior connections, crafts practice and trades include (not exclusively): thatchers and/or bricklayers. These work in unison with surrounding environment(s) and non-human species [13]. Thatching "uses straw or grasses as a building material" in roofing [14]. The 'craft' requires wetland conservation, reed agriculture, harvesting, and more. Because of this, straw/reed became despised as roofing for poor rural areas [15]. Thatching processes require high 'craft', from material cultivation (controlling waterways), cutting, transportation, placement, custom toolmaking and more.

In bricklaying terms, a crinkle-crankle wall "undulates in a series of serpentine curves" and is a perfect *Craft* $^{3.0}$ example [16]. It uses fewer bricks, does not require extensive foundations, but requires high craft skills. *Craft* $^{3.0}$ shares many social ambitions of the *Arts & Crafts Movement*. It echoes Ruskin's 'truth' – utilising honest displays of materials, creating outputs that respect culture(s) they have been developed from. *Craft* $^{3.0}$ operates on behalf of the planet. Our environmental safeguarding requires attributes of: skills, knowledge, frameworks, experience, economies, but also, policies in place. Recent *Right to Repair legislation* works for "consumers' legal rights to repair and modify products, pushing for the free availability of spare parts and manuals" [17]. Our material goods are shifting from ownership to custody models, the next step is natural world impact mitigation.

3 Methodology

Authors curated a public roundtable entitled: *'The Cost of Change: new trajectories of production and consumption'*. With an emphasis on design's responsibility in the climate crisis. Authors invited: *Prof. Tony Fry* (defuturing), *Timothée Parrique* (economics and politics of degrowth) and the *Crafts Council*'s Head of Research, *Julia Bennett* (social challenges and sustainable craft). Leading peers were also interviewed, and responses thematically analysed. They identified contemporary *Craft* $^{3.0}$ examples across art, design, and crafts practices, consolidating fresh trajectories. Interviewed peers contributed to frame *Craft* $^{3.0}$ by identifying boundaries, material relationships, methods, principles, and value(s). These elements challenge modern orthodoxies in which artefacts, systems and services are uncoupled from the environmental crisis. The authors

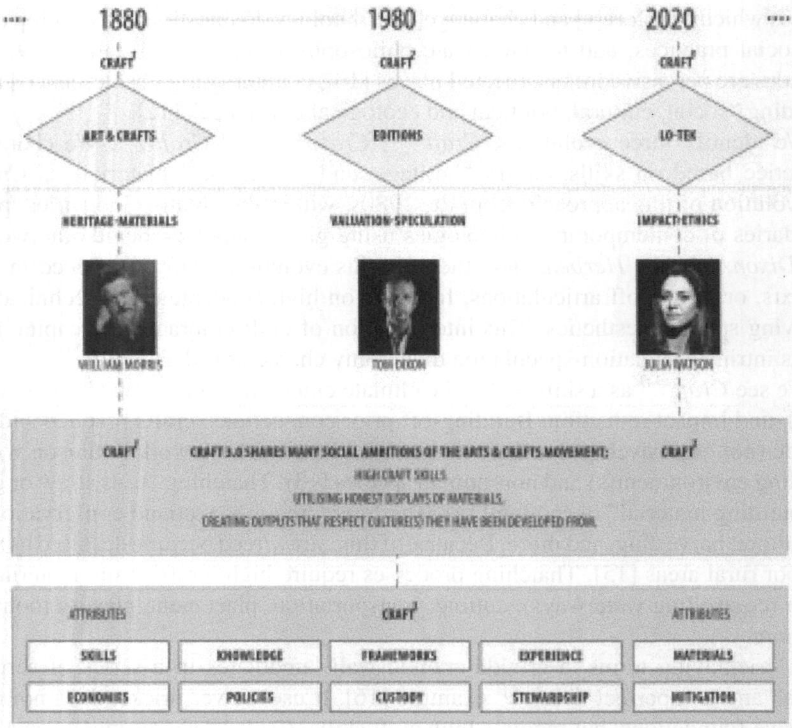

Fig. 1. From heritage to impact; framing the evolution of craft's main paradigms.

approach this evolution of craft from a position of 'guardianship' rather than financial optimisation. We aim to build a framework in which new economies with ecological visions can be grown through 'material reductionism through design'.

3.1 Data Collection

To frame the intended outcome, a progressive and systematic integrative review was conducted. It was decided to use this approach to insert flexibility into the cataloguing of the cases. The search criteria were articulated based on their relevance to the subject. This sourcing approach was preferred to papers, due to their immediacy and relevance. Papers (often) showcase a specific method, technique, or process, whereas a specialist blog offers a full view of cases in almost real-time. We also included reports from institutions to complement and expand data collection to provide a broader, more inclusive, and representative perspective. The criterion for inclusion was the relevance to the practice of regenerative design. The selection was conditioned by our searches; therefore, it was somewhat arbitrary. The cases collected in this paper represent a sample of data. In effect, we were interested in documenting a sample of data in real-time to extract high patterns of knowledge to build "knowledge for future actions" [19].

4 Data Analysis

We collected (38) cases, which we used as a foundation to build a set of categories and subcategories (gradients) to provide a preliminary understanding of the field and practises around regeneration. Here, we are not conducting research on archives or museums or collections. In using elicitation, we aim to capture cases as soon as the phenomenon occurs, to record examples of subtractive design as soon as they are found in the digital landscape. In following *Nigel Cross*, as design researchers we are concerned with extracting and identifying patterns of activity emerging from the collected experience of the material culture, and the collected body of experience, skill and understanding embodied in the arts of planning, inventing, making, and doing in the artificial world, to infer knowledge for future actions in the context of appropriateness.

As we have stated, we have documented subtractive practices as they are evolving. In this scenario, we have selected the cases from the point of view of design practice. Accounting for (38) cases in total, these interventions are a record of places, dates, embodiments and strategies and the proposed categorical structure operates as a type of framework system, which we have enabled by articulating a graphic organisational matrix. The classification of the interventions into categories emerged during the process of collecting. The classification process was executed in the moment, therefore was influenced by contextual elements and personal interpretations and judgements. Once we classified the cases and extracted and organised their main elements into categories, we could implement categorical analysis to underpin and subcategories in specific categories. This process presents a design-led, graphical, and visual alternative or complement to pure statistical models.

5 Results

5.1 Nature's Value Within *Craft* $^{3.0}$

Our nature connectedness, dependence upon it, and the responsibility we have for its health, is critical for ecological futures. Positive natural world interactions include: Nature's Health Service [20], community benefits, understanding of risk and wellbeing, connectedness [21], economics [22], play [23], emotional connection [24] and mental health [25], among others. The Natural Childhood depicts nature's effects: reduction in ADHD, and increased nature engagements could "sav[e] the health service £2.1 billion per annum" [20]. Natural England's medical advisor comments, "the outdoors is a great outpatient department, whose therapeutic value is not fully realised" [26]. The Office for National Statistics estimated the UK's 'nature capital' at "over £1.5 trillion" with rural tourism alone, "annually yielding £14 billion, with 17% of UK tourism involving wildlife watching" [27]. Presented literature demonstrates the value, complexity of locations to inhabitants and visitors alike. To comprehend future intervention points of material 'reductionism', we reviewed contextual and embedded knowledge domains. To unpick the concept of materially 'crafted' relationships aiming to embed environmental knowledge with positive interactions for its surrounding(s), we must deploy crafted interconnectedness of environmental 'causes/effects', when 'materials, food, and health' are seasonally reliant.

We summarise the natural world's importance, for human life on earth, as intertwined and co-dependent. We see these factors as fundamental. Human civilization is presently overshooting critical planetary boundaries, facing a multidimensional crisis of ecological breakdown, dangerous climate change, ocean acidification, deforestation, and biodiversity collapse. We have become detached from the environment, and materials we consume – potentially due to growing urbanisation, the neoliberal economy, our working roles, and our distance from materials, manufacture, in the context of repair and/or the waste we produce. Our culture "must transform from single use interventions to consequential mitigation in the first instance providing health back to the planet under regenerative design frameworks" [28].

The importance of safeguarding the planet is the provision of life, our ecosystem and biodiversity that sustains all life. Indigenous communities [29], bushcraft practices [30] and Inuit societies [31], mainly, consider their environment, material use, resource overuse and seasonality of ecosystems to reduce impactful consequences. Our connection with nature and its seasonal material dependencies, is becoming lost in our current everyday lives. The authors advocate for 'designed-in' accessible means to environmental agency and see this transition and understanding as a skill for enabling *Craft* $^{3.0}$. In the past three decades, multiyear ice, the thickest /oldest "type that supports the Arctic marine ecosystem, has declined by 95 percent" [32]. The "Inuit continue to be the human barometer of climate change; they have been saying to the international community for years that climate change is happening at a rapid pace" [33]. For example, in Canada shipping companies leverage local knowledge to inform haulier routes, i.e., the Arctic Corridors and Northern Voices (ACNV) project responded to this knowledge gap about local areas [34]. More recently Sami reindeer herders (protected by Norwegian law) are raising lawsuits against green energy turbines, "It's a paradox, really, you are squeezed between the impact of climate change and green energy, which is the answer to climate change" [33]. Some call ecological disappearance "solastalgia" coined by Albrecht addressing feelings of shock in Australia, after open-cut coal mining transformed the valley [33]. We must deploy interconnectedness of environment(s) causes and effects, especially when 'materials, food, and health' are seasonally reliant.

5.2 *Craft* $^{3.0}$ Contextual Examples

Contextual examples were highlighted and framed by interviewees and are unpacked. An industrial *Craft* $^{3.0}$ example includes lobster pots, reliant on material, and environment(s) knowledge that they operate within. Inkwell withy (lobster) pots are made from foraged willow sticks, becoming a symbol of their localities' distinctive maritime culture and identity. The craft historically was passed down from captain to crew. In the 20th Century their use declined, replaced by modern non-biodegradable fishing tools. The "U.K.'s *Heritage and Craft Association* lists Willow withy pots as 'critically endangered' believing just 11 withy pot makers remain" [35]. The pots vary regionally, and "only those with direct knowledge know the dialect, rationales for form variations and ways the pots are made" [36]. A second industrial *Craft* $^{3.0}$ is beekeeping. Beekeepers are stockholders of a completely wild and undomesticated creature, the honeybee. Bees can visit 1500 flowers, and fly up to 500 miles, during their lifetime. Their work equates to a large percentage of pollination for our food chain: "without bees, McDonald's would only have the buns to sell" [37]. Prior to *Langstroth Hives* (patented in 1852), beekeepers weaved 'straw skep hives' [38]. Skep hives were crafted: for local climates, catering to differing volumes, and built from contextual material knowledge. These longstanding design examples are situated within ecologies, systems, care, and materiality, i.e., embedded 'craft'.

Within architecture practices, low-tech (akin to *Craft* $^{3.0}$) seeks to re-balance the relationship between buildings and technologies. It concerns "leanness, fewer components, preferring natural, low-embodied carbon materials, reducing reliance on technology and mechanical servicing, robustness and flexibility" [39]. *Craft* $^{3.0}$ 'noun' is defined as design-led practices and approaches beyond immediate requirements, considering systemic interventions including the exit and termination of proposals. This approach is academically supported by authors of *Conceptualizing Sustainable Craft:* who communicate the "[c]oncept of sustainable craft is a multidisciplinary global phenomenon, applied as policies and practices, markets and economy, materials and life cycle in intertwined contexts of use, aiming to reconcile and revive traditions" [40]. In 2023 *the Crafts Council* hosted discussions on how traditional craft practices can help repair our ecosystems, i.e., craft and futures can be strategic collaborators [41]. Scholar *David Pye* defines craft as: "Workmanship using any kind of technique or apparatus, in which the quality of the result is not predetermined, but depends on the judgement, dexterity, and care which the maker exercises as [they] work" [42]. Authors align this to a designer's role by being an Ecological Citizen [43], benefiting each other, wider communities and the planet we rely on for all life.

Our last example of *Craft* $^{3.0}$ considers legacy/sustainable practice is the "dying art of hedge laying" [44]. Hedgerows have an aesthetic value in the landscape; whilst supporting conservation, providing corridors of shelter and food for birds and small mammals. Hedge laying (Fig. 1) is recognised as a heritage craft, with a history of 2,000 years. It embodies a direct link combining: knowledge of material environment, context, designing for decades (not trends), and is considerate towards its impacts, legacies, and construction. The UK Government attempts to cultivate this, paying farmers £9.40 a metre [45]. That cost barely meets the workers' minimum wage of £9.50 per hour, as of April 2022 [46]. This skill, when considered alongside the payment rate, indicates the lack of value it holds to the UK Government. The craft of hedge laying (Fig. 2) encompasses contextual understanding of: habituating species, accurate knowledge of land boundaries, live material knowledge, ecological value, seasonality (e.g., not harming nesting species) etc., tacit knowledge, tool knowledge and experience. A hedge expresses a 'long tail', as it stands for generations, and nurtures species and serves as a corridor. The created hedge is an ecosystem informed by all its constituents, i.e., a product with co-dependent relationships. These approaches are transferable to: designing outputs, questioning legacies and deconstruction.

An arts exemplar of *Craft* $^{3.0}$ is *Project for a Rift Valley Crossing*, where the artist *Simon Starling* constructed a canoe out of magnesium extracted from 1,900 L of lake water (Starling, 2015). The magnesium to create *Starling's* boat was extracted from the waters of the Dead Sea, the world's most concentrated source of magnesium. After exhibiting it, *Starling* returned it to its source, using the vessel to cross from Israel to

Fig. 2. Hedge Laying Process. Images care of *Richard Lewis Hedge laying.*

Jordan. The canoe's material was ethically sourced, leveraged new models of manufacture, and demonstrates a mastery of *Craft* $^{3.0}$. A design-led example of *Craft* $^{3.0}$ is Daniel Metcalfe's *Hannafore Project* that centres around a concrete outfall pipe on Hannafore Beach in Cornwall [47]. The project recrafted the pipe's surface, enhancing its ecological function (creating a habitat for marine species), maintaining the walkway, and primary functions as a sewage outfall pipe. A community-led *Craft* $^{3.0}$ example, *the Crochet Coral Reef*, offers an expression of coral reef knowledge [48]. The community fostering project unites people to crochet handmade sculptures, embodying millions of stitches on display. The project combines a deep knowledge of coral reefs, with manual skills, utilising biodegradable materials.

Local Works Studio (LWS) brings a circular economy approach to design and construction through craft. LWS focuses on the "creative reuse of site-based resources, local manufacture, and processing" [49]. Heavy Gardening (made by LWS) was a sculpture trail of artworks in Liverpool. *Something & Son* commissioned LWS to make an artificial reef, from waste oyster shells. The reef was cast with LWS's Shellcrete material, using waste shells from seafood restaurants. The artificial reef sits on the seabed, with anchoring ropes running vertically up to the surface, for mussels, oysters, and sea life to populate. Finally, Castlemaine (Australia) is a systems *Craft* $^{3.0}$ example. The town's artists launched their own clay currency as an economic experiment called *the Silver Wattle*. $10,000 worth of the currency was created in two denominations, from locally acquired clay [50]. This approach to material value/local production (that was hard to fraudulently copy) opens new economics of *Craft* $^{3.0}$ (Fig. 3).

6 Principles of *Craft* $^{3.0}$

Craft $^{3.0}$ sits in the context of degrowth, considerate craftsmanship, ecological consideration, and material reduction, as its fundamental pillars. The examples demonstrate *Craft* $^{3.0}$ attributes, they are envisioned and created by material custodians, seeking to be better ancestors, rather than evolutionary 'capital designers'. Contextual knowledge, *Craft* $^{3.0}$, and contemporary design approaches provide new opportunities for design that go well beyond materiality; starting with outputs 'end-focused' on cultural, community

Fig. 3. Construction of the 'Shellcrete' material created by *Local Works Studio*. Images Courtesy of Local Works Studio 2022.

and environmental contexts. We agree that the breakdown of: material knowledge, expertise, ecologic understanding, tacit knowledge, and locality is critical. As Fry [51] notes, understanding 'being in the world' with humans and others, in time, brings with it a reconceptualised responsibility of material ethics of care, for designers. Technology can aid ecological interventions and optimise processes. Boehnert emphasises, in relation to the acute denial of our interdependencies with nature, "humans have not yet learned how to use technology on scale in ecologically benign ways" [52]. *Craft* $^{3.0}$ as a discourse sheds light on how technological usages, work with nature and understanding planetary boundaries, may be emphasised and desirable.

6.1 Transferable Principles of *Craft* $^{3.0}$

Current crafts practices are very aware of 'tool marks' and legacies they leave within materials/created works. However, there is often a disconnect when considering the ecological impacts of those materials, requiring further investigation. Material provenance, "the history of ownership of valued objects", and legacy data are already used within craft practice of precious materials [54]. *Craft* $^{3.0}$ legacies consider the after effects of projects, initiatives and how they are financially, resource and environment sustaining. Precious metal hallmarking (a common practice) not only supplies provenance but can identify ethical sourcing. In time digital hallmarking could provide transparency, alongside material passports. Traditionally (for example) wheelwrights selected timbers for their role in assembly, with knowledge of sourcing, material seasonality and using appropriate species. Even exporting timber for firewood has ramifications, as "regulations to protect natural environments and plant-based industries of receiving countries from introductions of harmful organisms" [55]. Considerate material use is evident in thatching, beekeeping, and contemporary practices, and it needs updating for *Craft* $^{3.0}$ within a wider forum. These considerations in turn inform how the material(s) and environment(s) are left when practitioners exit. Not just from a material perspective but, assisting individuals/communities and not 'extracting' assets. These should be considered at the inception/undertaking of projects.

Craft $^{3.0}$ should reskill agents with new and future perspectives, rather than reviewing the future with today's tools. We should consider mechanisms to upskill and continually review with contemporary best practice for planetary challenges. *Craft* $^{3.0}$ reskilling, aligns makers with steps to make more ecological choices and inform contemporary best practices. The final element (that is often an afterthought) are the connections between materials, practices, their consumption and regeneration. We need to understand that this field is interconnected; practically, culturally, economically, and ecologically. All dependent on their context of operation and application. Designers should work closely with the locality (e.g., communities) and collaborate with ecological / environmental expertise. We have outlined a range of considerations vital for associated ecosystems, skills, and its economic implementation. We demonstrate that one of the core skills is mastering 'potential impacts', both environmental and ethically. This demonstrates an overarching concept that communicates the interdependencies of *Craft* $^{3.0}$.

- *Designing for Exit:* At every stage of the design and execution process, including communities, locations and subsequent opportunities for repair. - *Considerate Material Use:* This is all bound in the context of use, location and purpose. These decisions are often formed too quickly.
- *Contextual Ramifications:* The cultures, communities, and use cases are the most important points. We need to change the rhetoric that unsustainable practices are 'just bad', polarised or just 'cost saving' and support transitioning to appropriate carefully crafted choices as better custodians.
- *Place Based Knowledge:* Not everything has to 'scale', there will be rural, sub urban and international needs/perspectives that will be totally different. - *Seasonality:* Materials and the natural world have seasons of regrowth, optimisation, left fallow and deeper understanding is required. - *Impact Considerate:* Both in project, legacies, communities and construction. - *Building Contextual Tacit Knowledge:* Outcomes are not predetermined, but depend on the judgement, and care which *Craft* $^{3.0}$ makers exercise. - *Community Craft:* The care and consideration of collaboration, co-creating knowledge and the careful production of assets.

7 Conclusion

Building from *Glanville*, we see research in design about creating knowledge for future actions [19]. We depart from traditional notions of value, as associated exclusively with economic value. Instead, we are now understanding that value can also be social, or environmental, or a combination of these. In positioning *Craft* $^{3.0}$ at the intersection of research and practice, or applied research in the context of design, we see its implementation as creating knowledge for future actions, delivering environmental, social, and economical value. This structure changes traditional orthodoxies in which economical value is used to override any other possibility. We need a deep contextual understanding of designing within human/non-human settings; being critically informed, beyond our own perspective, and in consideration of caring for places and notions of othering. In time, legislation around new ecological laws should readdress the breakdown of materials from the onset. We are in times of commodity scarcity, with increasing comprehension of material impact over time and all its traits, i.e., centralised farming generating large (silage piles), soil erosion and acidification.

We should question avenues for these stockpiles of waste, and urban mining potentialities. Our responsibilities are tied to resource use and efficiencies of design, i.e., disposal, reclamation, urban mining (taking thatching and earth building as an example), material growth, skill in material selection, installation, and durability. There are legacies and complexities of materiality in different situations and contexts. Seasonality, not just considering weather, for example, may encompass certain work being scheduled to avoid the nesting season. The challenge is, which complexity 'element' do you make a priority? We must consider how design and manufacturing can offer regeneration in situations, so that we are not just working from a capitalist perspective, and in terms of extraction of resources, generating displacement and negative impact. Further considerations include: material end-of-life, comprehension of wear, conditioning, and material degradation over time. Also of importance is the question of designing for repair (whilst maintaining manufacturers warranties) and the interesting interplay between people's skill, knowledge, and trust. These challenges are intertwined, benefiting each other. It is within this expanded notion that we see 'nature-based design' and Ecological Citizenship [42] as a timely planetary-scale, multi-species contemporary craft. The authors have characterised this evolution of craft as *Craft* $^{3.0}$. The question now is, how we as practitioners, designers, and crafts people can evolve our practices to *Craft* $^{3.0}$?

Acknowledgements. Dr. Delfina Fantini van Ditmar for their collegial discussion, contribution(s) and criticality. Professor Tony Fry (University of Tasmania), Head of Research Crafts Council Julie Bennett. The Design Futures MDes, RCA colleagues specifically Alon Meron. Lifelong colleagues; Prof Stephen Boyd-Davis, James Tooze, Nick Gant, Prof Sharon Baurley, Prof Roberto Fraquelli, and Prof Ashley Hall. **Rights Retention Statement:** This work was part funded by a UKRI grant (EP/W020610/1). For the purpose of open access, the author has applied a Creative Commons Attribution (CC BY) licence to any Author Accepted Manuscript version arising. Dan Price for administration and proofing support. Ecological Citizen(s) is a cross-RCA research network led by Dr. Rob Phillips, with Professor Sharon Baurley, and Tom Simmonds, in partnership with Professor Sarah West of the Stockholm Environment Institute (SEI) at the University of York, and Professor Alec Shepley of the Faculty of Arts, Science and Technology at Wrexham University, and partners from industry, third sector, NGO. Ecological Citizens is a Digital Economy Network+ project funded by the UKRI Digital Economy Programme, that is focused on digital interventions that would create 'the conditions to make change' towards a sustainable post-industrial society. The work was supported by the EPSRC Network+ award (EP/W020610/1).

References

1. Richardson, K., et al.: Earth beyond six of nine planetary boundaries. Sci. Adv. **9**2023https://doi.org/10.1126/sciadv.adh2458
2. Smith, D., Blowers, A. Here today, There Tomorrow: The Politics of Hazardous Waste Transport and Disposal. Waste Location. Routledge, pp. 208–226 (2021)
3. Hickel, J.: What does degrowth mean? A few points of clarification. Globalizations **18**(7), 1105–1111 (2021)
4. Irwin, T., Kossoff, G., Tonkinwise, C.: Transition design provocation. Des. Philos. Papers **13**(1), 3–11 (2015)
5. Morris, W.: 'What is Arts and Crafts?', Homepage of William Morris Tile, https://williammorristile.com/articles/what_is_arts_and_crafts.html2022. Accessed 13 Dec 2022

6. Crafts Council: We are the national charity for craft. Craft Council, https://www.craftscouncil.org.uk/2022. Accessed 13 Dec 2022
7. Lovelace, J.: Craft: Seriously, What Does the Word Mean?. Craft Council, https://www.craftcouncil.org/magazine/article/craft-seriously-what-doesword-mean2022. Accessed 13 Dec 2022
8. Frayling, C.: On Craftsmanship, Towards a new Bauhaus. Oberon Books Ltd, Great Britain, Chippenham (2011)
9. Craft Research: 'Craft Research', Journal Website. In: Niedderer, K., Townsend, K., (eds.) Intellect Books https://www.intellectbooks.com/craft-research. Accessed 13 Dec (2022)
10. Macbeth, P., Barber, C.: Craft in unexpected places. Craft Res. **6**(2), 275–285 (2015)
11. Chatterjee, A., Alvelos, H.: Re-inscribing the value of craft in times of dictated obsolescence. Craft Res. **11**(2), 275–294 (2020)
12. Castelblanco Perez, S.: Craft as resistance: a case study of three Indigenous craft traditions. Craft Res. **13**(2), 387–409 (2022)
13. Ocejo, R.E.: Masters of Craft. Princeton University Press (2017)
14. Long, K. Introductory Guide to Thatching. 1st edn. NBRI/CSIR (1978)
15. Wichmann, S., Kobbing, J.F.: Common reed for thatching—A first review of the European market. Ind. Crops Prod. **77**, 1063–1073 (2015)
16. Pevsner, N.: [Architectural glossary]; Pevsner's Architectural Glossary. 1st edn. Yale University Press, London
17. Bradley, K., Persson, O.: Community repair in the circular economy – fixing more than stuff. Local Environ., 1–17 (2022)
18. Singh, A., Malhotra, S.: A researcher's guide to running diary studies. In: Proceedings of the 11th Asia Pacific Conference on Computer Human Interaction, pp. 296–300 (2013)
19. Glanville, R.: Design propositions. In: Belderbos, M., Verbeke, J. (eds). The Unthink-Able Doctorate: Brussels: Sint Lucas (2005)
20. Moss, S.M.: Natural Childhood, 1st edn. National Trust London, London (2012)
21. Lumber, R., Hunt, A., Richardson, M., Harvey, C.: Nature Connections 2016 conference report: implications for research and practice. University of Derby, Derby (2017)
22. Bockstael, N.E., Freeman, A.M., Kopp, R.J., Portney, P.R., Smith, V.K.: On measuring economic values for nature (2000)
23. Brown, J.M., Kaye, C.: Where do the children play? An investigation of the intersection of nature, early childhood education and play. Early Child Dev. Care **187**(5–6), 1028–1041 (2017)
24. Munoz, M.C., 'How can we all help conserve nature?', Homepage of frontiers for young mind. https://doi.org/10.3389/frym.2019.00084. Accessed 13 Dec 2022
25. Cox, D.T.C., Gaston, K.J.: Human-nature interactions and the consequences and drivers of provisioning wildlife. Philos. Trans. Royal Soc. London. Ser. B, Biol. Sci. **373**(1745) (2018). https://doi.org/10.1098/rstb.2017.0092
26. Holden, P., Abbott, G.: RSPB Handbook of Garden Wildlife, 1st edn. Bloomsbury Publishing, London (2018)
27. Juniper, T.: What has Nature Ever Done for Us? How Money Really Does Grow on Trees, 1st edn. Profile books, UK (2013)
28. Wahl, D.: Designing Regenerative Cultures, 1st edn. Triarchy Press, New York (2016)
29. Dawson, J., et al.: Infusing Inuit and local knowledge into the low impact shipping corridors: an adaptation to increased shipping activity and climate change in arctic Canada. Environ Sci Policy **105**, 19–36 (2020)
30. Mears, R., Lore of the wild - Ray mears, bush survival expert, talks to christian amo-deo about life in the wilderness. Geographical, 18–21 (2002)
31. Strzyzynska, W., 'Sámi reindeer herders file lawsuit against Norway windfarm', The Guardian, https://tinyurl.com/jzssddrj. Accessed 13 Dec 2022

32. Johnson, A.: 'As ice melts, the Inuit strive to keep their culture alive', National Geo-graphic. https://tinyurl.com/mry68es2. Accessed 13 Dec 2022
33. Cecco, L.: The face of climate change: how Inuit youth led the fight to save the Arctic. The Guardian, https://www.theguardian.com/environment/2015/dec/10/paris-climatechange-talks-arctic-inuit-youth-cop21. Accessed 13 Dec 2022
34. Richard, C.: This centuries-old British tradition might soon be lost', National Geographic. https://www.nationalgeographic.com/travel/article/this-centuries-old-britishtradition-might-soon-be-lost. Accessed 13 Dec 2022
35. Leach, H.: Sociophonetic variation in Stoke-on-Trent's pottery industry (Doctoral dissertation, University of Sheffield) (2018)
36. Langworthy, G., Henein, M., Mohser, H.: The Vanishing of the Bees. Hive Mentality Films, USA (2009)
37. Hughes, C.: Urban Beekeeping, A Guide to Keeping Bees in the City, 1st edn. The Good Life Press Ltd, Preston (2012)
38. Fowles, E.: Make low-tech our mantra and design clean and simple. The RIBA Journal. https://www.ribaj.com/intelligence/low-tech-resetclimate-emergency-feilden-fowles. Accessed 13 Dec 2022
39. Väänänen, N., Pöllänen. S., Conceptualising sustainable craft: concept analysis of literature. Des. J. **23**(2), 263–285 (2020)
40. Crafts Council: Gaining Ground: How ancient craft knowledge can shape our future. https://www.craftscouncil.org.uk/whatson/gaining-ground-how-ancient-craft-knowledge-can-shape-our-future. Accessed 13 Dec 2022
41. Pye, D.: The Art and Craft of Workmanship. Cambridge, Cambridge (1968)
42. Phillips, R., et al.: Defining ecological citizenship; case-studies, projects & perspectives; analysed through a design-led lens, positioning 'preferable future (s)'. In: Design for Adaptation Cumulus Conference Proceedings Detroit 2022, pp. 264–288. Cumulus Association (2023)
43. Ingram, S.: Country diary: the vanishing craft of hedge laying. The Guardian. https://www.theguardian.com/environment/2021/mar/19/countrydiary-the-vanishing-craft-of-hedge-laying. Accessed 13 Dec 2022
44. GOV UK: BN5: Hedgerow laying. https://www.gov.uk/countryside-stewardship-grants/hedgerow-layingbn5. Accessed 13 Dec 2022
45. GOV UK: National minimum wage and national living wage rates. https://www.gov.uk/national-minimum-wagerates2022. Accessed 13 Dec 2022
46. Metcalfe, D.: Hannafore project. Royal college of Art, United Kingdom.https://danimetcalfe.com/index.php/research/multispecies-design/. Accessed 13 Dec 2022
47. Wertheim, C.: An ever-evolving nature-culture hybrid. Homepage of Crochet Coral Reef. http://www.crochetcoralreef.org. Accessed 13 Dec 2022
48. Bosence, B.: Local works studio. https://localworksstudio.com/2022. Accessed 13 Dec 2022
49. Lawrence, S.: Castlemaine clay coin currency launched to strengthen town's economy. Homepage of ABC Central Victoria. https://www.abc.net.au/news/2022-05-09/clay-coins-currency-launchcastlemaine/1010477002022. Accessed 13 Dec 2022
50. Boehnert, J.: Design, Ecology Towards the Ecocene. Bloomsbury Publishing, Politics (2018)
51. Collecting & Provenance Research (Getty Research Institute). https://tinyurl.com/yc8azj6j. Accessed 15 May 2022
52. Forestry Commission: Importing and exporting wood and timber products. GOV.UK. https://www.gov.uk/government/collections/importing-andexporting-wood-and-timber-products. Accessed 15 May 2022

Strategic Design for Green Transition Inspirations

Prologue: Change by Design - Learning to Learn From and With Each Other

By Sabine Junginger, Professor of Design, Northumbria University, UK

The "What's around Design – Designing Nature and Humanity-Centered Futures" Conference brought together an impressive international group of engaged and concerned researchers. Moreover, it shone a light on a less salient location in the Algarves and its opportunities to contribute to such a future. According to the conference organizers, one of the reasons I was invited to speak was that my work has gotten the ear of politicians, policymakers and public managers. There is some truth to that. One of the projects I led with MA students in Switzerland with a local municipality has triggered action by local politicians. Together, we investigated the lack of participation by residents in local elections through a participatory design inquiry. We found that many long-term residents struggled with the Swiss Citizenship requirements for one reason or another. Yet, absent citizenship, they are not eligible to vote. After an immediate effort to call for a chance of local law that would allow non-Swiss long-term residents to vote on the local level was rejected, the initiative has been picked up again this year. And our findings continue to be central to this change.

This project – along with other work I have been able to conduct, for example, with the WHO Global Initiative for AI for Health and the Brazilian Ministry for Management and Innovation in Public Services – reinforce that human-centred design research enables us 'to move beyond dispute and debate towards action', one of the key challenges in a pluralist democratic society Richard McKeon (1957) so astutely observed. Together with John Dewey, McKeon provides the needed foundation for what is now re-labelled 'humanity'-centered design. Design inquiries that have leaned on both philosophers have never abandoned humanity but grounded their thinking and doing in the realization that we cannot go about designing in a 'human-centered' way when that means to pollute the air, eliminate bio-diversity, abandon nature and the planet (Junginger 2016, p. 18). Design research contributes by examining, visualizing, explaining and differentiating principles and methods of doing and making. Design research serves other disciplines by showing how outcomes connect to processes and approaches people employ. At the same time, it proposes co-creative and participatory paths forward. In doing so, design researchers facilitate dialogues on the value and limitations of specific approaches in particular problematic settings. They illustrate through examples and experiments how different approaches complement each other and explain product/outcome failure in terms of design principles, processes, practices and methods. This explorative, reflective and analytical mode invites people to experiment with human-centered design principles and approaches. In general, this research fosters theories and understanding of design itself – across disciplines and professions.

In our VUCA world, design research gained renewed recognition for its role in social innovation and in the development of new business models. Design thinking is understood in the context of wicked problems (Buchanan 1992). The EU New European

Bauhaus initiative underlines the links among engineering, architecture and design in the social, economic and environmental context. For design research, the current interest on the EU level generates new and exciting opportunities for international research collaborations where methods like user research, co-designing and cocreation will be central. In fact, all three streams of Horizon Europe recognize that the development of novel solutions to complex problems of business and society depend on new methodological approaches to design. The same can be said about the UN Sustainable Development Goals (SDGs) – each goal implicitly calls for new design thinking, processes and methods. Together, they refer to the need of new design capabilities that allow for new ways to conceive of, plan, develop, deliver and execute new products and services through the skilful and conscious employment of design methods and processes. Drawing on Dewey (1938), McKeon 1957, Simon (1969 [1996]), and Buchanan (1995), designing inherently involves learning, learning involves changing and changing often requires people designing to influence the conditions under which a design effort is taking place to allow for new ideas and novel solutions that can be acted upon. Policymakers and public managers in the Algarve have a lot to gain from design and ISMAT promises to have a main role here.

Refereces

Buchanan, R. (1992). Wicked Problems in Design Thinking, Design Issues, Vol. 8, No. 2 (Spring, 1992), pp. 5–21.

Buchanan, R. (1995). Branzi's Dilemma', in: Päivi Tahkokallio and Susan Vihma (Eds.) Design – Pleasure or Responsibility? University of Art and Design, Helsinki UIAH 1995.

Dewey, J. (1938). 'The Pattern of Inquiry', in Logic- The Theory of Inquiry, Holt, NY.

Dewey, J. (1934). Art as Experience, London.

Junginger, S. (2016). Transforming Public Services by Design: Re-orienting policies, organizations and services around people. Routledge. McKeon, R. (1957). 'Communication, Truth and Society', in McKeon, Z. K., ed. (1990), Freedom and History and Other Essays – An introduction to the thought of Richard McKeon, University of Chicago Press, IL. Simon, H.A. (1996). The Sciences of the Artificial, MIT Press, 3rd edition.

The New European Bauhaus: https://new-european-bauhaus.europa.eu/index_en.

The UN Sustainable Development Goals: https://www.un.org/sustainabledevelopment.

Introduction to Part 5: Strategic Design as a Catalyst: Case Studies

Mostafa Zekri, ISMAT, Portugal, and Onur Mengi, Izmir University, Turkey (Conference Track Chairs)

Strategic design plays a crucial role in driving the green transition by embedding circular economy principles across various sectors. This chapter highlights the importance of design in fostering environmental sustainability through implementation cases, examples, and best practices. Case studies emerge at the intersection of material innovation, educational reform, and systemic industrial change. Key strategies include developing bio-based and renewable materials, integrating circular principles into design education, and establishing collaborative networks that connect designers with stakeholders across sectors.

At the heart of the green transition lies the circular economy, a system focused on minimizing waste and maximizing resource efficiency. Strategic design acts as a bridge between innovation and sustainable practice, enabling the shift to more sustainable models. One example, "Circular Diversification and Design" shows how manufacturing companies can expand their product portfolios by adopting circular principles. This approach not only reduces environmental impact but also uncovers new market opportunities. By fostering circular innovation, design can drive systemic transformation in industrial contexts. In the textile industry, the "Circular Network of Textiles Biomaterials" study examines how biomaterials and circular design practices can revolutionize material selection. It emphasizes categorizing materials into biological and technical cycles based on their longevity and end-of-life considerations. The Circular Network, designed as an educational tool, illustrates how strategic design influences both production processes and educational practices, enabling designers to make sustainable choices that reduce environmental impact.

The educational role of strategic design is further explored in "Next Skills for Circular Design Profiles: Bio-Based Innovation Towards Climate-Neutral Cities." This initiative addresses educational gaps in circular product design and urban planning, fostering bio-based innovation while integrating circularity into the built environment. By linking academia, vocational education, and the broader socio-economic environment, this multidisciplinary approach accelerates the transfer of research into practical applications and supports green start-ups, driving systemic change.

Material innovation is a recurring theme in the green transition, with "Materials Design Trends Fostering Environmental Sustainability" highlighting the use of sustainable materials that balance economic growth with ecosystem health. By focusing on materials with reduced emissions, resource use, and enhanced biodiversity, designers can create environmentally conscious solutions. Mycelium, a fungi-based biomaterial, exemplifies this potential. The study "A Different Kind of Growth" discusses its versatility, biodegradability, and eco-friendliness, positioning mycelium as a promising alternative to conventional materials like plastic and synthetic textiles. Despite scalability challenges, mycelium offers the potential to transform material usage in various industries.

This chapter emphasizes the transformative power of strategic design in the green transition. By integrating Circular Economy principles, material innovation, and sustainable practices, design can enable systemic change across industries. The case studies presented, including bio-based materials like mycelium, demonstrate how strategic design not only supports environmental sustainability but also opens new market opportunities.

Circular Diversification and Design: The Contribution of the Design Discipline and Practice to Identify New Production Opportunities and Create a Closed Loop System

Noemi Emidi(✉), Eva Vanessa Bruno, and Beatrice Lerma

Politecnico di Torino, Torino, Italy
noemi.emidi@polito.it

ABSTRACT. The current global climate crisis demands an immediate shift towards new business models that operate in circular and non-linear logic. However, despite discussions about this transition from the late 1960s, achieving a circular economy remains lengthy and challenging. Manufacturing companies must acquire and develop new skills and expertise to implement this change successfully. Additionally, they must continually analyze and evaluate their performance from multiple perspectives over an extended period to obtain optimal results.

The main objective of this research is to delve into the approaches and tactics utilized by four companies based in Italy, representing manufacturing realities, including circular models in their portfolio diversification strategy. The research theorized this model as "Circular Diversification" (CD), and the study seeks to analyze how these companies have implemented it in practice, placing particular emphasis on the role of Design, actual or presumed, by going to find out which Design professions play a crucial role in CD.

The research uses qualitative case-based reasoning methods to extrapolate meta-knowledge; Issue-Concept-Form (ICF) is the semantic model chosen to break down information from case studies. Next, three parameters are crossed through an Alluvial Diagram to show intersections: problem/motivation (former *Issues*), strategy (former *Concept*), and form/solution (former *Form*), which Designers adopt are identified. Three relevant Design figures operating for a Circular Diversification are determined from the intersections: Systemic-Circular Designer, Product Designer, and Strategic Designer. Specific forms/solutions can be replicable and scalable recommendations in other manufacturing and territorial contexts, paving the way toward a Circular Diversification support service.

Keywords: Industrial circularity · Resource saving · Waste reduction · Portfolio management · Design strategies

1 Introduction

In the modern global, capitalistic, and profit-driven economy, the discipline and practice of Design play a crucial role in guiding companies towards balanced approaches that benefit profit, people, and the planet (Elkington, 1997). Given that almost everything in

A. Mateus et al. (Eds.): WAD 2024, What's Around Design?, pp. 247–261, 2026.
https://doi.org/10.1007/978-3-032-08163-6_18

our artificial world has been designed (Norman, 2023), designers have significant intervention potential. The urgency to address environmental and climate crises, highlighted in United Nations Climate Change Conferences, is pressing. The OECD (2019) predicts that if current trends continue, planetary resources could be depleted by 2060, exacerbating global environmental and social crises (Vezzoli & Manzini, 2007). According to the European Environment Agency (2023), achieving a circular-oriented economy—which aims to minimize pollution, maximize material and product value, and regenerate natural ecosystems (Ellen MacArthur Foundation, 2013)—requires doubling the circular material use rate by 2030, beyond mere recycling improvements.

A significant shift occurred at the turn of the century when products began to be designed for short lifespans, reducing their supply chain permanence (Maldonado, 1998), a trend intensified by technological advancements (Papanek, 1971). A circular-oriented economy, however, seeks to minimize waste and restore value chains, creating synergies among stakeholders. The ISO 59000 standards define Circular Economy (CE) as an economic system that maintains resource flow through recovery and value retention, contributing to sustainable development (ISO, 2024). Implementing CE requires radical changes in industrial structures and business models, including adopting closed-loop value chains, cleaner technologies, and increased supply chain interdependencies (Tan et al., 2022). However, existing business models and resource allocations can resist change (Asgar & Asgari, 2021). Firms must adapt organizational cultures and capabilities to capitalize on CE opportunities, facing challenges in transitioning to circular business models, product-service systems, reverse logistics, product redesign, and changing consumer preferences (Chizaryfard et al., 2020).

Circular diversification could be a way to overcome these challenges. The research aims to answer the following RQ: how can designers strategically lead circular innovation supporting manufacturing companies through a circular diversification? Circular Diversification (CD) is a strategy to upgrade and expand product portfolios through Product Diversification. CD aims to minimize waste by reusing internal production waste or waste from other chains, enabling companies to reach new customers and market niches by converting waste into valuable resources, thus reducing environmental impact. Analyzing the transition to CE through CD requires a multi-level perspective, addressing institutional, industrial, organizational, and societal norms and behaviors (Chizaryfard et al., 2020). Using vertical and multi-layer analysis of case studies, this research employs the Concept-Issue-Form framework (Zambelli, 2022) to extract meta-knowledge—objectives, problems, strategies, and features used to close the loop. The Alluvial Diagram visualizes semantic labels from meta-knowledge, forming recommendations for other companies to implement CD in their business models. Case studies demonstrate that embracing CD can offer benefits like reduced costs, increased competitiveness, and improved corporate image, fostering innovation across structure, network, and brand dimensions (Keeley et al., 1999).

2 Circular Diversification

The upcoming section introduces the concept of CD by delving into the meanings of the two terms constituting this neologism: Product Diversification and CE. By merging the two concepts, the authors define CD as a potential Design-oriented strategy for manufacturing companies to expand their product range by incorporating materials, processes, or strategies that lead to a self-sustaining system. The case studies presented show the implementation of this strategy.

2.1 Product Diversification Strategy

Product Diversification (PD) is a strategic approach where companies expand their product lines to increase revenue and reduce risk while maintaining their core business (Benito-Osorio et al., 2012). Identified by Igor Ansoff in the mid-1900s as a growth strategy for volatile markets (Ansoff, 1957), PD allows companies to strategically plan the introduction, enhancement, or elimination of products based on their life cycle stage and market trends. Ansoff categorizes diversification into three types: Horizontal, Vertical, and Lateral. Horizontal D involves developing new products within the company's existing technological, financial, and marketing expertise. Vertical D extends the company's production to include components, parts, and materials. Lateral D, which explores outside the company's industry boundaries, aims at opening new market structures. The correlation between existing and new businesses is crucial in diversification strategies. There are two main types of diversification based on product portfolio consistency: related and unrelated. Related diversification occurs when a company introduces new products or enters new markets that are related to its existing business activities, making it easier for consumers to access complementary goods or services. Unrelated diversification involves adding new, unrelated product lines or entering markets without a direct connection to the existing business.

In literature, PD is also referred to as an extension and is one of the four business model changes defined by Cavalcante, alongside business model creation, revision, and termination (Cavalcante et al., 2011).

2.2 Circular Economy Model

The Circular Economy (CE) is an economic system addressing current inefficient and linear production and behavioral models (Thackara, 2016; Papanek, 1971) that lead to global crises and environmental degradation (WCED, 1987). Emerging in the mid-20th century, CE advocates for both economic, social, and ecological development within the planet's biophysical limits, countering the negative environmental impacts of the Industrial Revolution (McDonough & Braungart, 2002). Key international strategies, such as the European Green Deal (European Commission, 2019) emphasize product, energy, and material circularity to enhance system efficiency.

Transitioning to a restorative and regenerative model requires firms to overhaul core assets, including facilities, infrastructure, capabilities, and local supply chains, to achieve sustainable growth and reduce environmental impact (Dumée, 2022). Europe has adopted this model to ensure the resilience of its member states (European Commission, 2020), promoting restorative economic practices (Ellen MacArthur Foundation, 2013) encouraging innovative applications and awareness of waste treatment and re-valorization (Rognoli et al., 2021). CE operates on "closed loop" or "cycle" concepts involving biological and technical cycles. Biological cycles reintroduce materials into the biosphere through biodegradability and composting, while technical cycles valorize inorganic or synthetic materials within an organized system, turning waste into new valuable resources (McDonough & Braungart, 2002). This system fosters cooperation among industrial stakeholders and new productive networks (Carta & Ronsivalle, 2015). In design, CE strategies often focus on closing product life cycles through "R-Imperatives" (Virtanena et al., 2017). Ecodesign guidelines facilitate cradle-to-cradle approaches (McDonough & Braungart, 2002), enhancing competitive advantages for companies through partnerships and collaborations (Carta & Ronsivalle, 2015) and transforming waste into viable solutions (Pellizzari & Genovesi, 2017). However, to date, companies prioritized value-retention without a defined CE roadmap, often distinguishing between short, medium, and long loops based on consumer proximity and product upgrading (Reike et al., 2018). Current end-of-life options for products involve upcycling, reconditioning, recycling, and heat recovery before landfill disposal, with progressive efficiency loss (Ashby & Johnson, 2009; Dumée, 2022).

2.3 The Role of Design and Designer in CD

Central to Circular Economy (CE) strategy is moving from traditional linear models to closed-loop systems, necessitating reduced dependence on virgin materials, increased use of bio-based materials, and elimination of energy-intensive materials. Circular design is crucial in this transition, promoting the development of reusable and energy-efficient products while redefining waste as a valuable resource (Dumée, 2022; Morone & Yilan, 2020). The European Commission's 2022 Ecodesign for Sustainable Products Regulation (ESPR) aims to standardize sustainable product design, encouraging circular and energy-efficient business strategies. This regulation sets requirements to improve the design of widely consumed products in Europe and includes measures to discourage the destruction of unsold products, applicable to small, medium, and large companies, excluding listed micro-enterprises. Addressing sustainability challenges requires more than technical management; it necessitates multi-scale, long-term strategies as outlined in Transition Design theories, which advocate for radical economic, social, and environmental transformation (Willis, 2015; Irwin, 2015). Designers' transdisciplinary skills are essential for tackling systemic issues by envisioning future-oriented strategies and fostering innovative, place-based solutions (Earley & Hornbuckle, 2023). These long-term visions catalyze innovative solutions at multiple levels, from materials to spaces, encompassing political, social, economic, ecological, and territorial domains, as well as temporal dimensions (Ceschin & Gazilously, 2016). In the context of industrial transformation towards CE, this multi-level perspective ranges from Macro to Micro levels (Chizaryfard et al., 2020). The inclusion of a Nano level further refines the analysis of the micro dimension, highlighting the design role through empirical case studies. This approach underscores the importance of comprehensive strategies in achieving a sustainable and circular economy.

3 Methodology of Analysis

Italy is the leading European country in Design infrastructure, with over 30,000 enterprises, primarily SMEs, integrating Design into their industrial fabric, leading to higher innovation and business success (Symbola, 2021; 2022). This research explores CD from a Design perspective through case studies. The methodology flow detailed in Fig. 1. The findings were validated in a workshop with 13 participants, including researchers, students, CE experts, business developers, civil society members, and policymakers.

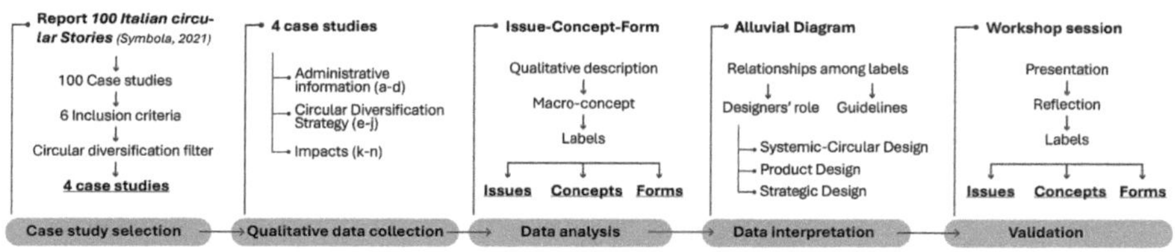

Fig. 1. Flow of methodology: from case studies to the two main research outcomes (recommendations and the role of Design in the CD process). © Authors

3.1 Data Gathering and Collection Protocol

This study analyzes four emblematic cases of Circular Diversification (CD) from the "100 Italian Circular Stories" report by Symbola (2021). The selected cases were chosen based on exclusion criteria (Yin, 2018) such as CD strategy, SMEs, Italian context, Product-Service-System, relevant years, and adherence to the Ecodesign for Sustainable Products Regulation. Data was gathered through a triangulation of sources including Symbola, company databases, and websites. Each case was individually described in a shared document, containing both concise and detailed descriptions. The analysis utilized a spreadsheet documenting company information, CD strategies, and business model impacts.

This initial set of data (1, in Table 1) provides a broad outline of the organization, starting from the localization and its founding year. The second set of data (2, in Table 1) provides a deep view of the CD Strategy adopted by the company to achieve shared goals and objectives. According to the levels depicted by Chizaryfard and colleagues (Chizaryfard et al., 2020), the case studies are one per level of production system (h) (*Nano, Micro, Meso, Macro*), considering various application fields—such as materials, products, services. Finally, the authors analyzed the Design contribution to CD to highlight how the discipline and the practice of Design can support solving complex issues, adopting the Celaschi's theory, which introduced the Value levels of design approaches (j) (Celaschi et al., 2011).

3.2 Analysis Approach

The study uses a qualitative case-based reasoning method (Zambelli, 2022) to analyze the role of design and designers on multiple scales. The authors derived an analysis

Table 1. Data sets used to analyze, compare and chose the selected companies

Administrative information (1)	CD strategy (2)	Impacts (3)
a. Location, foundation and diversification year	e. Internal/external	k. Motivations to implement CD
b. Size	f. Product scale	l. Motivations to choose the new sector
c. Sector (before and after)	g. Correlation with previous market (related/non-related)	m. New production sector chosen
d. ATECO (before and after)	h. Systems Level (*Nano-Macro*)	n. Consequences of CD
	i. Ecodesign for Sustainable Products Regulation (ESPR)	
	j. Value levels of designers' approaches	

method from the Issue-Concept-Form (ICF) model (Zambelli, 2022), which deconstructs case project stories into independent knowledge blocks. This method helps interpret the meanings and learnings of each project, which constitutes a "precedent" (Oxman, 1994). Macro-concepts from each qualitative description were summarized in a spreadsheet using keywords (labels). The ICF model breaks down cases into parts: Issues (questions, problems, or objectives), Concepts (design solutions), and Forms (morphologic types manifesting the Concepts). To interpret and reassemble this knowledge, the study used a graphic medium, the Alluvial Diagram, to display the relationships among labels. The diagram shows how categorical dimensions correlate, representing them as flows. It visually links three categories—Problem/Motivation, Strategy, and Form/Solution—with companies and identified items (Fig. 2). Each rectangle in the diagram represents a unique value in the selected dimension, with its height proportional to its value. Curved lines represent the correlations between dimensions, with the width of these lines proportional to their value. This visual representation facilitates the inquiry and exploration of the multidimensional knowledge warehouse of each case study.

The validation process was structured into three main parts. Initially, the study was presented to a heterogeneous group of participants during a workshop session, focused on CE issues to show the results obtained from applying the ICF model to the cases in question. Subsequently, an individual reflection about the problem/motivation and strategy was conducted on possible implementations of the results using stick notes. Finally, the participants returned and shared them collectively and the authors used the insights gathered to implement the findings of the analysis.

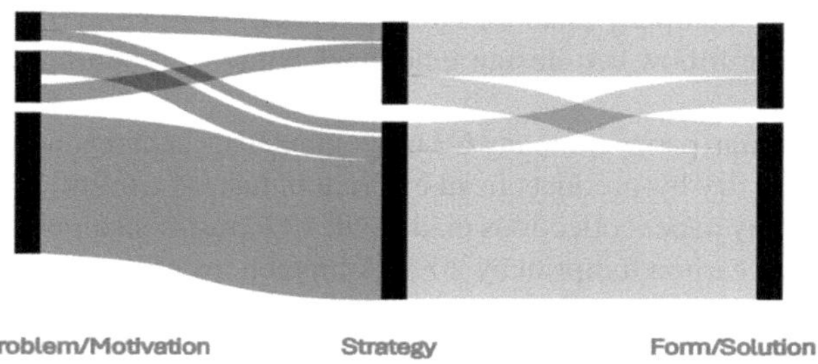

Fig. 2. The Alluvial Diagram visually represents the transversal correlation between the ICF model features. Each ICF category is represented by a node, the flows from one node to another indicate the relationships and interactions between different elements and their width corresponds to the strength of that feature. © Authors

3.3 Materials: Case Studies of Circular Diversification

The four case studies the research is going to present below are four Italian companies whose names have been hidden and will be called Company1 (C1), Company2 (C2), Company3 (C3), and Company4 (C4).

C1 is a start-up based in Milan which transforms spent coffee grounds from the food industry to develop new thermoplastic, recycled and bio-based materials. C2 is a medium-sized furniture company, which, over the years, has increasingly moved towards the principles of CE, putting responsibility toward the planet and people at the center of its business model. C3 Is an Italian paper mill specializing in producing innovative graphic papers based on natural raw materials; since the 1990s, the company has been constantly searching for more sustainable processes and raw materials. Finally, C4 engages in contract development and manufacturing, which began its sustainability journey in 2008 toward improving environmental, social and economic performance.

Founded in 2022, C1 is a Milan-based spin-off from a social cooperative involved in environmental services since 1996. The company's core business project involves transforming organic waste into durable, bio-based thermoplastic materials as an alternative to disposable plastics, blending coffee grounds with polymers to create a range of products suitable for different industries. C1 collaborates with local and international firms, academic experts, and social enterprises to repurpose industrial by-products. Additionally, C1 integrates fragile individuals into its logistics activities to promote social inclusion. The replicability of C1's model depends on mastering end-of-waste laws and partnering with plastic manufacturers.

The second company analyzed is a medium-sized, family-owned company founded in the 1990s, internationally recognized in interior design. Their vision, project of living, guides the dialogue between the lived environment and the natural habitat, involving recycled and eco-friendly materials, following eco-design principles, and conducting life cycle assessments. C2 also works on innovative solutions to improve sustainability performance through reuse and recycling policies, end-of-life management, and research collaborations. The company aims to reposition itself as green by substituting materials/components that follow eco-design guidelines while maintaining the same industry standards.

In 1992, the Italian paper company C3 implemented circular economy and upcycling principles by using dry by-products from external industrial controlled supply chains to produce high-quality paper collections made with 40% post-consumer recycled material, reducing its overall carbon footprint by 20% (compared to conventional papers). C3 uses industrial symbiosis to exchange energy through local infrastructure offsets emissions with Carbon Credit.

Established in 1979, C4 is a Milan-based company specializing in contract cosmetics. C4 commitment to sustainability with report publications since 2017, emphasizing water conservation, recyclable packaging, natural raw materials, and waste recovery. C4 has two notable CE projects including external diversification and internal CD. In terms of Corporate Social Responsibility, the company focuses on social, environmental, and economic pillars. Notably, the company engages in social projects and donates surplus stock to local associations in need.

4 Results

4.1 Interconnections Between Nano-Macro Levels

The semantic extrapolation of labels from the analyzed cases contributes to deconstructing projects into richer categorizations with minor complexity. For this purpose, we broke down projects' descriptions into more detailed and precise categories by extracting meaningful labels and terms from the analysis. The projects' knowledge was divided into meaningful units (semantic blocks). These units enable organizing and visualizing each case study's various elements (issues, concepts and forms) through maps (Zambelli, 2022). Maps show how different outcomes are interconnected across various levels of systems, from small-scale to large-scale, so far as nano to macro dimensions for each company. The visualization in Fig. 3 illustrates how different strategies are used to solve specific problems or fulfil particular motivations within the company. It shows the resulting solutions and forms and provides insights into where these strategies overlap or work together synergistically.

The authors designed the map through a bottom-up approach. Starting from systematically sorting the tangible and physical solutions implemented through different strategies (Fig. 4), the network map traces these solutions back to the original motivations or problems the company needed to address. Despite the adopted approach, the network can be understood and read without a strict hierarchy, meaning that it does not need to be interpreted in a top-down or bottom-up manner. It can be explored in various ways to find and freely explore connections and patterns. This approach aims to find practical solutions to common problems yet addressed by other projects, regardless of whether these solutions come from different areas or levels of the system (Zambelli, 2022). Similar paths or strategies might be used across different contexts. Moreover, the network can be further expanded to create more connections. It has the potential to be transformative, incorporating insights and approaches from various disciplines embracing interdisciplinary contaminations.

Finally, the author adopted a design lens to provide an in-depth perspective on the transformative potential of CD within industrial enterprises. The design lens allows for a detailed analysis of how CD strategies can revolutionize production systems, emphasizing both the tangible and conceptual aspects of these transformations, revealing potential opportunities and spaces for the Design agency, intervention and contribution. The CD strategy is designed to benefit from these new connections and relationships between different pieces of knowledge, making it more comprehensive, holistic-oriented and informed. This comprehensive understanding is crucial for evaluating the success and sustainability of CD strategies. It helps identify the broader implications of CD, such as its potential to reach new customers and market niches and its role in reducing the environmental impact of production processes. For this purpose, the contribution outlines key guiding recommendations derived from the CD of companies' case studies under analysis.

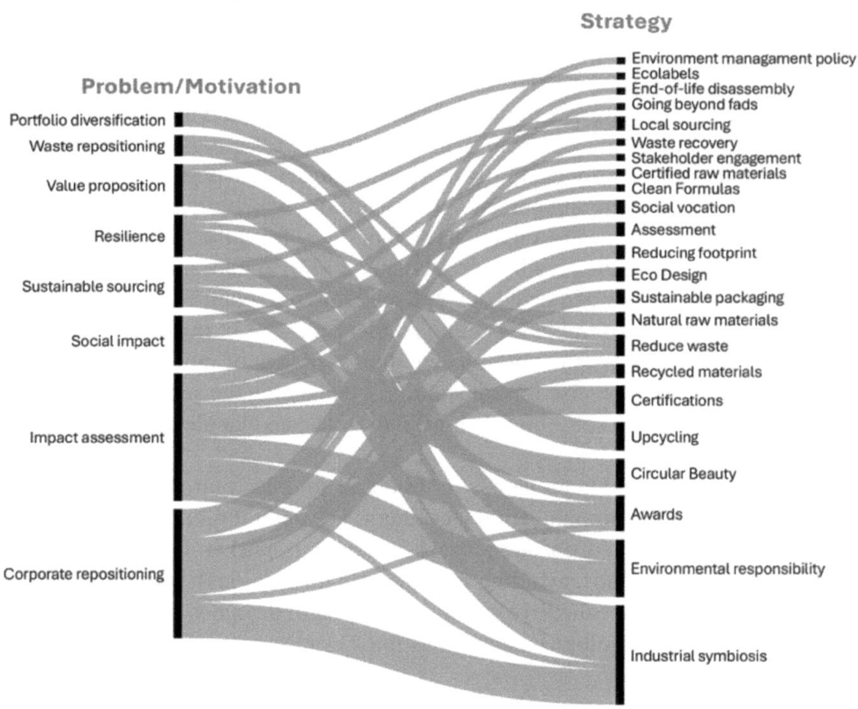

Fig. 3. First part of the diagram indicating relationships between issues and concepts, allowing to cross-compare commonality and alternative strategies to similar problems or objectives. © Authors

4.2 Design-Driven Strategies

The next step was to identify the pivotal role of Design and designers in the mentioned D processes (Fig. 4). Therefore, proceeding backward from the companies-problem-strategy-solutions diagram, three Design professions were identified that contribute most to CD. It should be specified, however, that the boundaries of these professions are blurred, because Design professions often intersect.

To better understand the role of designers within CD, each strategy was assigned to a disciplinary area of design research according to Perspectives on University Research in Design in Italy (Bertola & Maffei, 2008). Three were highlighted: Systemic-Circular Design, Product Design, and Strategic Design. These areas translate into three design professions.

Within CD, Systemic-Circular Designers can deal with managing the macro issue of policies and adapting circular products and materials to local matter in question (Hahladakis & Iacovidou, 2018), and thus create created guidelines that indicate ways of managing products, businesses and system sustainability (Bistagnino, 2009)—i.e. sustainable material selection and carbon footprint reduction as well as employing tools such as the sustainability report and consequent standards (such as ISO 14001). In addition, systemic designers oversee managing the supply chain, and thus the raw materials processed in manufacturing companies that, through CD actions, are raw secondary materials from a short supply chain, from a post-industrial or post-consumer, recycled and recyclable, with a swinging percentage of natural raw materials.

Product Designers intervene on a level farther away from the business organization and closer to the final product (Ashby & Johnson, 2009); in particular, CD relies on the design of products with low perceived obsolescence and zero planned obsolescence and is easily repairable and disassembled (Ogunmakinde et al., 2022). In addition, the Product Designer is the figure who can communicate the circularity of industrial strategies and material resources. With precise formal solutions, enhancing the sensory characteristics of these materials through meaningful signs, narratives, and tactile and visual stimuli is possible.

Finally, Strategic Designers can be considered middle figures between Systemic Designers and product/communication Designers. The central figure directs the product designer to avoid perceived obsolescence by analyzing market and consumer trends, identifying long-lived trends, avoiding fads (temporary fashions), formulating an innovative vision (Calabretta et al., 2016). Finally, the type of designer who deals with innovative materials and technologies contributes to CD success and the product portfolio's consequent expansion.

4.3 Circular Diversification Design Directions

Based on the results of the previous round of analysis, in the contribution authors highlight recommended manifold directions for CD improvements (Fig. 5). Designers and companies could cooperate in redefining corporate strategies, identifying relevant partners for strengthening systemic relationships, ensuring the reliability of the secondary resources' supply chain, and relying on valuable tools to assess their ecological performances. Moreover, efforts can be further focused on improving the brand visibility,

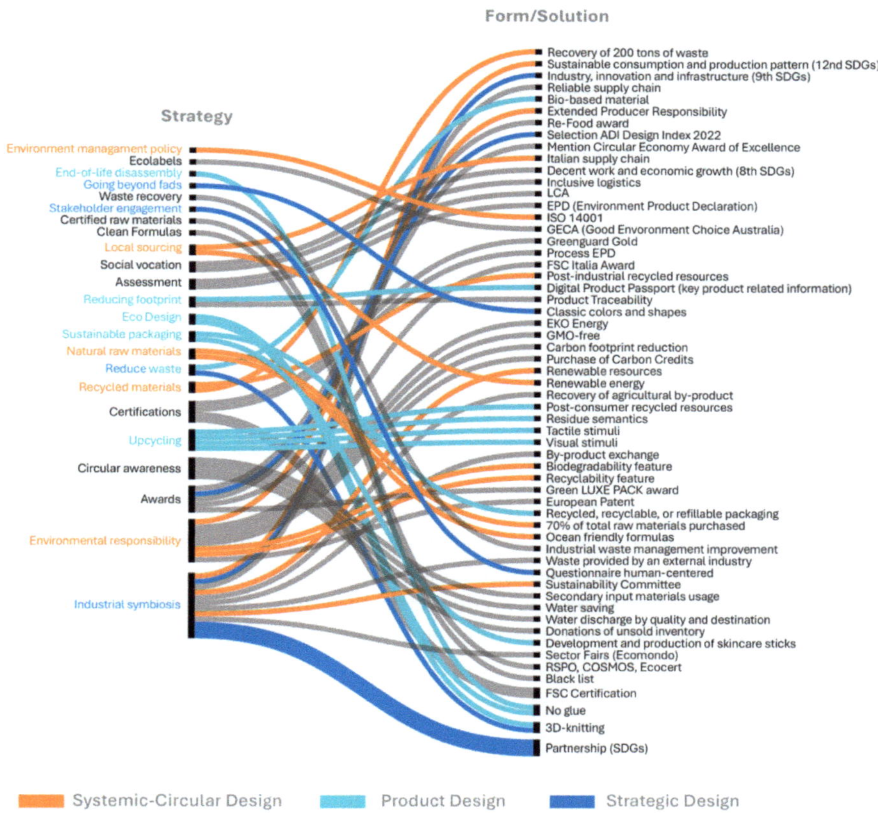

Fig. 4. Physical features from different strategies results in a more complex and polymorph diagram and design, implicitly or explicitly, fits into the strategies identified through precise solutions. © Authors

prompting its relevance at international events and awards, as well as designing social empowerment solutions with local positive impact.

These recommendations provide an ongoing framework for companies looking to implement CD, enhancing their sustainability and competitive edge. It also provides powerful insights for designers working with transitioning companies.

4.4 Discussing Circular Innovation

The research aimed to guide designers in the strategic support of manufacturing companies through a CD.

The circular changes resulting from the introduction of EC principles into current business models provide new design spaces for portfolio management and thus supporting company resilience. By means of the analysis of the CD strategies through business case studies, the contribution highlights how the different cross-disciplinary

Fig. 5. Recommendation for the CD strategic implementation in SMEs' organizations. © Authors

competences of experts in Design—Systemic-Circular Design, Product Design, and Strategic Design—can support the introduction and implementation of CD. Therefore, this study fits in the discussion of new circular business model development for restorative and regenerative transition, blurring the borders between Design and Innovation Management.

5 Conclusion

The research presented is an initial analysis of critical circular advancing directions addressed by manufacturing companies to implement CD updating and expanding the company's product portfolio to follow the CE trend. The research aims to create training action knowledge to engage a network of productive firms towards circular patterns through designed Diversification. CD presents significant challenges for companies, especially for SMEs. The panel within the workshop validated and integrated the meta-knowledge on problems/motivation and strategies extrapolated from the desk analysis. In addition, barriers to implementing CD emerged from the panel. The primary obstacles to CD could be cognitive inertia and switching costs, which may entail significant plant, equipment, and personnel training investments.

Additionally, access to circular technology can be difficult as it is often still in development and expensive. Companies may also need more skills and knowledge to implement circular solutions. However, despite these challenges, CD is an increasingly important and growing trend that companies should embrace to remain competitive and sustainable.

The first limitation that future implementations aim to overcome is the small number of case studies. In this research, four case studies, one for each level of impact (*Nano, Micro, Meso, Macro*), were chosen to exemplify the analysis methodology. Despite the limited number of cases under investigation, such examples can display virtuous and promising practices for reversing current production patterns. Since the *forms/solutions* depicted by the study are specific to the inquired manufacturing context, the study outlines strategies that are more generalized and scalable to different production sectors.

However, the research team is committed to selecting a single production sector and conducting case studies using a validated analysis method to identify more CD guidelines specific to that sector. This step is crucial to make the research increasingly concrete and use it as the basis for a consulting service for manufacturing companies that want to embark on a CD path. Once a substantial number of guidelines have been collected, they can be proposed to companies that have yet to implement CD strategies to boost solutions with positive impact. Additionally, Design's intervention at system, product and strategy levels involving CD recommendations may improve business performances in transitioning towards CE models, ensuring that their efforts are not only economically sustainable but also strategically aligned with broader social and environmental goals.

References

Asgar, A., Asgari, R.: How circular economy transforms business models in a transition towards circular ecosystem: the barriers and incentives. Sustainable Prod. Consumpt. **28**, 566–579 (2021)

Ashby, M., Johnson, K.: The Art and Science of Material Selection in Product Design. Elsevier (2009)

Benito-Osorio, D., Ángel Guerras-Martín, L., Ángel Zuñiga-Vicente, J.: Four decades of research on product diversification: a literature review. Manag. Decis. **50**(2), 325–344 (2012). https://doi.org/10.1108/00251741211203597

Bertola, P., Maffei, S.: RM: Design research maps: prospettive della ricerca universitaria in design in Italia = visions of academic design research in Italy: 2003–2007. Poli.design (2008)

Bistagnino, L.: Design sistemico. Progettare la sostenibilità produttiva e ambientale [Systemic design. Designing productive and environmental sustainability]. Slow. Food Editore (2009)

Calabretta, G., Gemser, G., Karpen, I.: Strategic Design: Eight Essential Practices Every Strategic Designer Must Master. Bis Publishers (2016)

Carta, M., Ronsivalle, D.: Territori Interni. La Pianificazione Integrata per lo sviluppo circolare: metodologie, approcci, applicazioni per nuovi cicli di vita [Inner Territories. Integrated Planning for Circular Development: Methodologies, Approaches, Applications for New Life Cycles]. ARACNE Editrice int.le (2015)

Cavalcante, S., Kesting, P., Ulhøi, J.: Business model dynamics and innovation:(re) establishing the missing linkages. Manage. Decis. 1327–1342 (2011)

Celaschi, F., Celi, M., García, L.M.: The extended value of design: an advanced design perspective. Des. Manage. J. **6**(1), 6–15 (2011). https://doi.org/10.1111/j.1948-7177.2011.00024.x

Ceschin, F., Gaziulusoy, İ: Evolution of design for sustainability: from product design to design for system innovations and transitions. Des. Stud. **47**, 118–163 (2016)

Chizaryfard, A., Trucco, P., Nuur, C.: The transformation to a circular economy: framing an evolutionary view. J. Evol. Econ. **31**, 475–504 (2020)

Dumée, L.F.: Circular materials and circular design—review on challenges towards sustainable manufacturing and recycling. Circ. Economy Sustainability **2**, 9–23 (2022). https://doi.org/10.1007/s43615-021-00085-2

Earley, R., Hornbuckle, R.: Design Materials and Making for Social Change. Routledge (2023).https://doi.org/10.4324/9781003260356

Elkington, J.: Cannibals with Forks. The Triple Bottom Line of 21st Century Business. Capstone (1997)

European Environment Agency. How Far is Europe From Reaching Its Ambition to Double the Circular Use of Materials? (2023). https://doi.org/10.2800/69671

Ellen MacArthur Foundation. Towards the Circular Economy. Economic and Business Rationale for an Accelerated Transition, vol. 1 (2013)

European Commission. The European Green Deal (2019). https://eur-lex.europa.eu/legal-content/EN/TXT/?uri=COM:2019:640:FIN

European Commission. Communication From the Commission. A New Action Plan for the Circular Economy. For a Cleaner and More Competitive Europe (2020). https://eur-lex.europa.eu/resource.html?uri=cellar:9903b325-6388-11ea-b735-01aa75ed71a1.0020.02/DOC_1&format=PDF

European Commission. Ecodesign for Sustainable Products Regulation (2022). https://environment.ec.europa.eu/document/download/11246a52-4be4-4266-95b1-a15dbf145f51_en

ISO 59010. Circular Economy – Guidance on the Transition of Business Models and Value Network. 3.1.1 (2024)

Irwin, T.: Transition design: a proposal for a new area of design practice, study, and research. Des. Cult. **7**(2), 229–246 (2015). https://doi.org/10.1080/17547075.2015.1051829

Hahladakis, J.N., Iacovidou, E.: Closing the loop on plastic packaging materials: what is quality and how does it affect their circularity? Sci. Total. Environ. **630**, 1394–1400 (2018)

Keeley, L., Pikkel, R., Quinn, B., Walters, H.: The Ten Types of Innovation. Doblin Inc. (1999)

Maldonado, T.: Reale e virtuale [Real and virtual]. Feltrinelli (1998)

McDonough, W., Braungart, M.: Cradle to Cradle: Remaking the Way We Make Things. North Point Press (2002)

Morone, P., Yilan, G.: A paradigm shift in sustainability: from lines to circles. Acta Innov. **36**(1), 17–32 (2020). https://doi.org/10.32933/ActaInnovations.36.1

Norman, D.A.: Design for a Better World: Meaningful, Sustainable, Humanity Centered. The MIT Press (2023)

OECD. Global Material Resources Outlook to 2060: Economic Drivers and Environmental Consequence. OECD Publishing (2019).https://doi.org/10.1787/9789264307452-en

Ogunmakinde, O.E., Egbelkin, T., Sher, W.: Contributions of the circular economy to the UN sustainable development goals through sustainable construction. Resour. Conserv. Recycl. **178**, 106023 (2022). https://doi.org/10.1016/j.resconrec.2021.106023

Oxman, R.E.: Precedents in design: a computational model for the organization of precedent knowledge. Des. Stud. **15**(2), 141–157 (1994). https://doi.org/10.1016/0142-694X(94)90021-3

Papanek, V.: Design for the Real World. Human Ecology and Social Change. Bantam Books (1971)

Pellizzari, A., Genovesi, E.: Neo materiali 2.0 nell'economia circolare [Neo materials 2.0 in the circular economy]. Edizioni Ambiente (2021)

Reike, D., Vermeulena, W., Witjesb, S.: The circular economy: new or refurbished as CE 3.0? – exploring controversies in the conceptualization of the circular economy through a focus on history and resource value retention options. Resour. Conserv. Recycl. **135**, 246–264 (2018). https://doi.org/10.1016/j.resconrec.2017.08.027

Rognoli, V., Pollini, B., Alessandrini, L.: Design materials for the transition towards post-Anthropocene. In: Camocini, B., Verganti, R. (eds) From Human-Centered To More-Than-Human Design. Exploring the Transition, pp. 101–130. FrancoAngeli (2021)

Symbola. Design Economy 2022. I quaderni di Symbola. [Design Economy 2022. Symbola Notebooks] (2022). https://symbola.net/ricerca/design-economy-2020/

Symbola. 100 Italian Circular Economy Stories (2021). https://symbola.net/wp-content/uploads/2021/12/Rapporto-100-CircularEconomy_pgDoppie.pdf

Tan, J., Tan, F.J., Ramakrishna, S.: Transitioning to a circular economy: systematic review of its drivers and barriers. Sustainability **14**, 1757 (2022). https://doi.org/10.3390/su14031757

Thackara, J.: Development, globalization, sustainability. In: Huppatz, D.J. (ed.), pp. 193–213. Bloomsbury, London
Vezzoli, C., Manzini, E.: Design per la sostenibilità ambientale [Design for environmental sustainability]. Zanichelli editore (2007)
Virtanena, M., Manskinena, K., Eerola, S.: Circular material library. an innovative tool to design circular economy. In: 12th EAD Conference Sapienza University of Rome (2017). https://doi.org/10.1080/14606925.2017.135268
Willis, A.: Transition design: the need to refuse discipline and transcend instrumentalism. Des. Philos. Pap. **13**(1), 69–74 (2015). https://doi.org/10.1080/14487136.2015.1085687
World Commission on Environment and Development (WCED). Report of the World Commission on Environment and Development: Our Common Future (1987). https://sustainabledevelopment.un.org/content/documents/5987our-common-future.pdf
Yin, R.K.: Case Study Research and Applications: Design and Methods, 6th ed. Sage (2018)
Zambelli, M.: La conoscenza per il progetto. Il case-based reasoning nell'architettura e nel design. Firenze University Press (2022). https://doi.org/10.36253/978-88-5518-654-4

Circular Network of Textiles Biomaterials: Green Strategies for Future Design Practices

Pedro Ferreira[1], Arlete Apolinário[2,3](✉), and Gabriela Forman[1]

[1] CIAUD, Research Centre for Architecture, Urbanism and Design, Lisbon School for Architecture, Universidade de Lisboa, Rua Sá Nogueira, Polo Universitário do Alto da Ajuda, 1349-063 Lisboa, Portugal
pedoc85f@gmail.com

[2] Departamento de Física e Astronomia, Faculdade de Ciência da Universidade do Porto, Rua do Campo Alegre s/n, IFIMUP – Institute of Physics for Advanced Materials, Nanotechnology and Photonics of the University of Porto, 4169-007 Porto, Portugal
arlete.apolinario@fc.up.pt

[3] LaPMET – Laboratory of Physics for Materials and Emergent Technologies, Rua do Campo Alegre s/n, 4169-007 Porto, Portugal

Abstract. The textile industry is increasingly seeking materials within a circular economy framework to mitigate waste and foster regenerative ecosystems. Design education is concurrently evolving, challenging designers to make progressive and sustainable choices in product selection. Integrating circular economy principles with material selection involvers categorizing materials into biological and technical cycles, each governed by specific paraments such as safe cycles, longevity, and end-of-life considerations. This study focuses on the principle of eco-design, often linked to reducing the ecological footprint. Sustainability and circular economy principles are examined, encompassing various aspects related to materials and processes, including upcycling, downcycling, cascade use, safe loops, durable products, natural cycles, bio-neutrality, plastic-free solutions, and lower CO_2 emissions. A comprehensive literature of scientific works on biomaterials for textile application is conducted, and insights are disseminated through interviews with Portuguese designers and researchers who specialize in these materials. Employing exploratory and generative methods, this research proposes an exercise termed the Circular Network, designed to be utilized in future design practice as a pedagogical tool to enhance material and circular economy education. While the network itself will not be implemented in this study, guidance is provided for its future testing and further research. The interviews yield approaches to understand the interplay between processes and materials within circular, sustainable pathways. The strategies connection to the proposed network, allows this research aim to facilitate and promote engagement with sustainability and CE in design, leading to teaching approaches and pedagogical practices. Thereby this work contributes to the ongoing discourse on sustainable design and circularity in the textile industry.

Keywords: Circular economy education · Biomaterials for textiles · Future design practice · Textile industry · Green strategies

A. Mateus et al. (Eds.): WAD 2024, What's Around Design?, pp. 262–276, 2026.
https://doi.org/10.1007/978-3-032-08163-6_19

1 Introduction

The European Green Deal and the Circular Economy (CE) Action Plan (European Commission, 2020) underscore that the transition to a CE can significantly reduce greenhouse gas emissions, foster a more circular economic model, enhance the management of natural resources, and shift reliance from fossil fuels to renewable sources. This study specifically addresses the latter objective: the substitution of traditional materials (fossil-based) with sustainable alternatives (plant-based) within the textile sector to facilitate this transition.

The textile value chain is widely subdivided, complex, and spans various sectors, including petrochemicals, forestry and agriculture, manufacturing, logistics, distribution and consumption processes, and services related to treatment, reuse, and recovery (Millward-Hopkins et al., 2023). The textile industry employs several strategies to mitigate the environmental impact of production. Among them the use of renewable energy and the reduction of chemicals in finishing processes are the most attractive to reduce the environmental footprint of fibres (Shirvanimoghaddam et al., 2020). While sustainable manufacturing aims to align with environmental considerations, the textile industry is still marked by unstable factors, such as escalating water pollution from chemicals and poor working conditions that undermine social sustainability practices (Warasthe et al., 2022). To mitigate these impacts, a crucial approach is the integration of product, process, and practice (Rosen & Kishawy, 2012). According to the European Environment Agency (2023), developing safe and sustainable products has the potential to prevent pollution and mitigate climate change by enabling a circular economy.

Recent data from European Bioplastics indicates a decrease in bioplastic production from 2020 to 2023, with bio-based, non-biodegradable bioplastics production decreasing from 48.5% to 47.9%, and biodegradable bio-based bioplastics production increasing from 51.5% to 52.1% (European Bioplastics, 2024). These fluctuations highlight the consumption trends of these materials. A key question purposed in this work is: *How can design promote the use of both types of bioplastics and textiles in the sense of a circular economy?*

Bioplastics offer a sustainable alternative to harmful materials commonly used in traditional textiles, such as cotton, polyester, nylon, leather, silk, and fur (Material Innovation Initiative, 2022; Textile Exchange, 2022a). Polyester is a widely used synthetic material in textiles and valued for its durability, waterproof properties, and recyclability but is derived from fossil raw materials. Polyester waste, both pre- and post-consumer, is prevalent in landfills and oceans. Seaqual Polyester, for instance, is a material sourced 50–70% from the Mediterranean Sea. Seaqual is a global European initiative addressing marine plastic pollution, producing yarn from upcycled marine plastics labelled by SEAQUAL LAB. This yarn also includes other materials like metals, glass, organic substances, and non-plastics (Seaqual, n.d.). Recycling marine plastics is particularly challenging due to environmental factors like UV radiation and salt water, which degrade the materials. Consequently, Seaqual yarn is blended with other fibers, such as recycled cotton, organic cotton, Refibra, and Tencel, to maintain quality (Seaqual, 2024). This initiative exemplifies how integrating recycled materials can support sustainable design practices and promote environmental stewardship in the textile industry.

Another example is the incorporation of bio-based contents in material composition. By integrating a small percentage of bio-based feedstock, materials can become more environmentally friendly in both production and use. Bio-PET, a potential substitute for PET in textile applications, exemplifies this approach. Bio-PET is a biosynthetic drop-in material consisting of 20–30% bio-based ingredients, with the remainder derived from fossil monomers (Storz & Vorlop, 2013; Lamberti et al., 2020). It can be produced using the same technology as PET and retains the same properties as the synthetic material (Polymerdatabase, 2015; Volanti et al., 2019). In a CE context, bio-PET aims to ensure the downcycling of materials through cascade use, allowing for recycling until the material no longer retains the necessary structural properties (Ferreira et al., 2023; Lamberti et al., 2020). However, recent years have seen a decline in bio-PET production due to emissions associated with downstream processes. Market value for these bio-based fibres has also diminished (Ali et al., 2023; Ibrahim et al., 2021).

The Ellen MacArthur Foundation (EMF), inspired by McDonough and Braungart's Cradle-to-Cradle concept, developed the butterfly diagram to illustrate CE. This diagram depicts two material cycles—biological and technical (Ellen MacArthur Foundation, 2019; McDonough and Braungart, 2002). Designers using drop-in materials like Bio-PET operate within the technical cycle, where understanding how to design for the R's (reduce, reuse, repair, recycle) is crucial (Metic et al., 2021). The technical cycle includes additional R's, such as reject, refocus, reintegrate, respect, undertake, and repost (Textile Exchange, 2022b). Materials like bio-PET belong in this cycle due to their recyclability and reusability. Designers must consider how to incorporate these materials into the loop, despite the complexity of accessing relevant information. Therefore, producing scientific works in this area is essential to capture and disseminate crucial knowledge in a practical and effective manner (Ferreira et al., 2023). One challenge lies in translating this information into accessible language for designers. Another challenge is selecting and applying these materials in design projects after understanding their properties. The purpose of our article is to contribute to the use of biomaterials in textile projects, considering CE and material flow (i.e. dual circular economy). In particular, this research aims:

to establish effective circular and sustainable strategies to future design practices in the transition to a greener economy.
to establish a relationship between biomaterials selection for textiles projects and the circular economy using a Circular Network.

The increasing use of biomaterials can be beneficial not only for designers but also for industry and consumers. Designers bear economic, ecological, and social responsibilities. Moreover, designers have easier access to social groups, as design is an interdisciplinary path that incorporates social dimensions. According to Forman et al. (2022), the connection between design and communities is integral to achieving a sustainable society, where the designer acts as a facilitator. Elvin Karana, a pioneer in the field of Design Methodology and Advanced Materials, developed the Material-Driven Design methodology, which places materials at the core of the design process (Camere & Karana, 2018; Karana et al., 2015). Frascara (2004) asserts that design methods do not aim to solve problems but to empower designers to engage in critical thinking, integrating the social environment into their creative outputs. In Karana's method, the use of case studies involving users helps to understand the connections between materials and consumers, particularly when designers create biomaterials through Biodesign approaches (Lin & Lin, 2022). Biodesign is an emerging field that intersects biology and design to experiment with materials based on natural principles—not by mimicking living organisms but by integrating them into the process (Collet, 2018; Myers, 2018). Myers, a pioneer in this field, revolutionised the way designers seek materials compatible with the planet, focusing on regenerative and compostable options (Myers, 2012). In other words, designers should consider CE in mind through material flows understanding their suitability for a green transition.

This work is divided into five parts. The introduction contextualises the research, outlining the main concepts, the problem, the research question, and the aim of the study. Section 1 reviews the literature on Design for Circularity to identify essential skills for working with circular design, focusing on key areas of knowledge and CE principles that navigate between biological, technical, and dual cycles (Dokter et al., 2022; Metic et al., 2021; Sumter et al., 2021). This enables the determination of material value and quality supporting the aim of this research. Section 3 underpins the statements from the literature review using exploratory, triangulation, and brainstorming methods to create an exercise called a Circular Network for designers, helping to provide answer for the research question through the development of green strategies. Additionally, a generative methodology is employed by conducting surveys with designers and specialists in textile biomaterials in Portugal. This section provides several insights focus on how designers can utilise materials for a sustainable and circular future. Section 4 discusses the findings of the work, contributing to designers utilise biomaterials in their textile projects. To conclude, Sect. 5 offers inputs to shift the mindset of stakeholders involved, facilitating the green transition through more conscious thinking in product development.

2 Theoretical Framework

2.1 Design for Circularity

The development of sustainable and ecological materials highlights society's growing concern for environmental protection, particularly in response to issues associated with fossil materials, energy consumption, greenhouse gas emissions, climate change, resource scarcity, ecosystem degradation, human health concerns, and overpopulation (Das et al., 2021; Mestre & Cooper, 2017; Savio et al., 2022). While these factors present significant challenges, design can develop capabilities that support designers in creating tighter and longer loops, reinforcing CE principles and acting as catalysts for problem-solving (Manzini & Vezzoli, 2003; Sumter et al., 2021).

Following the comprehensive literature review on different materials in the "Introduction" section, gaining knowledge about them can help to understand how they can be integrated into circularity. According to Sumter et al. (2021), this system integrates a set of operational factors between products and services and encompasses the biological and technical cycles of nature, which is known as the dual circularity. These three groups are fundamental for a natural system through the filtration of several interdisciplinary areas of knowledge (Metic et al., 2021). Thus, the CE process relies heavily on circular education practices that link and incorporate them into the main goals and principles to fulfil the field (see Table 1). This collaboration is crucial for achieving synergy and transdisciplinary in the designer's activity (Dokter et al., 2022).

Table 1. Main objectives and principles of the circular economy. Font: adapted by the work of (Dokter et al., 2022)

Main Objectives of the Circular Economy
The reuse of material waste for new products services and systems
Reducing the use of non-renewable materials
Disposal of waste
Extending the life cycle of products, services and systems
Decoupling economic growth and resource consumption
Satisfying the needs of end consumers
Taking inspiration from nature (biomimicry)
Looking at the supply chain and establishing links or co-operation between stakeholders
Considering the local context (in terms of materials, infrastructures and industries)
Assessing the environmental impact of the different concepts
Consideration of the end user
Main Principles of the Circular Economy
Tangible and Technical Solutions (Physical Products)
Intangible solutions that facilitate the CE (systems and services that share the principles)
Life cycle of the product in question (improvements in terms of circularity through designs with the need to utilise steps, activities and actors)
Design tasks to be carried out through systematic perspectives, conceptual life cycles and value chains

As it is a new field, reconciling materiality and CE can be a challenging and complex process (Ferreira et al., 2023). Acquiring knowledge about materials could therefore enable designers to design with CE in mind and form interdisciplinary teams so important to incorporate a range of parameters (such as, material flows, dematerialization, lifespan, and recycling) to contribute to a more sustainable and localised economy realised through environmental sustainability methods minimising the impact of materials (OECD, 2021; Dokter et al., 2022; Ellen MacArthur Foundation, 2017; Markevičiūtė & Varžinskas, 2022). Through these new practices, studies, and research in product design, it is possible to take a green and sustainable path.

3 Methodology

In this work, qualitative methods were used to analyse the integration of textile biomaterials into the circular economy. Based on the literature review, key elements of the study topic were identified to outline the problem, formulate the research question and define the objective of the study. Following the theoretical framework in Sect. 2, it was important to understand the CE and integrate a set of principles and objectives for working with circularity as a future design practice. In addition, we support some statements in the introduction by formulating an interaction between the acquisition of knowledge about materials and the design for circularity (a system that integrates several parameters related to material cycles – biological and technical). Including the fundamental aspect of acquiring knowledge about materials to understand in which cycles CE can be integrated, a complementary methodology was added to better understand materials in the CE, using 2 surveys with two experts in biomaterials for textile projects: Mónica Gonçalves (MG) and Marita Setas Ferro (MF). Both are designers with extensive experience in the field of circularity and sustainable textiles (Bento, 2022; Moreno, 2024). They are internationally active, both at academic and professional level. MG is a material designer, teacher, and researcher who founded the material brand Alfarrobatex. She has participated in various national and international projects, acquiring significant experience in product development through material characterization research. Her initial work focused on developing a cork yarn to replace wool yarn, leading to the Griji Project, which produces clothing from cork yarns and other local raw materials such as burel, linen, and vegetable silk. Through these projects, MG has gained skills beyond design, including understanding the microscopic morphology of materials. MF is the designer and creator of the brand Marita Moreno, an international footwear and accessories brand known for its eco-friendly and circular textiles. Her main objective is to highlight Portuguese culture and heritage through her designs. Additionally, MF is a textile sculptor whose projects combine artistic influences with practical fashion products. She was the first Portuguese designer to use burel in shoes, along with other emerging textile and non-textile materials. Despite her innovative use of materials, MF also values traditional animal leather for its strength and historical significance as a sustainable material.

At this point, the correlation between the use of these surveys and the theoretical framework described above helps to understand which key parameters are important for designers to recognise materials in the CE. Section 3.1 describes how we generate data by a software— MAXQDA. Section 3.2 conducts the thematic analysis based on the main keywords for the study and creates an action plan. Section 3.3 presents the Circular Network, a speculative exercise that could serve as a tool for designers to integrate materials into the CE, based on the key parameters identified through theoretical frameworks and verified by surveys. Both methods help us to support the work carried out and provide results and discussion that show how this interplay could create dialogues for future design practices by considering sustainable into technical and biological cycles.

3.1 Data Analysis and Thematic Analysis

The data obtained from the interviews were analysed using MAXQDA software to investigate the phenomena discussed in the subsequent section (GmbH, 2023). This analysis allowed for a deeper understanding of the role of circular design in the textile industry and provided insights into the practical application of these principles by experienced designers. Through a deductive approach, it was conduct a thematic analysis relate theoretical concepts guided by predetermined categories (Bingham & Witkowsky, 2022). This method facilitated the identification of connections and patterns based on seven explicit keywords relevant to the study (Bingham, A.J., 2023). Figure 1 illustrates the most frequently used terms by the interviewees, reflecting their perspectives and insights on the subject matter. According to the software, the associated standard deviation is 0.0723, while the average value is 1.12%. This shows the variation associated with the construction of the word frequency developed.

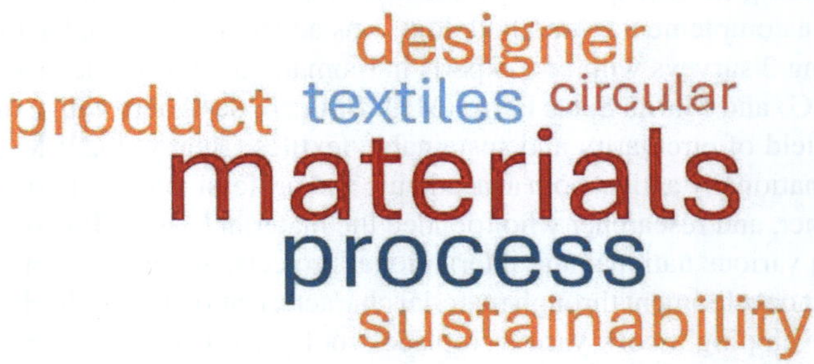

Fig. 1. Keywords Frequency developed using the MAXQDA software.

3.2 Thematic Analysis: Keywords and Action Plan

The keywords analyzed in this study concern several issues. The most important ones for this work are: *"Which concepts are most relevant for the use of sustainable and circular materials for textile applications? How do you link these concepts when developing a product or selecting materials? What results do you get from this process?* Based on the keyword "process", this can be related to several criteria in the optimization of knowledge in part of industrialisation, material production and end-of-life treatments. In this way, MG emphasised that understanding the entire process—from inception to completion—is crucial for designers. As she noted, "it is essential to recognise the process from A to Z and justify decision-making processes". In the context of sustainability and CE the term of "process" can be broken down into several key parameters:

- **Sustainable and Vegan Materials:** MG highlighted that materials should align with natural product cycles, emphasising the importance of understanding and respecting these cycles. She pointed out that "this condition is tried by the time of duration," referring to the longevity and sustainability of materials.
- **Sub-products of the Food Industry:** MF expressed a preference for animal leather, viewing it as a byproduct of the food industry that is reused in other sectors. She argued that "you do not throw anything away; you reuse, reorganize, and recycle," highlighting the importance of the "R's" in material management.
- **Material Properties:** Both designers emphasised critical attributes such as comfort, strength, UV resistance, toughness, durability, and weight (light/heavy) when evaluating materials.
- **Eco-design Benefits:** They discussed eco-design aspects such as energy and water consumption, CO_2 emissions, and land use, which are essential for assessing the environmental impact of materials.

- **Funding for Prototyping:** MF noted that funding for new product development and prototyping, especially in small companies, presents a challenge in the material selection process.

In response to these findings, an action plan was developed to address the materials-related challenges and integrate sustainable circular strategies into decision-making. Key elements of the action plan include:

- **Reuse and Repair:** MF emphasized the importance of maximizing the reuse of leftover materials to create new collections, highlighting the value of repurposing materials in CE.
- **Bio-neutral and Natural Circular:** MF advocated for selecting organic and natural materials with fewer harmful substances, citing examples such as animal leather and cork for their natural strength and durability.
- **End-of-Life Recovery:** Recovery, as described by MF, involves considering how products can be de-manufactured and recycled. Different materials require specific recycling processes, underscoring the need to reduce resource use and optimize recycling.

The complex decision-making process challenges designers to acquire in-depth knowledge about materials. According to MG, designers should be akin to "doctors of textiles," while MF believes that experimentation and prototyping are crucial for understanding material suitability. This exploratory approach allows designers to leverage new materials and test their viability, fostering a "win-win relationship" as MG described.

In summary, ongoing research and the emergence of biomaterials for textiles applications enable designers to recognise next-generation categories, technical information, and essential concepts related to circular materials. This enhanced understanding supports more informed decision-making processes and promotes the advancement of sustainable and circular design practices.

3.3 Circularity Network

The transition to a green economy necessitates the use of circular materials to restore natural systems and promote sustainable practices. Understanding material flow is a fundamental aspect of advancing a CE (OECD, 2008). Therefore, the study of biomaterials is essential for designers to determine how materials can integrate into biological or technical cycles (Ferreira et al., 2023). The butterfly diagram created by the EMF is instrumental in helping designers grasp these concepts within broader ecological frameworks.

In this context, a Circular Network has been developed (see Fig. 2) as a speculative tool for educators and designers. This network serves to illustrate how circular materials can be effectively integrated into design processes and selected according to their functional roles within the biological and technical cycles.

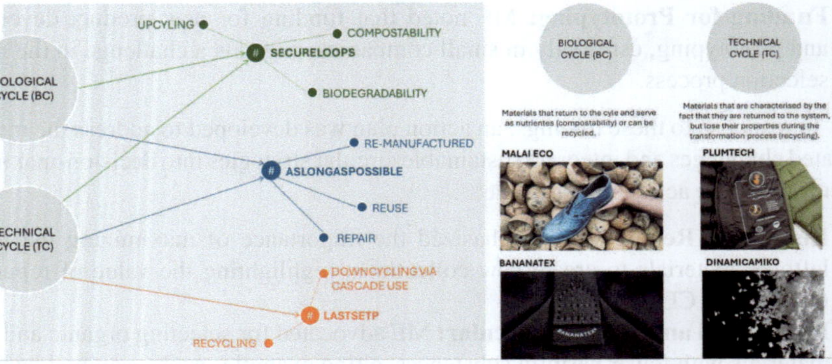

Fig. 2. The circular network. Font images: MALAI ECO (Raut, 2019), BANANATEX (Around the World in 80 Fabrics, 2022), PLUMTECH (Save the Duck, n.d.), DINAMICAMIKO (MAXCOM, n.d.).

The Circular Network categorizes materials into the following cycles:

#secureloops: Includes materials with upcycling value, biodegradability, and compostability. Examples include materials that can be safely cycled through natural systems or composted at the end of their life.
#aslongaspossible: Represents materials with potential for reuse, repair, and reprocessing. These materials are designed to extend their lifecycle through multiple uses and restorative practices.
#laststep: Consists of materials suitable for downcycling through cascading use and recycling processes. These materials are designed to be recycled or repurposed when they can no longer serve their initial function.

For instance, materials such as *Dinamica, Sequal Polyester, Recycled Plumtech,* which incorporate post-consumer PET and microfibers, are part of the technical cycle in CE (Material Innovation Initiative, 2022). These materials are designed for durability and recyclability within the technical system.

Conversely, materials like *Bananatex* (QWSTION International GmbH, 2022), *Malai Eco* (Malai.Eco, 2021)*, and Mirum* (Natural Fiber Welding, 2021) fall under the biological cycle, as they are derived from renewable resources and possess biodegradable or compostable properties. However, it is important to note that renewable resources alone do not guarantee environmental friendliness. *Mylo*, a mycelium-based material, for example, is renewable but partially biodegradable through chemical treatment steps. To ensure high resistance and durability, the composition of this material consists of 80% mycelium, some lyocell and a PU-based (Polyurethane) water solution (Bolt Threads, n.d.). This highlights the importance of understanding the complete production lifecycle of materials (i.e., knowing the process in various criteria as we point out in Sect. 3.2).

Additionally, according to Natural Fiber Welding (2022), evaluating whether a material is 0% plastic-free, part of a natural circular system, bio-neutral, and produces minimal CO_2 emissions is critical. Designers must consider these factors when working with biomaterials for textile projects. At this point, the Circular Network provides designers with a structured approach to assess materials, helping them make informed choices that align with CE principles and contribute to a more sustainable future.

4 Results and Discussion

The study underscores the critical importance of acquiring comprehensive knowledge about materials to integrate them in the CE using biological and technical cycles. Continuous research is pivotal for designers to understand the specificities of CE principles and effectively incorporate these materials into their work. The Circular Network exercise developed in this study serves as a tool for designers to recognize the roles and characteristics of various materials within the CE framework considering materials flows.

Designers are central to advancing the green transition by integrating circular practices into their designs. However, it is imperative for designers to be well-informed to avoid the pitfalls of greenwashing. Accurate knowledge about material production processes and lifecycle impacts is crucial. Designers must educate themselves on the production paths of materials and transparently communicate these details to consumers, from raw material sourcing to end-use. This aspect was achieved through the surveys conducted, as it is important to optimise knowledge during the process.

The textile industry faces complex challenges in different pathways. One of them is related to recycling and material degradation. According to Recover Textile System (2021), less than 1% of recycled garments are reused as raw materials, highlighting a significant issue in recycling efficacy. This problem is exacerbated by the deterioration of materials, which complicates standardisation efforts. Another significant challenge identified is the "Disappointment Curve," as reported by the Material Innovation Initiative (2024). Despite the readiness of the industry to introduce circular products, consumer reluctance poses a substantial barrier. This reluctance stems from the difficulty in shifting consumer preferences from durable fossil-based products to less familiar bio-based alternatives. Furthermore, social sustainability—often underrepresented in existing literature—emerges as a crucial yet challenging aspect of implementing CE. Addressing these social dimensions is essential for broader acceptance and effective implementation of circular practices.

The Circular Network, as depicted in Fig. 2, provides a structured exercise to understanding material flows within CE. It categorizes materials into three distinct direction that recognise the roles and characteristics of different materials within the CE framework. With this knowledge, designers can play an important role in the transition to a CE through material selection and circular practices. Understanding material flows based on #secureloops; #aslongaspossible; #laststep; allows designers to evaluate their decision-making process considering:

- **Environmental Impact:** Assess benefits related to water and energy consumption, CO_2 emissions, and land use.
- **Durability and Weight:** Consider the material's longevity and weight, which affect performance and sustainability.
- **Process Optimisation:** Align material choices with product requirements and client needs.

Designers must engage in critical thinking and dialogue about material choices, considering sustainability, functionality, and environmental impact. For example, if you show materials from the biological cycle, this cannot mean that these materials are better for the planet because renewable resources are used to produce the product. Designers should continuously question their material choices and their alignment with CE principles (Seixas, 2021). *Mylo* was ideally designed with nature in mind, but the production process and finishing treatments make *Mylo* part of the biological cycle designed for durability but partially biodegradable and not plastic-free (i.e. it is not a bio-neutral material). This approach ensures that designers can provide valid justifications for their selections, thereby enhancing the credibility and effectiveness of circular design practices.

Understanding and navigating material value chains are essential skills in eco-design. Designers must be equipped with detailed knowledge about materials and their lifecycle impacts to support circular and sustainable practices effectively. The Circular Network exercise and the insights gained from designer interviews highlight the importance of informed decision-making and continuous learning in advancing the circular economy.

5 Conclusion

The development of circular and sustainable strategies in design, as outlined in this work, emphasises the importance of several conceptualizations of teaching approaches:

– **Acquiring Deep Knowledge in Materials:** Designers must gain comprehensive knowledge about materials, including their properties, production processes, and end-of-life options. This understanding is fundamental for making informed decisions that align with circular economy principles.
– **Understanding Material Cycles:** Before incorporating materials into products, it is crucial to determine which cycle—biological or technical—they belong to. This step ensures that materials are utilized in a manner that supports their intended lifecycle, whether it involves reuse, recycling, or composting.
– **Developing a Sustainable Dialogue:** Engaging in a sustainable dialogue about materials and raising circular questions can foster a deeper understanding of their environmental and social impacts. This dialogue should involve discussions on how materials can be integrated into circular systems and how they contribute to sustainability goals.
– **Creating Social Environments:** Designers should collaborate with communities to facilitate access to information and insights. Utilizing methods such as the Material Driven Design (MDD) can help designers understand emotional connections to materials by using users as case studies. This approach not only enriches the design process but also supports the creation of social environments that encourage information exchange and community involvement.

In relation to some challenges identified in this work, we have looked to clarify how design can improve the use of textile biomaterials in their projects. Sound knowledge of materials helps designers to make informed decisions. By integrating the Circular Network explicitly presented in this work and using the acquired information, the designer can determine which cycle the material belongs to. The implemented method shows how designers can improve the materials in their projects with the help of CE by creating arguments to optimise the process of determining the selected material. In this way, designers can engage in a sustainable dialogue to get closer to the communities that share and democratise their knowledge. These strategies can enhance teaching approaches pedagogical practices enable a discussion between sustainability and CE in design. However, this can be challenged by the need for interdisciplinary education and continuous research in design. Integrating these instructional approaches into design education can strengthen it by promoting a more informed and responsible design practise in the education of designers. This can encourage the consideration of sustainability and circularity, ultimately leading to more effective and innovative solutions.

Acknowledgements. This research was funded in whole or in part by the Fundação para a Ciência e a Tecnologia, I.P. (FCT, https://ror.org/05qdjap63) under Grant of the Strategic Project with the references UID/04008/2025 and DOI https://doi.org/10.54499/UID/PRR/04008/2025, and through the individual research grant 2021.08190.BD.

A. Apolinário acknowledges FCT – Fundação para a Ciência e Tecnologia for her contract DL57/2016 (Ref. DL 57/2016/CP1454/CT0017) https://doi.org/10.54499/DL57/2016/CP1454/CT0017 (https://doi.org/10.54499/DL57/2016/CP1454/CT0017). AA also acknowledges support by FCT under project H2FlexiPEC's (ref. 2022.07332.PTDC), with https://doi.org/10.54499/2022.07332.PTDC (https://doi.org/10.54499/2022.07332.PTDC), Instituto de Física de Materiais Avançados, Nanotecnologia e Fotónica Universidade do Porto(IFIMUP)-UID/4968/2025 and Laboratório de Física para Materiais e Tecnologias Emergentes(LaPMET)-LA/P/0095/2020.

References

Ali, S.S., et al.: Bioplastic production in terms of life cycle assessment: a state-of-the-art review. Environ. Sci. Ecotechnol. **15**, 100254 (2023). https://doi.org/10.1016/j.ese.2023.100254

Around the world in 80 Fabrics. Bananatex. Bananatex (2022). https://images.squarespace-cdn.com/content/v1/61d329a936583d231b7bbb72/b9d4cda0-093b-409c-bf67-c4e42a65f153/Screen+Shot+2022-08-18+at+3.13.41+PM.png

Bento, D.: O inovador têxtil de alfarroba que foi criado por uma designer portuguesa. NIT (2022). https://www.nit.pt/compras/moda/o-inovador-textil-de-alfarroba-que-foi-criado-por-uma-designer-portuguesa

Bingham, A.: From data mangement to actionable findings: a five-phase process of qualitative data analysis. Int. J. Qual. Methods **22**(I–II), 11 (2023)

Bingham, A.J., Witkowsky, P.: Deductive and inductive approaches to qualitative data analysis. Analysing and Interpreting Qualitative Data: After the Interview (2022). https://www.andreajbingham.com/resources-tips-and-tricks/deductive-and-inductive-approaches-to-qualitative-analysis

Bolt Threads. Mylo Unleather. Materials (n.d.). https://mylo-unleather.com/

Camere, S., Karana, E.: Fabricating materials from living organisms: an emerging design practice (2018). https://doi.org/10.1016/j.jclepro.2018.03.081

Collet, C.: Biotextiles : Evolving Textile Design Practices for the Bioeconomy and the Emerging Organism Industry, pp. 87–99. Aalto University School of Arts, Design and Architecture, Helsinki, Finland (2018)

Das, O., et al.: Demystifying low-carbon materials. Mater. Circ. Economy **3**(1), 26 (2021). https://doi.org/10.1007/s42824-021-00044-0

Dokter, G., Jansen, B.W., Thuvander, L., Rahe, U., Duijghuisen, J.A.: Cards for Circularity (CFC): reflections on the use of a card-based circular design tool in design education. IOP Conf. Ser. Earth Environ. Sci. **1078**(1), 12057 (2022). https://doi.org/10.1088/1755-1315/1078/1/012057
Ellen MacArthur Foundation. Circular Economy Butterfly Diagram (2019). https://emf.thirdlight.com/link/bxqwo5kx53lq-2syjxg/@/preview/1?o
Ellen MacArthur Foundation. Materials Journey Mapping. Circular Design Guide (2017). https://www.circulardesignguide.com/post/journey-mapping
European Bioplastics. Bioplastics Market Development Update 2023. European Bioplastic Market (2024). https://www.european-bioplastics.org/market/
European Commission. A New Circular Economy Action Plan For a Cleaner and More Competitive Europe (2020). https://eur-lex.europa.eu/legal-content/EN/TXT/PDF/?uri=CELEX:52020DC0098
European Environment Agency. Designing Safe and Sustainable Products Requires a New Approach for Chemicals (2023). https://www.eea.europa.eu/publications/designing-safe-and-sustainable-products-1
Ferreira, P., Apolinário, A., Forman, G.: Optimising textile biomaterial selection for sustainable product and circular design: practical guidelines for a greener future. Mater. Circ. Economy **5**(1), 14 (2023). https://doi.org/10.1007/s42824-023-00086-6
Forman, G., Santos, M., Ferreira, P., Bandoni, A.: Biomateriality bridging design and the community. In: Human Dynamics and Design for the Development of Contemporarry Societies, vol. 25, pp. 186–194 (2022)
Frascara, J.: Communication Design: Principles, Methods, and Practice. Allworth Press (2004). https://books.google.pt/books?id=b40C60hN0KcC
GmbH, V.: MAXQDA (2023). https://www.maxqda.com/licenses
Ibrahim, N.I., et al.: Overview of bioplastic introduction and its applications in product packaging. Coatings **11**(11) (2021). https://doi.org/10.3390/coatings11111423
Karana, E., Barati, B., Rognoli, V., Zeeuw van der Laan, A.: Material driven design (MDD): a method to design for material experiences. Int. J. Des. **9**(2), 35–54 (2015)
Lamberti, F.M., Román-Ramírez, L.A., Wood, J.: Recycling of bioplastics: routes and benefits. J. Polym. Environ. **28**(10), 2551–2571 (2020). https://doi.org/10.1007/s10924-020-01795-8
Lin, Y.-S., Lin, M.-H.: Exploring indigenous craft materials and sustainable design – a case study based on Taiwan Kavalan Banana Fibre. Sustainability **14**(13) (2022). https://doi.org/10.3390/su14137872
Malai.Eco. Malai Material. About Malai Material (2021). https://malai.eco/blogs/news/about-malai-material
Manzini, E., Vezzoli, C.: Product-service systems and sustainability. Opportunities for sustainable solutions. J. Clean. Prod. (2003). https://doi.org/10.1039/c3tc00949a
Markevičiūtė, Z., Varžinskas, V.: Smart material choice: the importance of circular design strategy applications for bio-based food packaging preproduction and end-of-life life cycle stages. Sustainability **14**(10) (2022). https://doi.org/10.3390/su14106366
Material Innovation Initiative. Brand Engagement with Next-Gen Materials: 2022 Landscape (2022). https://materialinnovation.org/reports/
Material Innovation Initiative. 2023 – State of the Industry Report. Nex-Gen Materials (2024). www.materialinnovation.org
MAXCOM. DINAMICA (n.d.). https://www.maxcom-leather.net/media/wysiwyg/dinamica-poliestere.jpeg
McDonough, W., Braungart, M.: Cradle to Cradle: Remaking the Way We Make Things, 1st ed. North Points Press (2002)
Mestre, A., Cooper, T.: Circular product design. A multiple loops life cycle design approach for the circular economy. Des. J. **20**(suppl. 1), S1620–S1635 (2017). https://doi.org/10.1080/14606925.2017.1352686

Metic, J., McAloone, T.C., Pigosso, D.C.A.: A look into circular economy research: exploring the bio and techno cycles and the need for dual circularity. Proc. Des. Soc. **1**, 121–130 (2021). https://doi.org/10.1017/pds.2021.13

Millward-Hopkins, J., Purnell, P., Baurley, S.: A material flow analysis of the UK clothing economy. J. Clean. Prod. **407**, 137158 (2023). https://doi.org/10.1016/j.jclepro.2023.137158

Moreno, M.: About Us (2024). https://maritamoreno.com/pages/about-us

Myers, W.: Bio Design: Nature, Science, Creativity. Thames & Hudson Ltd. (2012)

Myers, W.: Bio Design: Nature Science Creativity – Revised and Expanded Edition. 304 (2018)

Natural Fiber Welding. How MIRUM is Made (2021). https://blog.naturalfiberwelding.com/how-mirum-is-made

Natural Fiber Welding. Manufacturing for Nature's Circularity: How Industry Circularity Definitions Differ from Nature and Why It Matters (2022). https://blog.naturalfiberwelding.com/circularity-definitions-white-paper

OECD. Measuring Material Flows and Resource Productivity (2008)

OECD. Sustainable design of plastics: problem formulation, objectives and scope. In: A Chemicals Perspective on Designing with Sustainable Plastics, pp. 15–21 (2021). https://doi.org/10.1787/f2ba8ff3-en

Polymerdatabase. Polyester Fibers (2015). https://blog.polymerdatabase.com/Fibers/Polyester.html

QWSTION International GmbH. Bananatex (2022). https://www.bananatex.info/

Raut, A.: Malai: A Sustainable, Vegan Alternative to Leather. Architectual Digest (2019). https://assets.architecturaldigest.in/photos/6008342e88d5cb634511e2c8/16:9/w_2560%2Cc_limit/malai-main-1366x768.jpg

Recover Textile System. Design for Cyclability (2021). https://recoverfiber.com/newsroom/design-for-cyclability

Rosen, M.A., Kishawy, H.A.: Sustainable manufacturing and design: concepts, practices and needs. Sustainability **4**(2), 154–174 (2012). https://doi.org/10.3390/su4020154

Save the Duck. Plumtech (n.d.). https://a.allegroimg.com/original/11a913/1030f12d4ee2a04d5db032d3e41b/Save-The-Duck-Plumtech-Kurtka-Puchowa-S-Wypelnienie-syntetyczne

Savio, L., Pennacchio, R., Patrucco, A., Manni, V., Bosia, D.: Natural fibre insulation materials: use of textile and agri-food waste in a circular economy perspective. Mater. Circ. Economy **4**(1), 6 (2022). https://doi.org/10.1007/s42824-021-00043-1

Seaqual. SEAQUAL YARN. A Starting Point for a Cleaner Ocean (n.d.). https://www.seaqual.org/seaqual-yarn/

Seaqual. Together for a Clean Ocean (2024). https://www.seaqual.org/

Seixas, S.: The Emergence of Sustainability and the Textile and Fashion Design Education. Pages of Arts and Design, pp. 19–38 (2021)

Shirvanimoghaddam, K., Motamed, B., Ramakrishna, S., Naebe, M.: Death by waste: fashion and textile circular economy case. Sci. Total Environ. **718**, 137317 (2020). https://doi.org/10.1016/j.scitotenv.2020.137317

Storz, H., Vorlop, K.: Bio-based plastics: status, challenges and trends. Landbauforsch – Appl. Agric. Forestry Res. **2013**(63), 321–332 (2013)

Sumter, D., de Koning, J., Bakker, C., Balkenende, R.: Key competencies for design in a circular economy: exploring gaps in design knowledge and skills for a circular economy. Sustainability (Switzerland) **13**(2), 1–15 (2021). https://doi.org/10.3390/su13020776

Textile Exchange. Preferred Fiber & Materials Market Report (2022)

Textile Exchange. The Sustainability of Biosynthetics (2022)

Volanti, M., et al.: Terephthalic acid from renewable sources: early-stage sustainability analysis of a bio-PET precursor. Green Chem. **21**(4), 885–896 (2019). https://doi.org/10.1039/C8GC03666G

Warasthe, R., Brandenburg, M., Seuring, S.: Sustainability, risk and performance in textile and apparel supply chains. Clean. Logist. Supply Chain **5**, 100069 (2022). https://doi.org/10.1016/j.clscn.2022.100069

Next Skills for Circular Design Profiles: Bio-Based Innovation Towards Climate-Neutral Cities

Federica Delprino(✉), Silvia Pericu, and Chiara Olivastri

Università Di Genova, Genova, Italy
federica.delprino@edu.unige.it

Abstract. Design, in its broadest sense, plays a pivotal role in achieving climate-neutral environments, with cities serving as key arenas for a green and resilient transition. The circular economy presents innovative pathways for creative design sectors, which are not only essential for this transition but also crucial for fostering the cultural change and social inclusion necessary for sustainable development. Through a multidisciplinary approach that views cities as ecosystems, the Innovation Alliance Circular Design (CiD) addresses the educational gaps in skills and knowledge among product and service designers, as well as architects and urban planners. The Alliance offers innovative models that connect design to circularity and urban transformation. CiD fosters innovation in circular design education by promoting entrepreneurship within academia. It provides educational programmes aimed at upskilling individuals for emerging labor market profiles and supports the establishment of green start-ups. This paper outlines the process of identifying skill gaps in circular product design education, with the ultimate goal of advancing climate-neutral cities. It activates a dialogue and brings together relevant stakeholders from academia and vocational education and training (VET), as well as the broader socio-economic environment—including research and technology organizations (RTOs), municipalities, civil society, and businesses. This collaborative effort aims to create an innovation ecosystem for the co-creation of knowledge and to accelerate the transfer of research and innovation into educational practices.

Keywords: Circular economy · Bio-based materials · Innovation ecosystem · Green skills

1 Introduction

To effectively support carbon-neutral cities, designing for a circular economy requires significant scaling and a collaborative and multidisciplinary approach to create regional innovative circular ecosystems for the built environment that prioritize bio-based and renewable materials. Currently, many products and systems in the construction and design sectors are structured in ways that lead to most materials being discarded through landfill, incineration, or unintended environmental release. This waste involves not only the materials themselves but also the associated energy, resources, labor, and creativity that are wasted.

A. Mateus et al. (Eds.): WAD 2024, What's Around Design?, pp. 277–290, 2026.
https://doi.org/10.1007/978-3-032-08163-6_20

Achieving community resilience necessitates a focus on not just energy and resource efficiency (IRP 2020) but also on the complete production and transformation processes that characterize built environments and urban areas. In Europe, 36% of total waste results from building construction and demolition (Eurostat 2020), which introduces a threefold challenge to attaining carbon neutrality: wasted energy in production and demolition, reliance on finite, carbon-based resources, and energy loss during transportation. By concentrating on bio-based materials in architecture, product design, and interior design and targeting embodied carbon emissions, we uncover opportunities to minimize our environmental impact. The principles of circularity and the circular economy provide various innovative routes, particularly in rethinking the creative fields of architecture, urban design, and product and service design. These sectors should shift their focus away from fossil raw materials and embrace bio-innovation, as they are vital in transforming the construction industry and fostering a cultural transition towards sustainable lifestyles, thus enhancing the attractiveness and benefits of sustainable, healthy, and resilient cities.

The shift to a circular economy in design and architecture relies not only on adopting circular business models and practices by both private and public sectors but also on contributing to systemic changes across different areas and advancing urban sustainability. The Innovation Alliance on Circular Design (CiD) aims to address the gap in green, resilience, and digital skills within creative professions by emphasizing academic education in product and service design in collaboration with architecture and urban planning. CiD introduces an innovative framework that links design to circularity, urban transition, and expertise from cultural, economic, and natural sciences, as well as the wider socio-economic landscape. Its focus on co-creation and knowledge sharing promotes a practical model that integrates new design methods and concepts with the principles of circularity and the circular economy, ultimately facilitating urban transformation toward sustainability.

2 The Alliance

The CiD Erasmus + Innovation Alliance, operational since November 2023, seeks to create an innovation ecosystem that develops new educational models with a focus on entrepreneurial intent and heightened social responsibility. It also aims to ensure the application and promotion of its outcomes beyond the project's boundaries. Grounded in the belief that "circular economy is driven by design" (Ellen MacArthur Foundation 2024), the innovation strategy is directed along three key pathways: first, to fundamentally reconsider design principles for a circular economy; second, to steer various design sectors towards achieving carbon-neutral cities; and third, to emphasize bio-based innovation in the built environment (Fig. 1).

To fulfill these objectives, the alliance includes three academic partners dedicated to educational innovation: University of Hannover (LUH) specializing in Urban Design and Planning, the Institute for Advanced Architecture of Catalonia (IAAC), and University of Genova (UniGe) with its group focused on Product and Service Design. For the purpose of enhancing its efficacy, CiD unites a diverse array of stakeholders—including vocational education and training providers (VETs), research and technology organizations (RTOs), businesses, sector and civil society organizations, as well as incubators

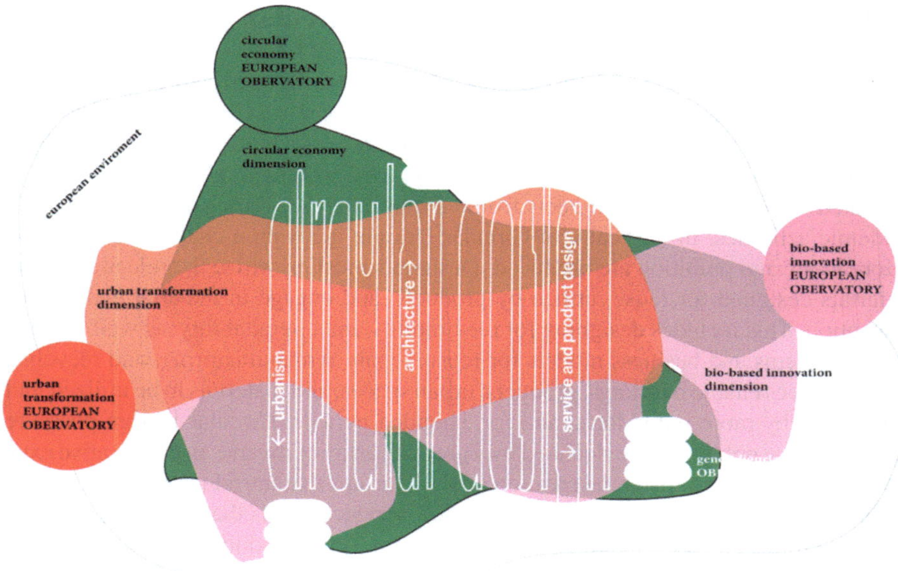

Fig. 1. CiD innovation alliance scheme, credits https://www.cid-innovationalliance.eu/

and investors—within a cohesive platform. This strategic synergy aims to develop an Innovation Ecosystem that facilitates collaborative knowledge generation and expedites the incorporation of research and innovation into educational frameworks.

The Innovation Alliance employs a distinct transnational strategy, featuring 11 partners from 8 different European countries. These collaborators include training providers, business entities involved in the labor market, and specialists in green skills, who participate throughout the project's lifecycle, from developing innovative educational models to their assessment and validation. This effort extends to establishing observatories to broaden the innovation ecosystems and to manage the exploitation, dissemination, and communication of the results.[1]

[1] Other partners are ARCES, an experienced Italian vocational training provider, ACE (Architects' Council of Europe), Materiom, a platform for SMEs focused on using bio-materials in design, Architekturos Fondas in Vilnius, which supports lithuanian SMEs in architecture, Business Inkubaator (TEI), business development center for SMEs in the creative and cultural industries in Tallinn. Additional partners with specialized expertise include: ALDA, Consultancy and evaluation expert, ERSILIA, consultancy and evaluation expert, and Biology Department of the National Research Council Italy (CNR), leading research and technology organization. As an associated partner, the Ellen MacArthur Foundation supports the activities of the alliance.

3 Transforming Design Education for a Circular Economy

The radical transformation of design education and practices necessitates not only a willingness to change but also creativity, collaboration across various sectors, and a diverse range of approaches. In this context, the process adopted by the CiD (Circular Innovation Design) initiative involves challenge-based methodologies that will be presented to designers and architects from various countries during the initial phase of academic activities. In the realm of Circular Design, innovation challenges arise even before waste and pollution are produced. Designers must prioritize the selection of safe materials designed for reuse, utilize by-products, and engage in material and product innovation. This includes designing for repairability and upgradability, as well as establishing systems and business models for reuse, repair, remanufacturing, and recycling. Such practices enable buildings, products, urban spaces, and materials to be utilized multiple times, by more individuals, and for extended periods. Circular Design possesses transformative potential, as evidenced by a growing array of diverse and innovative experiences. The goal is to foster the creation of new startups that introduce biomaterials into the building industry while also developing new service models that promote circularity and scale innovations to neighborhood and urban levels. In addition to academic education, vocational education and training (VET) for the circular economy must also be considered, as it plays a crucial role in bridging the skills gap through re-skilling and up-skilling. The objective is to increase the number of professionals proficient in green technologies, including digital solutions, who can develop green products, services, and business models, thereby minimizing the environmental impact of their activities. A key question arises: what skills are necessary for architects and product designers to achieve these goals? From the outset, when designing innovative curricula for academic and continuous education programmes, this question has been posed to local and European Observatories within the CiD Innovation Ecosystem. These observatories provide support, assessment, and sustainability perspectives for impact at various levels, extending to additional sectors and facilitating the project's exploitation and dissemination through multiplier events. This paper will outline this process, emphasizing findings that identify the skills gaps among product and service designers who aspire to become innovators in the circular economy, particularly in the transition away from fossil raw materials toward bio-innovation.

4 Integrating Skills in Circular Design Education

Profiles were constructed on the basis of local needs and a reflection upon the skills needed for new circular designers. Within the CiD project, a comprehensive reflection based on a Skills Matrix is presented, drawing from various sources (ACE, 2019; Sumter et al., 2021; Wiek et al., 2011; NEBC, 2021; Schröder et al., 2023). This Skills Matrix offers a foundational overview of the skills required for Circular Design, which include green skills, resilience skills, entrepreneurial skills, which can be considered in a perspective of an integrative approach.

In particular, green skills are essential for fostering a circular and sustainable built environment. To navigate this complex transition, individuals require a comprehensive skill set encompassing cognitive, interpersonal, intrapersonal, and technical abilities. Cognitive skills involve critical thinking, systems analysis, and innovation for addressing environmental challenges (Pavlova, 2018). Interpersonal skills, including adaptability, entrepreneurship, and a willingness to embrace new technologies, also emphasize collaboration, communication, and negotiation to balance economic, social, and ecological objectives; technical skills focus on quantifying, monitoring, and managing environmental impacts, such as waste, energy, and water (Pavlova, 2018).

Moreover, resilience skills, crucial for adapting to climate change, demands a blend of networking, problem-solving, and strategic thinking (Miceli et al., 2021). They also involve understanding urban and regional dynamics, fostering inclusivity, and implementing green features (Miceli et al., 2021). The communicability and realizability of these things is supported by digital and technical capabilities. Digital proficiency is paramount in contemporary design and sustainability. From digital design and fabrication to stakeholder engagement and marketing, these skills are essential for green initiatives (Soegaard, 2023).

Finally, but vital for upholding a circular economy mindset, entrepreneurial attitude is vital for driving innovation and market adaptation in the green economy. This includes critical thinking, networking, financial management, and risk assessment (Kickul, Lyons, 2020; Lyons et al., 2021). Linking green skills with entrepreneurial acumen can foster circular systems and startup creation within the construction industry (ACE, 2019; NEBC, 2021). To cultivate a holistic approach, the CiD project integrates green, resilience, digital, and entrepreneurial skills across multiple disciplines and scales. By fostering entrepreneurship and knowledge exchange, the project aims to accelerate the transition to a circular and sustainable built environment.

5 From a Skills Matrix to Circular Designers Profiles

In order to carry out a transformation from a generalist designer to one devoted to the circular economy at the academic training level, it was indeed necessary to first analyze the training offer, and the basic skills already acquired during the three-year degree by the students, and then compare them with the professional skills potentially useful in the world of work. Deganello (2019) proposes a different school, one that prepares new generations to claim a different design commission and allows 21st century design to contribute-through "aesthetic consensus"-to a radical "ecological conversion" of commodities. A new vision for the future that unites human activity with natural processes to form a regenerative, coevolutionary model for sustainable design (Fleming, Roberts, Isaac, 2024).

The profiling process involves identifying the needs of local stakeholders and regional demands regarding a tailored skillset designers should have to be hired and help local companies in circular processes, so that these challenges can be incorporated into the academic and training pathways of future professionals.

So, by integrating these demands to existing skills taught at design courses in bachelor's and master's degrees with those needed for circular design, comprehensive profiles are developed. These profiles combine green, resilience, digital and entrepreneurial skills to train designers equipped for sustainable practices in various sectors (Fig. 2).

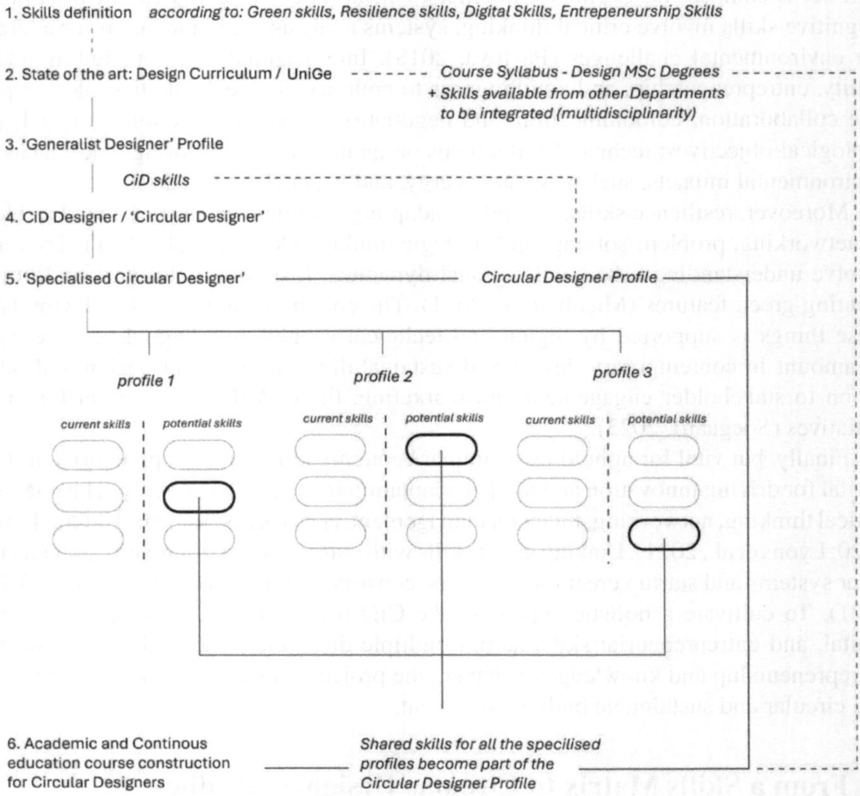

Fig. 2. Conceptual framework from the construction of skills and profiles to the research-based definition of academic and continuing education courses

Specifically, three reference profiles have been created, which will be a starting point for building the academic programmes, and also a point of departure and comparison with local stakeholders. The three profiles are considered complementary and collectively contribute to a comprehensive skill set that informs the development of an updated educational curriculum.

Each specialization falls within the broader domain of circular design, requiring a generalist foundation coupled with specific soft and hard skills. These professionals will operate within diverse contexts, including corporate, freelance, and startup environments, demonstrating social and ecological responsibility.

The UniGe team has proposed to the alliance, to be developed in their educational strategy, three distinct roles: the 'Biomaker', specializing in biomaterials; the 'Reprocesser', adept at systems thinking and process optimization; and the 'Community Maker', focused on service design and community engagement (Fig. 3).

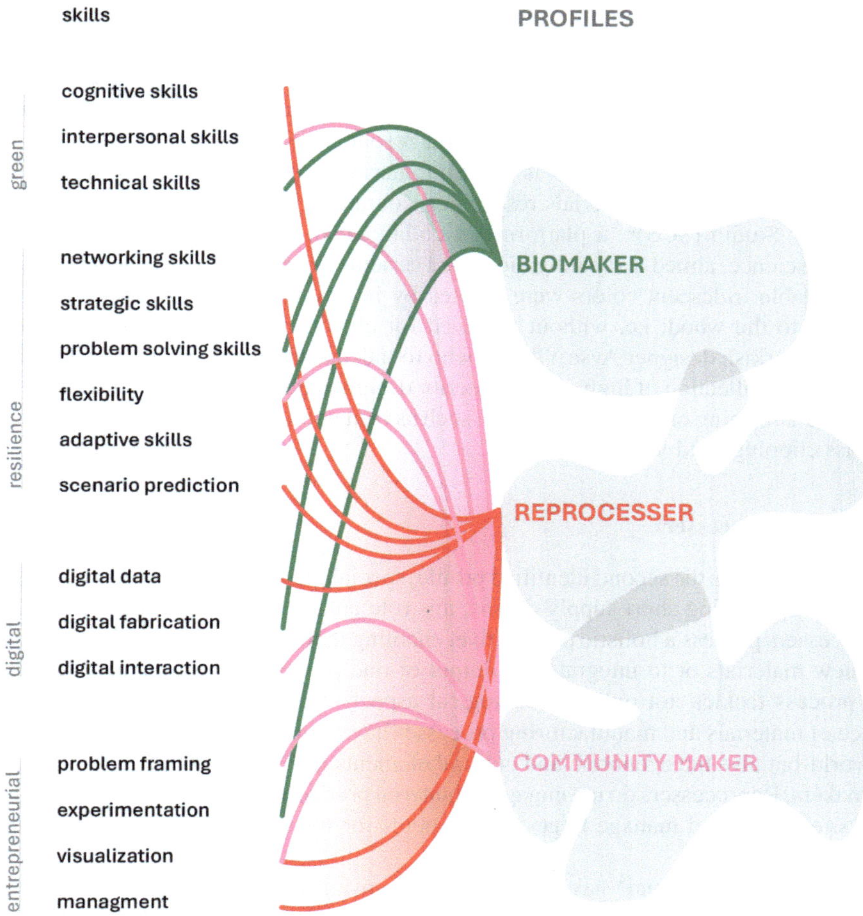

Fig. 3. Summary diagram of the main skills building up each circular designer profile

5.1 The Biomaker

The Biomaker is a hybrid role encompassing both design and fabrication. This individual experiments on a small scale with bio-based and recycled materials. These materials primarily originate from biodegradable living matter, such as plant fibers, using a biomimicry (Benyus, 1997) approach that comes from the Greek "bio" (life) and "mimesis" (imitation) and also from upcycled food or agricultural waste. Additionally, the Biomaker explores bio-fabricated materials produced by living organisms. As an explorer, the Biomaker seeks out discarded materials and identifies potential applications. They possess a designer's ability to envision new possibilities within natural elements, coupled with a hacker mentality to adapt materials for unconventional uses. Collaborating closely with engineers, biologists, and chemists in multidisciplinary teams, Biomakers are at the forefront of developing alternatives to fossil-based materials, aiming to reduce our reliance on these non-renewable resources.

The first interesting example is the Shimmering Wood, was created by Finnish designer Noora Yau and materials researcher Konrad Klockars, founders of the Structural Color Studio (SCS)[2], a platform for collaborative projects between design and materials science, aimed at creating bio-based structural colors, at Aalto University. The biodegradable iridescent colors were created by inserting microscopic nanostructures directly into the wood, i.e. without the need for chemical compounds. An upcycling project of Turkish designer Ayse Yilmaz, who founded a production start-up and created Ottan[3],*** a collection of high-quality interior design materials from an innovative process of transforming organic green waste, such as fruit peels, expired vegetable products, or grass clippings and leaves.

5.2 The Reprocesser

The Reprocesser is the second identified profile, specializing in waste recovery and upcycling. By optimizing short supply chains, this role enhances overall system efficiency. Reprocessers possess a holistic perspective, enabling them to transform perceived waste into new materials or to integrate the output of one process into the input of another. The process tackles not only the ecological aspects of sustainable design-designers' choice of materials and manufacturing processes have a tremendous impact on the natural world-but also the economic and cultural elements involved (Thorpe, 2007). Unlike Biomakers, Reprocessers do not engage in material production but collaborate with companies to access and manage large-scale systems for waste collection, transformation, and distribution.

The FabBRICK brand[4] has introduced an innovative solution: transforming waste fabrics into robust and versatile bricks for eco-friendly, i.e. environmentally friendly, construction. Founded in 2019 by French architect Clarisse Merlet, the idea for FabBRICK was born in 2017, during her university days. The student was looking to explore alternative construction methods and had a clear awareness of the environmental impact and energy-intensive nature of the construction industry. Therefore, she began to question the unexpressed potential of materials considered as 'waste' and thought to focus on the textile industry. Realizing the poor waste management in the industry, Clarisse began to imagine a new life for textile materials destined for landfill. Hence the revolutionary idea of turning waste textiles into building bricks.

Somewhat opposite to the previous case, but with a circular approach is the case of Brick Texiles, textiles made by the Danish Natural Material Studio[5] in collaboration with

---

[2] https://structuralcolourstudio.com/

[3] https://www.ottanstudio.com/

[4] https://www.fab-brick.com/

[5] https://naturalmaterialstudio.com/

Polish designer and researcher Zuzanna Skurka[6]. The textiles are made of overproduced or damaged bricks, but mostly recovered from demolition projects, which have been transformed into soft drapes for home furnishing. To make it, in addition to pulverized bricks, a bioplastic created by Natural Material Studio and composed of protein, natural softener and pigments was used. The company describes it as "a bio-textile made of ingredients and pigments obtained directly from nature. It is a living material: texture, color and shape are reactive to the material's environment and change with temperature, humidity and time".

5.3 The Community Maker

The Community Maker profile represents the transition from product-focused to service and system-oriented approaches. This role demands a comprehensive understanding of regional contexts, including cultural and ecological factors, with a particular emphasis on identifying underutilized or undervalued assets. The Community Maker recognizes the intrinsic value not only of the final product but also of the collaborative processes (Manzini, 2015) and experimentation involved in its creation. This co-design approach may serve as a countermeasure to what Armiero (2021) describes in Wasteocene as "the imposition of wasteful relationships on subordinate and more-than-human communities involves the construction of toxic ecologies made of contaminating substances and narratives." Integrating open-source principles and a participatory approach is essential for fostering broader ecosystem engagement and driving transformative change. To effectively communicate the project's narrative, along with its human and process-centric dimensions, it is crucial to disseminate knowledge through publications, exhibitions, conferences, and workshops.

A very good example is the Atelier LUMA[7] is the design research program of LUMA Arles, they carry out their bioregional design practice in their laboratories and workshops, and through collaborations and partnerships. Atelier LUMA develops materials, processes, and objects that address the economic, environmental, and human challenges of today's world. Through their work, they build productive and resilient networks that facilitate the exchange of knowledge and skills between communities, artisans, industries, scientists, students, and institutions. These references are a starting point for seeking confirmation and using collaborative activities to actually confirm whether these skill needs correspond to those of companies and organizations in the area.

6 Co-design and Participatory Meetings: 'Next Skills for Circular Design'

The definitive skills and profiles indicated were also constructed through collaborative work with local stakeholders, members of the observatories for the CiD project. Each partner, in fact, at a local and European level, has created a network of contacts that can support the academic and continuing education programmes through their direct

[6] https://zuzannaskurka.com/brick

[7] https://weever.atelier-luma.org/en

involvement. This mapping provides an overview of the key players involved in various sectors related to urban development, architecture, sustainability, and industry in the local territorial area. Each stakeholder is categorized based on their level of influence (local, regional, national) and their specific domain of activity. The diagram indicates a total of 15 stakeholders, categorized as local (red), regional (blue), and national (purple). Additionally, some entities are marked as "In Search of Stakeholder" (red circle), to denote a possible expansion of the network in relation to the category (Fig. 4).

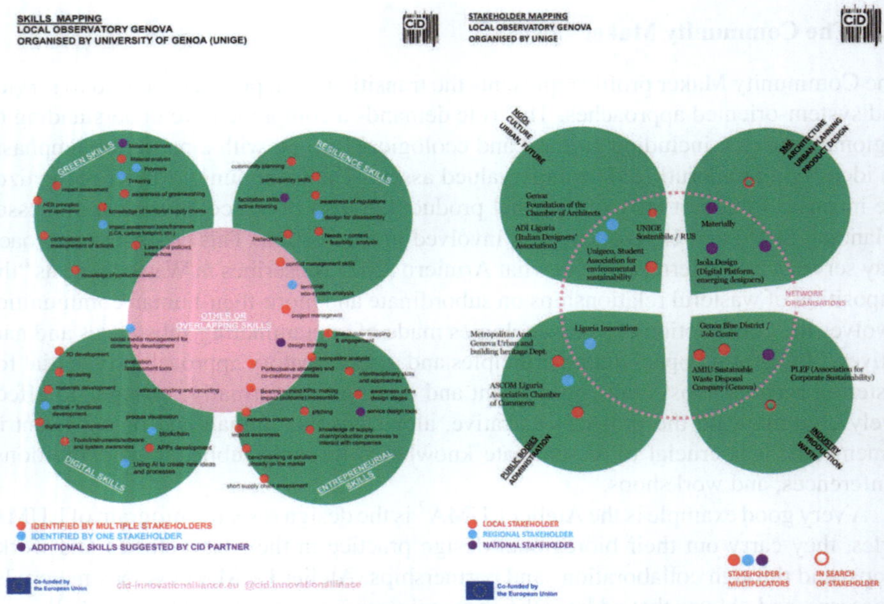

Fig. 4. Skills mapping scheme and stakeholder mapping (CiD graphic template)

During the meeting, the three designer profiles were presented and the participants assessed the current courses and skills that students acquire to enter the workforce and discussed which additional skills are necessary to align with circular needs, in a participative way. This was accomplished through a collaborative profile-based activity, including the use of participative personas through paper canvas and miro for the blended session, that led to the identification of skill gaps. Several specific skills emerged as essential for circular designers. For green skills, there is a need for awareness of certifications and impact measurement tools. Resilience skills require the ability to design for disassembly, ethical recovery, and knowledge of rules for interacting with evaluation criteria. Digital skills include the ability to visualize process data, create internal and external storytelling, and raise awareness within one's company or team. Entrepreneurship skills involve network creation, social media management, communication skills for community development, KPI-based thinking, and knowledge of the supply chain and production processes (Fig. 5).

Fig. 5. Some photos of the stakeholder meeting at University of genova (UniGe) – department of architecture and design

The collaborative activity was crucial in activating and solidifying the network of stakeholders, fostering individual contacts, and exploring four macro categories of expertise related to different fields and supply chains. The event highlighted the importance of creating a common basis within academic programmes for assessing and visualizing impact, enabling future circular designers to effectively intervene within networks and supply chains.

As a future working goal, it is essential to emphasize the relationship between the three designer profiles and their respective curricula, derived from a consolidated corpus of Master's Degree Design. The integration of the identified skills is vital for training circular designers. A key decision to be made is whether to develop all three paths through simulated teaching modules to shape each figure or to focus on a single path as a joint result of the master's courses, deploying specific skills within modules or integrative activities. This strategic approach will help in building a comprehensive and updated educational offering for future circular designers.

7 Skills Applied to Academic and Continuous Education Course Profiles and Curricula

These capacities were then incorporated into an academic and continuous training programme, which will start in 2024/2025 and 2025/2026, following an evaluation both on a local scale of its observatory stakeholders and through a participatory event organised online with the members of the European observatories throughout the project.

Regarding the Academic Program and its course construction, the Circular Product and Service Design curriculum at University of Genova (UniGe) focuses on equipping designers with the skills necessary for the circular economy, emphasizing sustainability and minimizing environmental impact. The course develops essential skills in green practices, resilience, digital proficiency, and entrepreneurship. These include understanding certifications, designing for disassembly, visualizing data, and building networks. These skills enable designers to integrate into circular economy processes, manage impacts, and foster climate-neutral behaviors. The program prepares students to innovate in product, service, and event design, utilizing recycled materials and improving sustainability across supply chains. Through methodologies like challenge-based learning and design labs, students engage in practical, real-world problem-solving. The curriculum aims to produce versatile designers capable of making informed decisions and leading initiatives in various scales and contexts of circular design.

On the other hand, the "Circular Product Design Processes and Practices" continuous education program held by University of Genova (UniGe) focuses on developing essential skills in green practices, resilience, digital technology, and entrepreneurship to foster innovation and sustainability in the circular economy, with a deep focus on startup-driven strategies. These skills enable professionals to design products and services with minimal waste and maximal resource efficiency. The program covers material selection, life cycle assessment, product life extension, recycling, policy frameworks, and the role of digital technologies in circularity. Participants, including designers and entrepreneurs, will learn to integrate circular design principles, develop startup capacities, and communicate sustainability challenges effectively. The methodologies include blended learning, challenge-based learning sessions, and hands-on workshops to promote international mobility and multi-actor networking. Through practical projects and networking, participants will gain the skills to innovate and lead in creating sustainable and circular design solutions.

The evaluation was done through guided online questionnaire compilation according to the indicated questions, email exchange and inputs, telephone and in-presence interviews to discuss questions related to programme evaluation and possible suggestions.

Regarding the academic programme for the curriculum 'Circular Product and Service Design,' a five-member evaluation panel confirmed its consistency and alignment with the learning objectives and competencies, particularly commending the inclusion of entrepreneurial skills. This ensures that graduates are well-prepared for circular design roles across various sectors. Applying the same evaluation strategy, the University of Genova (UniGe)'s continuous education curriculum was also assessed. In this case, the reviewers highlighted the importance of disseminating information on current processes and actively engaging the community to raise awareness of circular economy practices and sustainability. The curriculum was found to meet expectations, with a recommendation to integrate case studies into the course.

8 Conclusions

The transition towards carbon-neutral cities needs innovative pathways for creative design sectors, which are crucial for fostering the cultural change and social inclusion necessary for sustainable development in the building sector. The CiD project addresses a critical gap in design education for a circular economy, focusing on bio-based innovation. This paper has underscored the pivotal role of circular economy principles and bio-based innovation in achieving this goal. By fostering a collaborative ecosystem that integrates academia, experts from cultural, economic, and natural sciences, as well as the wider socio-economic landscape, the Innovation Alliance on Circular Design develops a process to address the challenges of architecture, urban design, product and service design in creating regional circular ecosystems that prioritize bio and renewable materials.

The integration of circular economy concepts into design education is crucial for cultivating a new generation of professionals equipped to tackle the complexities of sustainable urban development. By emphasizing entrepreneurial intent and social responsibility, CID aims to empower students to become catalysts for change with a focus on bio-based materials as an opportunity to reduce embodied carbon emissions, minimize waste, and create resilient, healthy urban environments.

The goal is to train design students and young professionals with systemic skills and specific knowledge of bio innovation, capable of integrating environmental requirements into the product and service development process to meet the demand of enterprises and public agencies for innovative practices in creating truly circular product-service systems. The process presented in this essay starts from the identifications of skill gaps in the curricula currently offered to design students to the identification of emerging profiles of designers and startuppers in order to be able to identify not only their profiles but also the skills needed. By identifying essential skills and developing three distinct designer profiles, the project fosters a new generation of professionals equipped to create climate-neutral cities. The integration of these skills into academic and continuous education programmes will be essential for driving the transition to a circular economy in the built environment. Collaborative partnerships and ongoing research are vital for the project's success and broader impact, also making the educational programmes more interesting, as they really have a dialogue and a connection to the demands of the local area, and thus respond concretely to the needs of a supply chain.

References

ACE Architects' Council of Europe. Designing for a Circular Economy: Statement of the Architects' Council of Europe. World Circular Economy Forum, 4–5 June 2019, Helsinki (2019). https://www.ace-cae.eu/uploads/tx_jidocumentsview/ACE_EWCF_Statement_FINAL.pdf

Armiero, M.: Wasteocene: Stories from the Global Dump. Cambridge University Press (2021)

Benyus, J.M.: Biomimicry: Innovation Inspired by Nature. HarperCollins (2002)

Carlgren, L., Rauth, I., Elmquist, M.: Design thinking: exploring values and effects from an innovation capability perspective. Des. J. **19**(3), 403–423 (2016)

Deganello, P.: Design politico: Il progetto critico, ecologico e rigenerativo: Per una scuola del design del XXI secolo. Altreconomia (2019)

Ellen MacArthur Foundation. Building Prosperity: Unlocking the Potential of a Nature-Positive, Circular Economy for Europe (2024). https://www.ellenmacarthurfoundation.org/building-prosperity

EUROSTAT. Recovery Rate of Construction and Demolition Waste (2020). https://ec.europa.eu/eurostat/databrowser/view/cei_wm040/default/table?lang=en

EUROSTAT. Waste Statistics – Statistics Explained (2020). https://ec.europa.eu/eurostat/statistics-explained/index.php?title=Waste_statistics#Total_waste_generation

Erichsen, P.G., Christensen, P.R.: The evolution of the creative process. Creat. Res. J. **25**(4), 351–360 (2013)

Fleming, R., Roberts, S.H., Isaac, C.: Sustainable Design for the Built Environment. Taylor & Francis (2024)

Genovesi, E., Pellizzari, A.: Neomateriali 2.0 nell'economia circolare. Edizioni Ambiente (2021)

Hertwich, E., Lifset, R., Pauliuk, S., Heeren, N.: Resource Efficiency and Climate Change: Material Efficiency Strategies for a Low-Carbon Future. United Nations Environment Programme (2020)

Kickul, J., Lyons, T.S.: Understanding Social Entrepreneurship: The Relentless Pursuit of Mission in an Ever-Changing World, 3rd ed. Routledge (2020)

Lyons, T.S., Lyons, J.S., Samson, J.A.: Entrepreneurship Skill Building: Focusing Entrepreneurship Education on Skills Assessment and Development. Springer International Publishing (2021)

Manzini, E.: Design, When Everybody Designs: An Introduction to Design for Social Innovation. MIT Press (2015)

Miceli, A., Hagen, B., Riccardi, M.P., Sotti, F., Settembre-Blundo, D.: Thriving, not just surviving in changing times: how sustainability, agility, and digitalization intertwine with organizational resilience. Sustainability **13**(4), 2052 (2021). https://doi.org/10.3390/su13042052

NEBC New European Bauhaus Collective. Common Ground: Making the Renovation Wave a Cultural Project (Conference Report) (2021). https://www.ace-cae.eu/fileadmin/user_upload/NEB_report_v3.pdf

Pavlova, M.: Fostering inclusive, sustainable economic growth and 'green' skills development in learning cities through partnerships. Int. Rev. Educ. J. Lifelong Learn. **64**(3), 339–354 (2018)

Schröder, J., Diesch, A., Cappeller, R., Scaffidi, F.: Circular design: towards regenerative territories. Jovis (2023). https://doi.org/10.1515/9783868598360

Soegaard, M.: How to Become a Digital Designer: The Complete Guide. Interaction Design Foundation (2023). https://www.interaction-design.org/literature/article/how-to-become-a-digital-designer-the-complete-guide

Sumter, D., de Koning, J., Bakker, C., Balkenende, R.: Key competencies for design in a circular economy: exploring gaps in design knowledge and skills for a circular economy. Sustainability **13**(2), 776 (2021). https://doi.org/10.3390/su13020776

Thorpe, A.: The Designer's Atlas of Sustainability. Island Press (2007)

Wiek, A., Withycombe, L., Redman, C.L.: Key competencies in sustainability: a reference framework for academic program development. Sustain. Sci. **2011**(6), 203–218 (2011). https://doi.org/10.1007/s11625-011-0132-6

Materials Design Trends Fostering Environmental Sustainability

Sofia Soledad Duarte Poblete(✉), Laura Anselmi, and Valentina Rognoli

Politecnico di Milano, Milan, Italy
sofiasoledad.duarte@polimi.it

Abstract. Humanity faces serious environmental challenges such as climate change, biodiversity loss and resource depletion. The urgency to move towards environmentally sustainable and circular design practices has become more pressing. In this context, the role of materials design practices in promoting environmental sustainability is increasingly recognised. However, despite this awareness, there is an urgent need to develop a critical perspective on these approaches. Materials designers must be able to address real-life issues and meet today's challenges to contribute in the transition towards sustainability and regeneration. With the aim to equip designers to make crucial decisions, this paper reviews and analyses 43 empirical material design projects collected over a decade (2014–2024). By capturing their characteristic features and industry applications, practical trends and patterns implemented to foster environmental sustainability are identified. The study presents emerging directions, principles and challenges in contemporary materials design. These directions aim to encourage a responsible economy that embraces design solutions with lower negative emissions, reduced and conscious use of resources, and promotion of biodiversity. By highlighting trends, principles, and challenges, this paper contributes to the strategic materials design discourse, offering insights and considerations for advancing ecologically conscious practices that support economic growth and reduce the deterioration of the ecosystem.

Keywords: Materials designers · Sustainability · Transition · Design practice · Principles

1 Introduction

Extreme climate change, waste pollution and biodiversity loss threaten ecological balance and human well-being. Low awareness of the life cycle and impacts of materials, products and production systems has accelerated these issues, highlighting the urgent need to rethink and redesign business systems and operational processes (Scranton, 2015). In these circumstances, the role of design is critical. As it is often argued design decisions significantly influence the environmental impact of products and services (Thackara, 2005; Vezzoli, 2018). Within this context, the contemporary materials design practice has evolved to offer more than performative, appealing, and distinctive materials design, but also design that could minimise environmental impact, fostering nature-based and circular solutions to harmonise the environment (Rognoli et al., 2021).

A. Mateus et al. (Eds.): WAD 2024, What's Around Design?, pp. 291–306, 2026.
https://doi.org/10.1007/978-3-032-08163-6_21

Despite the growing recognition of sustainable practices in materials and design (Ashby, 2009; Barati & Karana, 2019; Camere & Karana, 2018; Duarte Poblete et al., 2024), there remains a significant knowledge gap that hampers effective and wider application of the materials design practice in the real world. Designers often lack viable strategies to effectively work with materials in a sustainable way (Bak-Andersen, 2021). This issue underlines the need for greater awareness in the field. This paper address this gap by presenting information on contemporary materials design practices and principles that aim to foster environmental sustainability. The paper presents insights by reviewing 43 cases from a repertoire of materials design projects that were collected over a ten-year period (2014–2024). The analysis, which focuses on material attributes and industry sectors, identified practical trends in material design to promote environmental sustainability.

The remainder of the paper elaborates as follows: in Sect. 2, the background of the study. Section 3 outlines the methodological approach. Section 4 presents the findings and their analysis. Section 5 outlines the environmental sustainability trends in materials design. To finalise, conclusions are summarised.

2 Theoretical Background

2.1 The Materials Design Practice

Materials design has evolved from its foundational roots in materials science and engineering into a recognised industrial and product design field, characterised by a broad spectrum of research and creative practices (Rognoli & Levi, 2005). This expansion is transforming the role of *materials designers* as transdisciplinary agents that can influence new functionalities, experiential qualities, and material advancements (Pedgley et al., 2021).

The emergence of materials design can be traced back to the 1970s, when the influence of a design movement called *Design Primario* put the focus on the qualitative dimension of materials, emphasising their expressive potential, exploring aspects like chromatic, acoustic, and tactile (Rognoli, 2004; Squatrito & Ferrara, 2023). Renowned designers such as Clino Trini Castelli, Antonio Petrillo, Ezio Manzini, and Frida Doveil were pivotal in defining materials design as a distinct field, with a strong emphasis on material qualities leading to the CMF discipline (Colour, Materials, and Finish) (Becerra, 2016). By the 2000s, this area of research had started enlarging, and new tools, such as material libraries, emerged to connect designers with materials knowledge further (Ashby & Johnson, 2003). At this time, research focused on material design innovation for product application, following principles of sensory experience, authenticity as well as material meaning and cultural expression (Ferrara, 2004; Karana et al., 2009; Rognoli & Levi, 2005; Zuo, 2010). Since then, the number of research studies concerning material interactions and product design has grown considerably (Karana et al., 2015a; Karana et al., 2015b).

Around 2014, concepts like *materials experience* (Karana et al., 2014), the experience designers and people have through physical, aesthetic, sensory responses, emotions, meanings and interactions with materials, gradually began to be recognised in the field. Concurrently, material experimentation and self-manufacturing of materials becomes popular (Rognoli et al., 2015). This attitude led to the emergence of *DIY-Materials* and *materials-driven design* (Ayala-Garcia et al., 2017; Karana et al., 2015a; Karana et al., 2015b), where designers, through material *tinkering*, playing, and iteratively investigating ingredients and combinations, created novel materials, features, and experiences. Thereafter, principles such as experimentation, iteration, and materials experiences have been increasingly followed.

From 2015 to 2020, the role of materials designers evolved to integrate identities of craftspeople, chefs, and scientists, pushing the boundaries of materials design toward a circular economy and sustainable solutions (De Giorgi et al., 2020; Clèries & Rognoli, 2021, 2020; Vélez et al., 2022). This shift in materials design has blurred the lines between material and product design and forged new connections with fields like biology and nanotechnology, leading to the creation of hybrid products that merge design, art, engineering, and science (Oxman et al., 2015). Materials design is a field in continuous expansion and definition, with increased recognition in the sector yet to be professionally established (Symbola, 2024). Indeed, the principles of practice now emphasise promoting circularity, sustainability and regeneration, reflecting a major step forward towards environmental responsibility.

2.2 Materials and Design for Environmental Sustainability

The concepts of design for environmental sustainability and ecodesign have been developed to recognise the necessity of considering ecological impacts at all stages of a product's life cycle. From the production of raw materials and components to the distribution, use and disposal of the finished product (Vezzoli, 2018). This concept examines methods of reducing environmental impact to maintain ecological balance within the planet's natural environment and to conserve natural resources. While this concept focuses on environmental considerations, it always acknowledges the social and economic aspects, which must also be balanced in any sustainability development perspective.

Materials are so crucial in design that ecodesign strategies have undoubtedly considered them in many stages of the design process to reduce the environmental footprint of objects and services (Bovea & Gallardo, 2006). The designer's knowledge and decisions regarding materials and processes are pivotal in determining the impact of the design throughout the product life cycle.

As Pollini and Rognoli (2021) have shown, if the phases of a product's life cycle are observed within the ecodesign strategy wheel (Brezet & van Hemel, 1997), it can be recognised that materials have the potential to reduce the environmental impact of products and should be considered in all design phases. Figure 1 summarises the various stages and considerations of product design for environmental sustainability, highlighting those that relate to materials.

The elements highlighted in the ecodesign strategy wheel offer helpful insights to relate materials design practices with environmental considerations. Furthermore, this evidence suggests that integrating a deeper understanding of materials at the beginning of the design process, using a materials-centric or materials-driven approach (Bak-Andersen, 2018; Karana et al., 2015a; Karana et al., 2015b), can clarify the environmental impacts of products. This paper argues that guiding the design process with a thorough

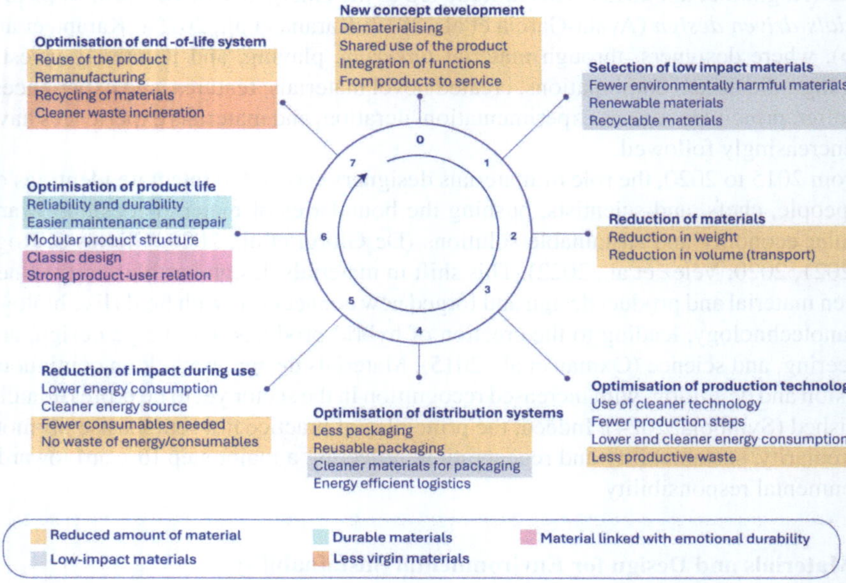

Fig. 1. Adapted from the ecodesign strategy wheel graphic from Brezet and van Hemel, 1997.

knowledge of materials and their potential implications in the environment and society can lead to more environmentally responsible design.

3 Methodological Approach

Collecting cases and examples of previous projects is a typical act of designers. Donald Schön (2017) refers to these cases as the designer's *repertoire.* It can consist of examples, images, understandings, and actions that practitioners use to deal with situations in their work. The repertoire used for this study is a collection of design cases that can be considered as material design projects and are related to material design practices. This repertoire was used as a tool to collect case examples from the web. The cases were retrieved between 2023 and 2024 through online sources such as the design magazines DeZeen, Designboom and Design Milk and social platforms such as Instagram and LinkedIn. Keywords used to find the projects were, for instance, 'new materials', 'materials experience', 'innovative materials', 'circular materials', 'sustainable materials' combined with 'design'. The search prioritised projects that were material-centred, material-driven, and material-based, i.e. that placed materials as the primary element and driver of the design process and demonstrated a deep connection to the design discipline and the designers, not just to materials science or engineering.

The project selection and exclusion criteria for the analysis were defined as follows: They must be presented after 2014 to ensure they reflect on advancements in materials, techniques, and design philosophies, capturing trends and recent developments.

They should differ from traditional materials known for their extended historical use and conventional properties. Projects should emphasise distinctive perceptions, experiences, and novelty in production processes, technologies, or techniques.
They should express low environmental impact decisions, like using renewable resources, enhancing recyclability, increasing durability, reducing waste, and controlling the carbon footprint.
They should achieve at least a technological readiness level (TRL) 5, indicating validation in a relevant environment and suitability for testing in simulated conditions. Preference was given to projects present on the market.
The information should be accessible, with descriptive details and clear images to understand the material practices.

Projects limited to initial prototypes, proofs-of-concept without subsequent development, and exclusively personal or highly speculative were excluded. Additionally, student projects and crafts produced on-demand without follow-up or evaluation and speculative or DIY-Materials lacking concrete design applications or potential for expansion were omitted. These criteria ensure that the selected cases demonstrate viability in real-world professional contexts.

3.1 Data Normalisation, Organisation, and Analysis

The *Notion* platform was employed to organise and categorise collected data, enabling precise classification through informative cards and labelling for each case. These cards included project name, authors, launch year, industrial sector, materials used, key features, TRL, and challenges, supplemented by geographic and workforce. To ensure accuracy, information was cross-verified against multiple online sources. The qualitative analysis, as suggested by Azungah (2018) was conducted by reviewing the detailed project descriptions. The process was supported by normalising labels with thematic relevance and clarity employing tables and graphs to identify patterns and trends.

4 Results and Discussion

4.1 General Analysis

A total of 43 projects were identified for analysis. A number determined by the observation of recurring patterns and deemed sufficient for a comprehensive study. The projects listed in Table 1 were selected for their relevance to the materials design field. Furthermore, they demonstrated commitment to foster environmental sustainability in different ways.

A review of this data, as shown in Fig. 2A, reveals a notable propensity for micro and small enterprises to pursue this kind of projects. Figure 2B shows thirty-eight of the evaluated companies are in Europe. This insight reflects a favourable ecosystem for materials design development, possibly supported by government policies and a solid technological infrastructure. Figure 3A shows the launch year of each project, highlighting a notable increase in initiatives between 2018 and 2020. This increase seems to be due to recent technological developments and a growing focus on solutions

Table 1. Provides key information on each of the materials design projects analysed

Case Name	Launch year	Author enterprise	N. Employees	Location
S-1500	2019	Snøhetta	201–500	Norway
Forite	2022	Snøhetta	201–500	Norway
Mylo™	2020	Bold Threads	51–200	USA
REISHI™	2020	Mycoworks	51–200	USA
Mogu PLUMA Wall Panels	2022	Mogu srl	11–50	Italy
ORB	2020	Biohm	11–50	UK
Pendant Lamp B-Wise	2021	Myceen	2–10	Estonia
Scoby Packaging Material (Film)	2023	MakeGrowLab	2–10	Poland
Weducer Cup	2022	Kaffeeform	2–10	Germany
Charlie Chair	2018	EcoBirdy	2–10	Belgium
Ohmie, the orange lamp	2021	Krill Design	2–10	Italy
ReKrill-Ment	2024	Krill Design	2–10	Italy
Coffeefrom Bio	2019	Coffeefrom	2–10	Italy
Nazena Material	2019	Nazena	2–10	Italy
Side Table	2020	Clandestino	2–10	Panama
Brassé	2020	Instead mobilier	2–10	France
Wood-Based Family	2020	Mixcycling	11–50	Italy
Espresso cup and saucer	2022	Coffeefrom	2–10	Italy
OTO (One to One) Chair	2023	One To One	2–10	Italy
Acoustic Pulp	2019	Baux	11–50	Sweden
Orca Material	2015	Smile Plastics	11–50	UK
EchoBoards®	2016	Echojazz	11–50	Switzerland
Ignorance is Bliss	2016	Agne Kucerenkaite	2–10	The Netherlands

(*continued*)

Table 1. (*continued*)

Case Name	Launch year	Author enterprise	N. Employees	Location
The light collection – Polygood®	2018	The Good Plastic Company	51–200	The Netherlands
Black Rock	2018	Plasticiet	2–10	The Netherlands
FabBRICK material	2019	FabBRICK	2–10	France
SCALITE®	2020	Scale vision	2–10	France
Cocoboards	2020	Naturloop	2–10	Switzerland
Blanc Marbre	2021	Pierreplume	2–10	France
Piñatex®	2014	Ananas Anam	11–50	UK
Fruit Leather	2018	Fruit Leather Rotterdam	2–10	The Netherlands
Bananatex®	2018	Qwstion	2–10	Switzerland
Desserto®	2019	Adriano di Marti	11–50	Mexico
Ananasse™	2020	Verabuccia	2–10	Italy
MyHelmet	2023	Studio MOM	2–10	The Netherlands
PEEL Chair	2023	PROWL Studio	2–10	USA
Kartell A.I. Chair	2019	Kartell	51–200	Italy
Fibre 12	2023	Caimi Brevetti	11–50	Italy
Carifa Carta	2024	Arper	201–500	Italy
Mat chair	2024	Normann Copenhagen	11–50	Denmark
Ooho	2017	Notpla	51–200	UK
Notpla Food Containers	2022	Notpla	51–200	UK
Sulapac premium	2018	Sulapac	11–50	Finland

that minimise environmental impact. Figure 3B presents the year of foundation of the companies behind these projects, with an increase between 2015–2020.

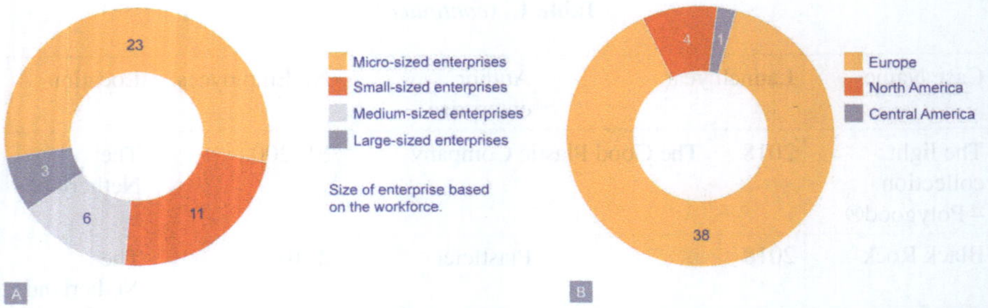

Fig. 2. General analysis. (a) Number of projects analysed in each category. (b) Number of projects per continent.

Fig. 3. General analysis. (a) Lunch of projects per year. (b) Company foundational year.

4.2 Practical Trends in Materials Design Projects

To identify practical trends and principles in materials design and their application in promoting environmental sustainability, an analysis of the selected projects was conducted, focusing on the following aspects:

The distinctive and novel attributes of the materials and sources in the projects.
The industrial sectors and application fields.

_Distinct and novel attributes

The particular and novel attributes of the materials in the projects are distinguished using diverse resources, categorised into two principal domains: natural and artificial, with some classified as mixed or borderline-origin due to significant human processing as in Duarte Poblete et al. (2024).

The natural domain includes resources derived directly from natural processes with minimal human modification. Examples include renewable resources, such as plant biomass, microorganisms like bacterial cellulose, cactus biomass, and repurposed organic food waste, such as coffee grounds and unused fruit. On the other hand, the artificial domain encompasses resources that have undergone substantial anthropogenic alteration, including engineered materials and the repurposing of non-organic industrial waste. Examples are mechanically reprocessed discarded postconsumer polymers or chemically recycled textiles transformed into new fibres. Lastly, the mixed/borderline-origin domain includes materials like industrial biopolymers from natural sources and recycled natural fibres (e.g., wool, cotton) that, although derived from natural sources, undergo extensive human processing to be used.

In summary, 28 projects utilise natural sources, 15 use artificial ones, and 22 involve natural sources with high intervention. This distribution underscores a solid commitment to waste valorisation principles through recycling and repurposing, highlighting circular practices across the projects (Fig. 4).

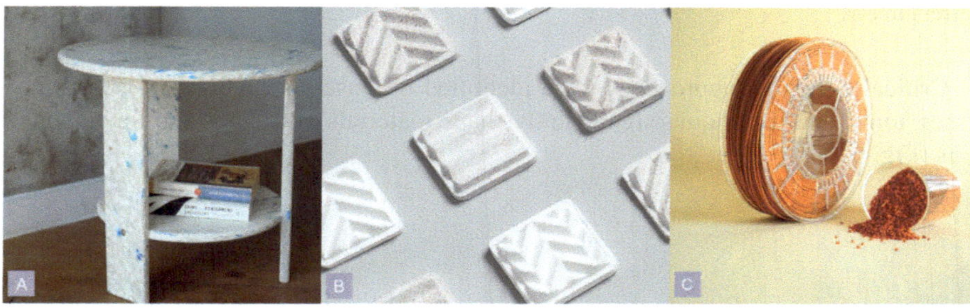

Fig. 4. Characteristic sources. (a) Artificial: postconsumer plastic. Photo credits Clandestino lab. (b) Natural: spruce, pine and wheat bran. Photo credits: Baux. (c) Mixed: pha biobased polymer with orange peel waste, photo credits Krill Design.

Regarding the aesthetic and perceptual distinctive aspects of the materials in the projects, there is a discernible inclination towards giving value to imperfections or the unconventional resources from which they originate. This move involves highlighting the unique qualities of these imperfections and weaving engaging narratives about their sources.

Two main trends are continuously present (Karana et al., 2015a; Karana et al., 2015b). On the one hand, the 'natural feel' is a trend that involves using natural resources, textures, colours, and finishes that emphasise the authenticity and origin of the materials, celebrating their natural attributes and intrinsic aesthetic values. On the other hand, a more 'artificial impression' is given using recycled materials of non-biological origins, where irregular colours, combinations, and textures provide the materials with a distinctive identity. This design approach evokes a positive sensory and emotional response, enhancing appreciation for the organic and imperfect while searching less environmental impact through the thoughtful use of recycled elements (Fig. 5).

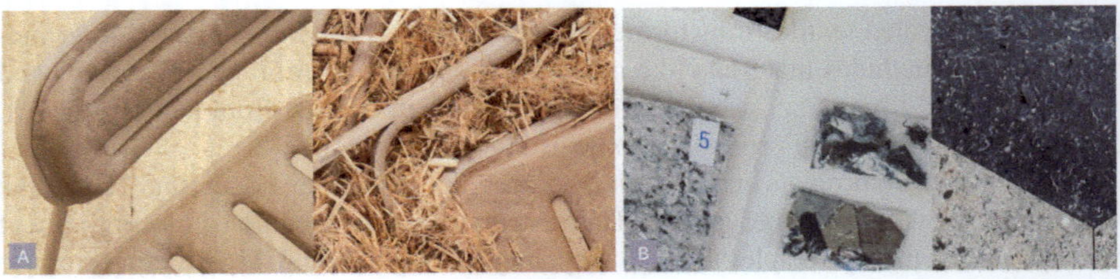

Fig. 5. Examples of aesthetic trends. (a) Natural feel, PEEL chair by PROWL Studio. Photo credits Noah Webb & Prowl. (b)Artificial impression, materials, by Pierreplume®. Photo credits Pierreplume®.

Critical emerging approaches were identified because production and design processes, tools, and techniques significantly shape materials' attributes and final applications (Fig. 6).

Fig. 6. This project utilises biofabrication with mycelium. The images depict the material process from growth to the final quality of finishing. Project and photo credits: ©MOGU s.r.l.

One of them is *biofabrication*, particularly natural by growing and self-bonding, which exemplifies the integration of biological sciences with design, directing to biodesign (Myers, 2022). These processes follow the principles of ecological balance and a closer link between humans and nature as explored by ©Mogu srl.

Two other examples are *traditional craft techniques* and digital tools, particularly *AI*. Both are exploring different ways to find innovative aesthetics, energy and resource efficiency.

The projects collected also exhibit unique or enhanced properties, including breathability, fire resistance, and acoustic solutions. Although these characteristics are found in traditional materials, they are notably being explored in projects using emerging natural and artificial resources. These enhancements search for materials that combine environmental sustainability with superior performance.

_The industrial sectors and application fields

The indicative industrial sectors from the design projects are diverse, encompassing eight distinct categories. These include 'Circular Materials and Product Design', 'Circular Materials Design for Architecture and Interiors', 'Biotechnology, Materials, and Product Design Research', 'Circular Textile/Leather Design and Manufacturing',

'Furniture and Home/Office Furnishings', 'Sustainable Packaging and Container Manufacturing', 'Service, Brand, and Product Design', and 'Architecture, Interior Design, and Planning'.

These examples showcase applications from decorative tiles and acoustic panels to chairs and lamps, highlighting the expansion of emerging materials across diverse products and sectors. This insight emphasises the versatility and adaptability principle in most projects.

Twenty-one companies from the first two highlighted sectors explicitly state their foundational purpose of promoting environmental sustainability through innovative material design. These companies prioritise materials in their operational and strategic discourse, embracing circular economy strategies like recycling and upcycling organic and inorganic waste. All these companies were established considering unconventional resources to develop new materials.

Five companies are working in a sector that intersects with biological sciences. Materials designers in these projects have a particular role in interpreting and translating scientific knowledge into possible applications.

Other five companies show interest in the fashion and textiles sector in response to a growing awareness of environmental impact associated with these industries. Emerging textile and leather propose performant and appealing materials for the fashion industry, starting with recycling and upcycling unconventional resources.

Four well-established companies from 'Furniture and Home/Office Furnishings' sector demonstrate their recent interest in using and developing materials through collaborations to foster the circular economy. They promote performance and beauty in their products through recovered and recycled resources and materials. While oil-based packaging is becoming a problem, just two projects fall in the industrial sector related to packaging. These projects demonstrate a way to generate packaging from alternative resources like algae and wood by-products. They use renewable resources, and biodegradation as the End-of-Life (EoL) cycle.

Two other projects come from the 'Service, Brand, and Product Design' sector. While these two companies are not primarily designing or developing materials, both are showing great attention towards renewables and biofabricated materials design, using mycelium and hemp by-products to innovate in aesthetics, perception and user connections to reduce the environmental impact, fostering a biodegradable and compostable EoL, aiming to foster regenerative solutions.

Finally, just one company with two projects falls under the 'Architecture, Interior and Planning' category. Like those in the 'Furniture and Home/Office Furnishings' sector, this established company, long dedicated to responsible design and innovation, is now increasingly committed to exploring and promoting sustainable practices by designing and exploring new materials.

5 Environmental Sustainability Trends in Contemporary Materials Design

Analysing the data and thematically separating the insights, eight trending strategies for reducing environmental impact in material design, alongside notable challenges, were identified. These strategies and their adoption rates are depicted in Fig. 7. Further explanation is discussed in this section.

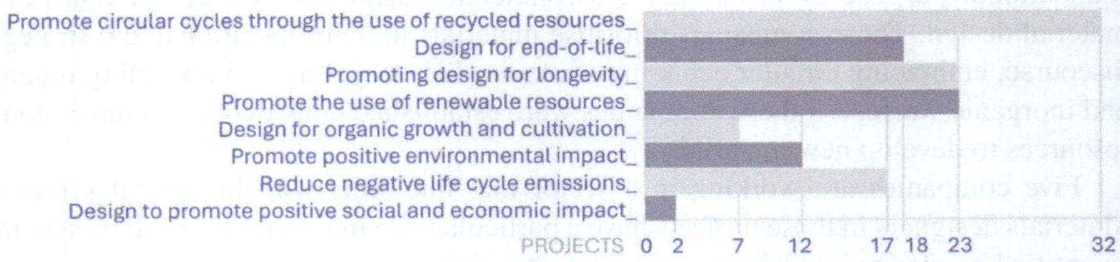

Fig. 7. Materials design practices fosters eight trending aspects of environmental sustainability.

Firstly, using recycled, upcycled, and waste materials to decrease reliance on virgin or petroleum-based resources and reduce landfill waste is growing. Standard routes to deal with these materials already exist, as explained by Singh et al. (2022). Designers employ these materials to extend the lifespan of virgin resources and lessen the need for dyes or finishes (Sauerwein et al., 2017). Significant challenges include material quality variability, waste traceability, market resistance to 'waste-derived' products, and inadequate waste management systems. Regulatory and safety issues also impede implementation, affecting scalability and market acceptance, as Polyportis et al. (2022) describe.

The physical and emotional durability of materials are popular strategies. Some projects use physical resistant materials such as recycled plastics or textiles that allow for a long service life. However, these materials pose management problems at the end of their useful life due to the difficulty of closing the cycle once the new product has been discarded (Vogt et al., 2021). On the other hand, in line with Haines-Gadd et al. (2018), emotional durability is based on fostering connections that extend materials' life by increasing user attachment through sensory qualities and aesthetic appreciation. However, widespread adoption of this strategy is limited, more research must be developed in this regard.

Using renewable resources in material design to minimise environmental impact is also popular. These projects prioritise biobased materials that aid carbon sequestration, supporting a circular bioeconomy (Sherwood, 2020). However, as Weiss et al. (2012) noted, challenges such as variability in resource quality, dependency on chemical processes for material durability, impacts, and ethical concerns over using potential food sources as raw materials still complicate their integration into existing systems and market acceptance.

There is increasing attention to detailing the EoL disposal of these materials and products. The design projects show the life cycle design through graphics and descriptions. However, it is still necessary to make clear management plans beyond labelling as 'biodegradable' or 'recyclable'. Veelaert et al. (2017) argue that more effective EoL design and management strategies are critical to mitigate environmental impacts at the disposal stage.

Strategies to reduce toxic emissions and minimise overall lifecycle emissions are well-known among the analysed cases. These initiatives prioritise designing materials and processes that are less energy-intensive and incorporate methods to capture and reuse carbon dioxide, as Gür (2022) suggested. They also aim to reduce or eliminate the generation of hazardous substances during material synthesis, biodegradability and compostability. However, maintaining the durability and functionality of these materials presents a significant challenge, often increasing research and development costs to meet environmental standards without compromising material quality. In this context, the knowledge of processes and engineering issues is essential, as life cycle analysis is required to assess the real emissions. Materials designers need to expand their knowledge and work in an interdisciplinary way to meet these challenges.

The strategies for contributing to positive environmental impacts and ecosystem regeneration is still scarce. Minimising harm by actively enhancing environmental health includes carbon-negative practices, CO2 sequestration, supporting biodiversity, and reducing water acidity (Badhoutiya et al., 2023). However, effectively implementing these positive strategies through materials design is challenging. Despite their benefits, the high costs associated R&D may hinder broad adoption. Similar challenges are faced by projects highlighting strategies for biological growth and cultivation, focusing on biofabricated and living materials and cultivated sources. The scarcity of successful projects using this strategy suggests it is an emerging area that requires development to effectively drive regenerative ecologies (Karana et al., 2023).

Finally, it is interesting to highlight that two projects of material and product design enhance social responsibility and boost local economies from a materials design perspective. These initiatives, create local jobs and support the regional circular economy, demonstrating concrete benefits by utilising local resources and skills. However, scaling these efforts requires further engagement and business change.

Many of these trends highlight several challenges that are still difficult to overcome. However, by following principles such as '*resource reclamation and utilisation*', '*valorising second-life resources*', '*low-impact technological adoption*', and '*performance and environmental compatibility*', designers are searching for ways to overcome these issues. Practical strategies that follow these principles include developing recovery programs and enhancing recycling systems. There are strategies to improve aesthetics, valorising imperfections, creating storytelling and meaningful experiences. Materials designers are reviewing new and traditional technologies and techniques that can lower the environmental impact of applications.

Furthermore, these designers are expanding their technical skills to assess and characterise emerging materials for meaningful implementation. Contemporary materials designers (Clèries & Rognoli, 2021; Duarte Poblete et al., 2024) should still increment their knowledge in mechanical, functional and environmental aspects, and assessment of materials, as well as in emerging technologies, crafts, biodesig and regenerative ecologies. Incorporating and integrating knowledge will make strategies not just to maintain ecological balance but foster regenerative practices. This is a challenge for the new generation of materials designers.

Furthermore, while forming interdisciplinary teams is becoming common, this paper advocates for comprehensive transdisciplinary training for materials designers. By adopting this principle, designers are encouraged to develop a broad understanding from 'hard' technical features to 'soft' experiential qualities of materials, integrating and translating knowledge across fields from biology to technology. This principle supports the idea that materials designers should be equipped to navigate across disciplines, upgrading their abilities. These designers must be fully aware of the *Materials for Transition* and of the quest to re-establish a healthy and balanced relationship between design and nature, integrating ecology principles acting as mediators for the whole of society to do so as well.

6 Conclusions

This study provides an in-depth review of 43 contemporary material design projects, identifying key trends and challenges in promoting environmental sustainability across various industrial sectors. The analysis highlights eight strategies that are currently being adopted to reduce environmental impact, including the use of recycled materials, enhancing physical and emotional durability, and leveraging renewable resources. However, significant barriers remain, such as market resistance, quality variability, and the need for better end-of-life management strategies.

In addition to these strategies, emerging areas like regenerative design, biofabrication, and local economic support are gaining traction, though they face high research and development costs and scalability issues. The findings emphasise the importance of interdisciplinary and transdisciplinary collaboration, suggesting that materials designers need to expand their expertise to include not only technical and mechanical aspects but also ecological and experiential dimensions.

By exploring these trends and challenges, this study offers valuable insights for practitioners aiming to incorporate sustainability into their materials design processes. It encourages the adoption of principles such as resource reclamation, valorisation of second-life materials, and low-impact technological adoption, guiding designers toward a more sustainable and regenerative practice. This work also underscores the necessity of continuous learning and collaboration to foster innovative approaches in materials design, ultimately contributing to the broader goal of achieving a more balanced relationship between design and nature.

References

Ashby, M.F.: Materials and the Environment: Eco-informed Material Choice. Butterworth-Heinemanan (2009)

Ashby, M., Johnson, K.: The art of materials selection. Mater. Today **6**, 24–35 (2003). https://doi.org/10.1016/S1369-7021(03)01223-9

Ayala-Garcia, C., Rognoli, V., Karana, E.: Five kingdoms of DIY-materials for design. In: Karana, E., Giaccardi, E., Nimkulrat, N., Niedderer, K., Camere, S. (eds.) Alive Active Adaptive: International Conference on Experiential Knowledge and Emerging Materials EKSIG 2017, pp. 222–234. TU Delft OPEN Publishing (2017)

Azungah, T.: Qualitative research: deductive and inductive approaches to data analysis. Qual. Res. J. **18**(4), 383–400 (2018). https://doi.org/10.1108/QRJ-D-18-00035

Bak-Andersen, M.: When matter leads to form: material driven design for sustainability. Temes de Disseny (2018). https://doi.org/10.46467/TdD34.2018.10-33

Bak-Andersen, M.: Reintroducing Materials for Sustainable Design: Design Process and Educational Practice. Routledge (2021)

Barati, B., Karana, E.: Affordances as materials potential: what design can do for materials development. Int. J. Des. **13**(3), 105–123 (2019)

Badhoutiya, A., etal.: Regenerative manufacturing: crafting a sustainable future through design and production. E3S Web Conf. 453, 01038 (2023). https://doi.org/10.1051/e3sconf/202345301038

Becerra, L.: CMF Design: The Fundamental Principles of Colour, Material and Finish Design. Frame Publishers (2016)

Bovea, M., Gallardo, A.: The influence of impact assessment methods on materials selection for eco-design. Mater. Des. **27**, 209–215 (2006). https://doi.org/10.1016/j.matdes.2004.10.015

Brezet, H., & van Hemel, C. (1997). Ecodesign: A Promising Approach to Sustainable Production and Consumption. United Nations Environment Programme

Camere, S., Karana, E.: Fabricating materials from living organisms: an emerging design practice. J. Clean. Prod. **186**, 570–584 (2018). https://doi.org/10.1016/j.jclepro.2018.03.081

Clèries, L., Rognoli, V.: Materials designers: a new design discipline. In: Material Designers. Boosting Talent Towards Circular Economies. Elisava, Barcelona

De Giorgi, C., Lerma, B., Dal Palù, D.: The Materials Side Of Design. The Future Material Design Cultures. Politecnico di Torino (2020)

Duarte Poblete, S.S., Anselmi, L., Rognoli, V.: Materials designers and the translational approach: a case from a product design company. In: Gray, C., Ciliotta Chehade, E., Hekkert, P., Forlano, L., Ciuccarelli, P., Lloyd, P. (eds.) DRS2024, Boston, USA (2024). https://doi.org/10.21606/drs.2024.1080

Duarte Poblete, S.S., Romani, A., Rognoli, V.: Emerging materials for transition: a taxonomy proposal from a design perspective. Sustainable Futures **7**, 100155 (2024). https://doi.org/10.1016/j.sftr.2024.100155

Ferrara, M.: Materiali e innovazione nel design. Gangemi (2004)

Gür, T.M.: Carbon dioxide emissions, capture, storage and utilization: review of materials, processes and technologies. Prog. Energy Combust. Sci. **89**, 100965 (2022). https://doi.org/10.1016/j.pecs.2021.100965

Haines-Gadd, M., Chapman, J., Lloyd, P., Mason, J., Aliakseyeu, D.: Emotional durability design nine—a tool for product longevity. Sustainability **10**(6), Article 6 (2018). https://doi.org/10.3390/su10061948

Karana, E.: Meanings of Materials. Ph.D Thesis. Faculty of Industrial Design Engineering. Delft University of Technology (2009)

Karana, E., Pedgley, O., Rognoli, V.: On materials experience. Des. Issues **31**(3), 16–27 (2015). https://doi.org/10.1162/DESI_a_00335

Karana, E., Barati, B., Rognoli, V., Zeeuw van der Laan, A.: Material driven design (MDD): a method to design for material experiences. Int. J. Des. **9**(2), 35–54 (2015)

Karana, E., Hekkert, P., Kandachar, P.: Material considerations in product design: a survey on crucial material aspects used by product designers. Mater. Des. **29**(6), 1081–1089 (2009). https://doi.org/10.1016/j.matdes.2007.06.002

Karana, E., McQuillan, H., Rognoli, V., Giaccardi, E.: Living artefacts for regenerative ecologies. Res. Direct. Biotechnol. Des. **1**, e16 (2023). https://doi.org/10.1017/btd.2023.10
Karana, E., Pedgley, O., Rognoli, V. (eds.): Materials Experience. Fundamentals of Materials and Design. Butterworth-Heinemann (2014)
Sauerwein, M., Karana, E., Rognoli, V.: Revived beauty: research into aesthetic appreciation of materials to valorise materials from waste. Sustainability, **9**(4), Article 4 (2017). https://doi.org/10.3390/su9040529
Myers, W.: Biodesign: Nature + Science + Creativity. Thames & Hudson (2022)
Oxman, N., Ortiz, C., Gramazio, F., Kohler, M.: Material ecology. Comput. Aided Des. **60**, 1–2 (2015). https://doi.org/10.1016/j.cad.2014.05.009
Pedgley, V.R., Karana, E. (eds.): Materials Experience 2: Expanding Territories of Materials and Design. Butterworth-Heinemann (2021)
Pollini, B., Rognoli, V.: Early-stage material selection based on life cycle approach: tools, obstacles and opportunities for design. Sustainable Prod. Consumpt. **28**, 1130–1139 (2021). https://doi.org/10.1016/j.spc.2021.07.014
Polyportis, A., Mugge, R., Magnier, L.: Consumer acceptance of products made from recycled materials (2022). https://doi.org/10.1016/j.resconrec.2022.106533
Rognoli, V.: The Expressive-Sensorial Characterisation of Materials for Design. Unpublished Ph.D. Thesis. Faculty of Design. Politecnico di Milano (2004)
Rognoli, V., Bianchini, M., Maffei, S., Karana, E.: DIY Materials. Mater. Des. **86**, 692–702 (2015). https://doi.org/10.1016/j.matdes.2015.07.020
Rognoli, V., Levi, M.: Materiali per il design: Espressività e sensorialità. Polipress (2005)
Rognoli, V., Pollini, B., Alessandrini, L.: Design materials for the transition towards post-Anthropocene. In: From Human-Centered to More-Than-Human Design. Exploring the Transition, pp. 101–130. FrancoAngeli Open Access (2021)
Schön, D.A.: The Reflective Practitioner. Routledge (2017)
Scranton, R.: Learning to Die in the Anthropocene: Reflections on the End of a Civilisation. City Lights Publishers (2015)
Sherwood, J.: The significance of biomass in a circular economy. Biores. Technol. **300**, 122755 (2020). https://doi.org/10.1016/j.biortech.2020.122755
Singh, N., Ogunseitan, O.A., Wong, M.H., Tang, Y.: Sustainable materials alternative to petro-chemical plastics pollution: a review analysis. Sustainable Horiz. **2**, 100016 (2022). https://doi.org/10.1016/j.horiz.2022.100016
Squatrito, A., Ferrara, M.: The emerging role of design in the circular materials field. Matériaux Techniques **111**(3), Article 3 (2023). https://doi.org/10.1051/mattech/2023019
Symbola. Design Economy 2024 (2024). https://symbola.net/wp-content/uploads/2024/03/Design24_WEB.pdf
Thackara, J.: In the Bubble: Designing in a Complex World. MIT Press (2005)
veelaert, L., Els, D.B., Sara, H., Karen, V.K., Kim, R.: EKSIG2017: alive, active, adaptive. In: International Conference on Experiential Knowledge, pp. 1–15. Rotterdam, The Netherlands (2017)
Builes Vélez, A.E., et al.: Education for Sustainability Approaching SDG 4 and Target 4.7. Pontifical Bolivarian University, Medellín, Colombia (2022). ISBN: 9786285000775
Vezzoli, C.A.: Design for Environmental Sustainability. Springer, London (2018)
Vogt, B.D., Stokes, K.K., Kumar, S.K.: Why is recycling of postconsumer plastics so challenging? ACS Appl. Polym. Mater. **3**(9), 4325–4346 (2021). https://doi.org/10.1021/acsapm.1c00648
Weiss, M., et al.: A review of the environmental impacts of biobased materials. J. Ind. Ecol. **16**(s1), S169–S181 (2012). https://doi.org/10.1111/j.1530-9290.2012.00468.x
Zuo, H.: The selection of materials to match human sensory adaptation and aesthetic expectation in industrial design. ODTÜ Mimarlık Fakültesi Dergisi **27**(2), 301–319 (2010). http://jfa.arch.metu.edu.tr/archive/0258-5316/2010/cilt27/sayi_2/301-319.pdf

A Different Kind of Growth: Challenges and Opportunities of Mycelium for Product Design

Shah Noor Shafqat[1(✉)], Bilge Merve Aktaş[2], and Hazal Gumus Ciftci[1]

[1] Arizona State University, Tempe, USA
sshafqat@asu.edu
[2] University of Miami, Coral Gables, USA

Abstract. Mycelium is a promising biomaterial that has recently gained popularity. This fungi-based material has been developed as an alternative solution to global environmental issues caused by non-renewable conventional materials, such as plastic. Often such environmental problems arise from material extraction, unsustainable manufacturing, pollution, and design practices. The study in this paper focuses on the perception and experiences of design students and designers to explore the potential and challenges of Mycelium for product design. Building on two workshops, survey answers of participants, and researchers' observations, the paper studies how first-time users characterize this material to develop solutions for expanding its boundaries. The aim is to highlight and explore mycelium's potential to revolutionize the intersectionality of sustainable design practices and manufacturing processes. The study also addresses the challenges of the material being functional at the mass production level. The findings provide insights into how the new material can be utilized in industrial product design and design education, opening pathways for novel, nature-inspired solutions to contemporary global issues.

Keywords: Mycelium · Hands-on explorations · Biomaterials · Industrial design · Design education

1 Introduction

New material development is a crucial strategy to reduce the negative environmental impact of industries and improve environmental sustainability. One reason materials gaining this much significance is due to their strong and close tie to industrial manufacturing as well as shaping new manufacturing technologies (Ashby, 2022). As such, research and development projects have been exploring new, emerging, and novel materials as alternatives to the ones that are fossil based or generate irreversible changes in environmental or social realms. Such emerging materials can be categorized as renewable, waste-based, and bio fabricated materials Poblete et al, (2024).

Mycelium is one of those recently emerging materials that can fall into the intersection of the above mentioned three categories. Mycelium, a foundational root structure of

A. Mateus et al. (Eds.): WAD 2024, What's Around Design?, pp. 307–321, 2026.
https://doi.org/10.1007/978-3-032-08163-6_22

mushrooms, comprises a complex network of numerous soft white threads that fungi use to absorb and provide nutrients to their surrounding ecosystem, benefitting many organisms. Due to it being versatile, biodegradable, and eco-friendly, Mycelium has become promising in improving and addressing environmental concerns and sustainability in the design field.

Designers and scientists have been experimenting with Mycelium and have suggested it can solve many contemporary real-life, global, and extraterrestrial (Doyle, 2024) problems. It can offer options to traditional materials used in industrial product design, such as plastic packaging, textiles, foams, cement, clay-based bricks used in architecture, and other unsustainable practices that contribute to waste in industrial manufacturing processes. With the introduction of Mycelium, a biomaterial alternative to the existing nonrenewable and fossil based ones, it could revolutionize how we approach sustainability and planetary health.

Mycelium is mainly based on manipulating or inoculating the fungi's growth by feeding it with organic matter. Once the mushroom species are activated under the right conditions, they release enzymes to break down organic matter like flour, hemp, and so on, and convert it into nutrients to grow (Fig. 1). The fungi then begin to colonize, eventually taking on a specific shape, preferably forming into a particular mold.

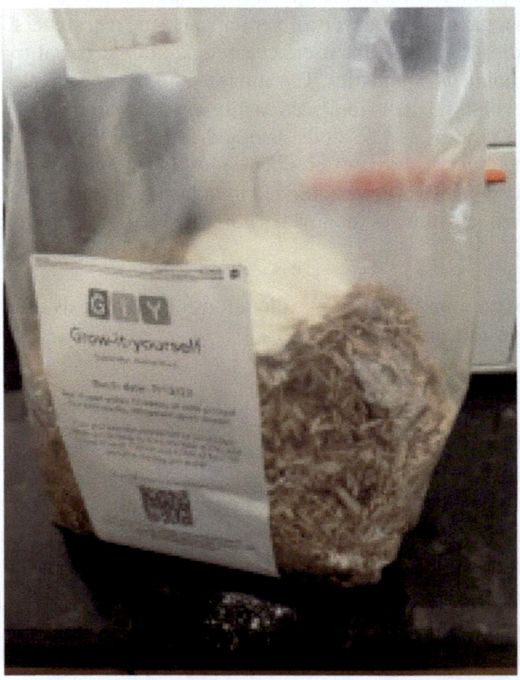

Fig. 1. Grow-it-yourself bags preparation before the workshops (Photos: Authors, 2024)

The properties like being zero-waste, easy to grow, and flexible to give shape have caused Mycelium to gain the attention of the researchers and creatives. The latest examples show that Mycelium can significantly reduce environmental impact while becoming a mainstream material. Our review of existing products made of Mycelium shows that the shift has already been happening in various industries while some other industries point out a potential shift to the use of Mycelium (Table 1, for a detailed review see also Sağlam & Özgünler, 2022). The applications of Mycelium can be categorized into two groups: small-scale products and large scale architectural projects. Small-scale products include everyday consumer goods like textiles, furniture, and various objects. Large-scale architectural projects encompass pavilions, bricks, and insulation materials (Bitting, et al. 2022). These examples have been explored by fashion designer Stella McCartney, computer lead Dell, NASA, and automotive giant Cadillac, as well as by artistic and experimental designers.

These applications of Mycelium demonstrate potential ways of integrating it into existing industrial processes. As a further step, continued research also shows potential for technical advancement of this material. Researchers have created Mycelium-Based Composites (MBCs) by growing Mycelium on agricultural waste or other organic substrates. MBCs have shown promising properties in terms of thermal insulation, acoustic absorption, and fire resistance (Sidhu et al., 2024). Moreover, research with MBCs created with different substrates utilized 3D printing as a rapid production method for textiles, furniture, packaging, electronics and building components (Bitting, et al. 2022). Recent studies have also explored the integration of nanomaterials with Mycelium, such as carbon nanotubes and halloysite nanotubes, for applications in water treatment and agriculture (Sidhu et al., 2024). Finally, recent advances in biotechnology, particularly CRISPR-Cas9 gene editing, have opened up possibilities for enhancing mycelium properties. Studies have demonstrated that deleting specific genes, such as the hydrophobin gene SC3 in Schizophyllum commune, can significantly improve the mechanical properties of mycelium materials (Sidhu et al., 2024)..

Over years, researchers and designers have made big improvements in growing and processing Mycelium. Better techniques like fermentation processes and controlled growth methods have been used to grow Mycelium in larger quantities and ways to optimize Myclium's growth to make it stronger and more useful. Other advancements include post-growth treatments that enhance material properties, for instance flexibility, durability, and use of protective coatings like bioplastics to increase functionalization of the material.

As such, Mycelium is at the forefront of exploring biomaterials to assist designers in creating sustainable products and services (Camere & Karana, 2018). One of its strengths is its versatility. From packaging, interiors, and architecture to fashion and industrial products, Mycelium can help designers explore various forms, shapes, and material properties and potentially revolutionize how designers perceive, interact, and adapt different materials (Alemu et al., 2022).

Another strength of Mycelium is its potential to be cost-effective compared to existing materials. For example, Mycelium-based blocks used in architecture have been proven to be cheaper than traditional cement-based bricks (Alemu et al., 2022). Another example is how customizable material property convinces designers to substitute it with other

Table 1. Selected project examples that use Mycelium. Click on examples to visit webpages.

Industry	Example	Use area	Impact
Interior	Mykor	MykoFoam: acoustic and thermal mycelium insulation sheet	Claims to omit 31% of global carbon emissions with their product
	Mycoflex	Foam like property	Being bio-based, Mykoflex doesn't require petroleum products for its production, helping to decrease demand for non-renewable resources
Packaging	Dell	Computer cushion in boxes	Transformed its packaging in 2011 by growing mycelium (cotton seed hull) based cushions that serve as an organic, compostable alternative to foams for their larger shipments
Textile & Fashion	Mycoworks, Bolt Threads, Mycotech Lab, Reishi, Stella McCarthney, Adidas	Myco-leather	A sustainable option for animal and plastic leather in textiles
Automotive	Cadillac	Car interior	To produce high-performance car interiors using red reishi mycelium based leather
Food	Atlast™	Mushroom based meat	Sustainable food alternative
Architecture	Shell Mycelium Pavillion	Temporary coverage	Alternative to conscious design through temporary structures
Healthcare	Biomedical Scaffolds (Antinori, et al., 2021)	Biomedical applications	Wound care materials and tissue scaffolds

(*continued*)

Table 1. (*continued*)

Industry	Example	Use area	Impact
Agriculture	Mycelium Based Composites (Madusanka, et al. 2024)	Agricultural waste	Growing mycelium on agricultural waste as a feedstock or other organic substrates reduces waste while expediting composting
Space Architecture	Myco Architecture (Rothschild, 2018)	Housing Construction, Radiation Prevention	Enabling long-term space exploration and reducing environmental impact by minimizing earth's resource use

materials. Additionally, it is affordable and accessible for anyone to experiment at a low-cost laboratory. The availability of the "raw material" facilitates and supports designer-led explorations as bottom up initiatives. Finally, living materials like Mycelium can allow designers to customize the growth conditions. It opens new pathways for the designer to explore innovative ways of product design and development (Epstein, 2021) while minimizing the amount of waste material generation.

All the notable advantages of this biomaterial contribute to its popularity and confidence in nature-based materials. On the one hand, this is opening new ways for designers to shift their design practices to more sustainable ways. On the other hand, it also brings several challenges that need to be tackled to offer further opportunities. Explorations and research show that despite all the potentials, Mycelium also has challenges, such as water absorption, consistency, contamination, cost, and scalability. Despite the recognition that Mycelium gains, it is still, to a certain degree, an experimental material. Research-based and creative studies explore ways of tailoring Mycelium properties for specific use areas as well as enhancing their capacities for scaling up.

To contribute to the ongoing explorations, we – three designers/artists and researchers – are presenting insights from a series of workshops based on Mycelium explorations. During the Spring of 2024, we organized three workshops to introduce this material to industrial design students and faculty. The aim of the workshops was twofold: to bring experiential learning to two sustainability-based design courses and to collectively explore the potentials and shortcomings of this material for designers. Thus, our research question was: How do designers who have limited experience working with mycelium perceive its material properties and potential applications?

In this paper, we discuss these early perceptions as proposed by workshop participants via surveys, and our observations emerging from our own experiences. Studying the survey answers and our observations pointed out six challenges and shortcomings for Mycelium to be used more widely. Before we present these blockades and our suggestions to surmount them, we will introduce our study setting. Then, we will share our findings based on survey answers and our observations from hands-on explorations.

Finally, we will discuss our suggestions for overcoming these challenges and the next steps.

2 Study Setting: Workshops & Surveys

In the Spring semester of 2024, we offered three Mycelium workshops at Arizona State University. We decided to conduct this study in the workshop format since it is an established method for creatives to explore together. It is also a research method for collecting qualitative data. Following a typical workshop setting, we designed the sessions in a pre-defined but flexible way (Ørngreen & Levinsen, 2017, p. 72). This structure gave freedom to participants to explore while ensuring that we fulfill the learning outcomes (Ørngreen & Levinsen, 2017, p. 72). Accordingly, each session started with demonstrations and was followed by participants' work. The sessions lasted for about an hour.

The first two workshops were offered to industrial design students from the undergraduate course "Design Ecology and Social Equity", offered by Shah Noor and Bilge, and from the graduate course "Design for Sustainable Development," offered by Hazal. Overall, around fifty students participated in both workshops. The professors of these courses, who are also the authors of this paper, all have a background in art and design with different degrees of working with Mycelium. Shah Noor is an interdisciplinary artist who has been working with Mycelium for three years. For this reason, she was also the lead convener. Bilge and Hazal both have backgrounds in industrial design and craft research. The latter assisted Shah Noor during the workshop planning, but it was also first-time for them both to work with this material. Accordingly, the conversation between the three of us vastly facilitated our understanding of the study.

The first workshop focused on bioremediation. The experiment was based on the utilization of mycelium as a solution for waste and its interaction with different types of disposed materials. Students were tasked with experimenting using various species of liquid mycelium (i.e., blue oyster, pearl oyster, and red reishi mycelium) to comprehend their interaction (Fig. 2). Students worked in groups and started by collecting four types of trash they could find around and initiating their Mycelium and waste experimentation. Subsequently, each group monitored the growth of mycelium on their collected waste within Petri dishes to assess its ability to degrade the materials over a time of 5 to 10 days. In the end, students reported that plastics such as tape and bags did not respond to Mycelium while cigarette butts started dissolving.

The second workshop focused on how Mycelium can be used for sculpting different forms and the possibilities for shaping this living organism. Students were given activated Mycelium Grow-It-Yourself (GIY) bags with hemp hurd substrates as the feedstock. They were tasked with using different molds that they created or selected to experiment and discover how they could work with this natural material. Working in groups, the students prepared the activated material for form-giving (Fig. 3). Then, each student filled their molds with the activated substrate to take care of Mycelium and its growth at home. The molds mainly were found objects, such as cake molds, cups, and 3D-printed ones. After the workshops, students filled out a survey about their perceptions of the future of Mycelium.

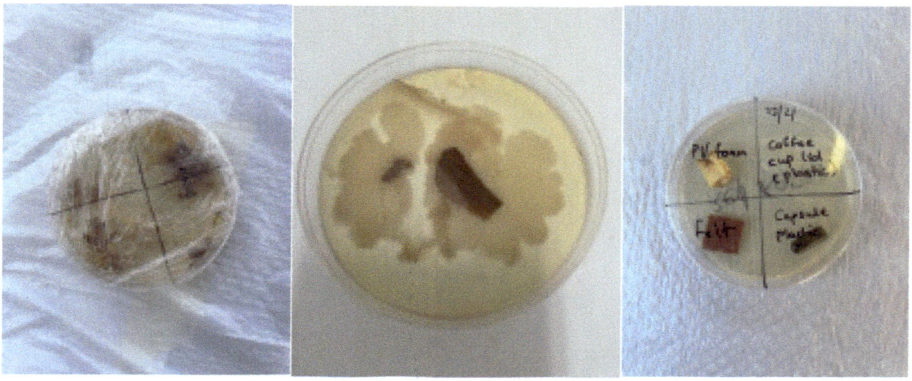

Fig. 2. Stages of Mycelium growth by feeding it with most common waste materials. Photos 1 & 2: various materials when they first come into contact with Mycelium. Photo 3: Experiment with a single material after it is completely overtaken by the Mycelium colony. Photos: Authors, 2024.

Fig. 3. Some mold examples from the second workshop. Photo: Author, 2024.

The last workshop invited faculty from the Design School and followed the same structure as the second workshop. At first, when we sent a survey to understand if there was demand from the faculty, the interest from faculty was high; however, due to the timing of the workshop, attendance was low. We had two professors who are teaching in the Design School about sustainability and social innovation who have not experimented with Mycelium before. The educators explored Mycelium with molds that they brought to see whether there is potential for them to include it in their curriculums. At the end of this workshop, they filled out a survey tailored to educators about Mycelium's potential future in design education.

In this session, as the three researchers, we also worked with Mycelium and combined it with wool, cotton, and synthetic fabrics to observe their reactions (Fig. 4). These explorations also helped us formulate the research question and discussion of this paper.

Fig. 4. Early tests of Mycelium with wool and loofah Photo: Author & Participant, 2024.

3 Initial Findings and Discussion

During the workshops, most of the students were actively engaging with Mycelium and asking questions to learn more about this material (Fig. 5). As the workshops also involved a level of peer-working, they proceeded as lively sessions. Some students also mentioned that this material was not interesting to them, and though they were happy to learn about it as creative professionals, they were not genuinely intrigued.

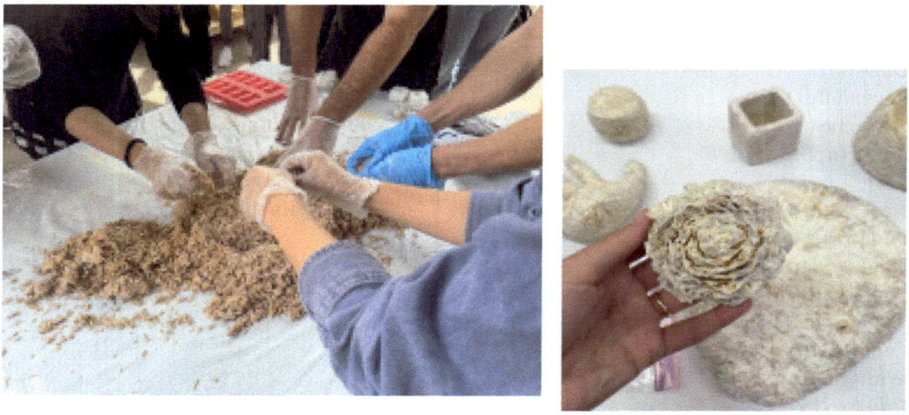

Fig. 5. Left: Mixing mycelium after activation can be a communal practice. Right: Form explorations with mycelium. Photos: Authors, 2024.

Seventeen students and two educators filled out the survey; however, the educator's answers were not included in the data analysis due to low participation. The questions in the student survey inquired about initial impressions of Mycelium, how they perceived it, and how this material might develop in the future. Most of the students (n: 11), were not familiar with this material at all before the workshop. When we asked them how they would describe it with three words, most of the students preferred different descriptions, yet some identifiers regarding the environmental characteristics were repeated. The descriptors could be grouped into five (i.e., creative characteristics, environmental characteristics, physical characteristics, compared to other materials, and production process) based on what they refer to (Table 2).

Table 2 shows that most of the descriptors were positive. This hints that there is a potential for these students to continue working with Mycelium in the future. One student claimed, "With love and care you can save the earth and make beautiful products"(Student 1), showing that future designers were interested in taking responsibilities. The fact that students referred to several aspects of this material can also propose the workshop was inspiring enough for them to think about it from different perspectives. One of them said, "It's very easy to work with and put into molds so it's surprising to see this concept not widely adopted" (Student 3).

Table 2. The adjectives and phrases students preferred to describe Mycelium and the categories these words relate to.

Creative characteristics	**Environmental characteristics**	**Physical characteristics**	**Compared to other materials**	**Production process**
Innovative (S6,12) Interesting (S8,14) Unorthodox (S8) Simple (S6) Dull (S5) Potentially awesome (S15) Multipurpose (S4) Neat (S17) Fun (S16,17) Unique (S10) Solution (S9) Applicable (S10) Funky (S14) Wit[t]y (S14) Crude (S15)	Organic (S1,9,10,11) Sustainable (S2,8,12,16) Eco-friendly (S6,13,16) Natural (S1,12) Biodegradable (S16) Recycling (S5,11)	Mossy (S5) Straw (S5) Smelly (S17)	Alternative (S2,8) Plastic (S2,5,8) Similar to woodchips (S3) Eating material (S4) Rubbery (S7) Chicken skin (S7)	Limitless (S1) Time-intensive (S1,7) Future (6,11,13) Easy (S3,9,15,16)

Some less positive identifiers were about it being time-intensive, generating negative multi-sensorial experiences (scent and tactility of it), and being dull. There were comments such as, "Mycelium takes a lot of technical care and time to work with" (Student 7).

Students were unable to propose potential industries or objects that could benefit from Mycelium, besides the already existing ones such as everyday consumer goods, objects, construction, and fashion. Though for some students this material has a long way to become mainstream (n: 4), most of them also mentioned potentially working with it in the future.

The fact that industrial design students mentioned care in the context of material development shows that they are willing to show additional attention to make their practice more environmentally-friendly and thoughtful. This willingness shows that the new generation of designers is ready to change existing production/consumption behaviors and become more careful in their practices.

Additionally, on our personal reflections and observations from the workshops and what students mentioned, we identified six key challenges that working with Mycelium posed.

Working with a living organism: As Mycelium is a living organism, working with it requires care and attention to keep it alive while creating the desired product. Some caring practices are related to the working environment, such as temperature, sterilization, and humidity. The designer can experiment with these to find the optimum environment for

Mycelium growth. This can also become challenging at times as it mandates experiments and trials.

Though Mycelium itself is a completely natural material, working with it requires several supportive tools that are currently mostly fossil-based single-use plastics, such as plastic bags to cover the workstation, latex gloves, surgical masks, molds – silicone, 3D printed or existing plastic vessels-, and conventional cleaning supplies. This simply means that working with Mycelium comes with the cost of creating more plastic trash than 'natural' products. This shows that considering that this is an emerging field, as designers we need to simultaneously streamline the production process along with tools and supplies so that the entire life cycle of Mycelium-based products is sustainable with the lowest possible environmental impact.

Finally, working with a living organism raises ethical questions. On the one hand, growing the material into the desired shape means not generating any waste material and creating the precise form the designer intends to create. On the other hand, from an ethical perspective, growing a living organism later to be baked to stop its growth raises questions. Though there are methods of reviving dormant Mycelium, which is a biological process similar to the fermentation process in yeast, once it is baked the Mycelium stops growing and for utilitarian functions, baking is commonly preferred. Considering the increasing number of debates around animal rights and welfare, growing mycelium later to be killed might not sound like the most conscientious story, raising several new questions regarding control and domination, what is alive, and what is dead? Can we sacrifice some life forms to protect others to the extent of tipping the ecological balance? This interesting ethical dilemma can become a limitation in some contexts but can also lead to fruitful philosophical conversations.

Consistency: As another side effect of being a living organism, complete control over Mycelium growth and shape formation may not always be possible. Depending on the environment, Mycelium can behave as it pleases, ultimately leading to results that may differ from the intended ones. While this uniqueness can be seen as a strength from an aesthetic standpoint, it also presents a significant challenge in terms of standardization. For example, after the workshops, some students observed one of their pieces becoming moldy while the other one remained healthy despite following near-identical steps for both pieces. For large-scale production, the challenges related to consistency should be addressed. Mold growth and environmental conditions is one of the ongoing issues in working with Mycelium, and researchers have been looking into ways for streamlining this process (see for example, Sağlam & Özgünler, 2022).

Water absorption: Currently, most products made of Mycelium are not very water-resistant. The water absorption depends on the substrate used as an activator, and innovators continue to experiment with it. However, the existing products made of Mycelium mostly are sensitive to outdoor or wet-environment use, which can limit the product range greatly. Alternative coating that can allow using Mycelium-made products in wet environments is already being developed. For example, in 2022, Steven Davies, a product designer produced a surfboard made of Mycelium and used beeswax and linseed oil to make it more water resistant. The early prototypes were successful and Davies is still developing this concept further (Davies, 2022). For large-scale production, the water-absorption requires special attention and should be developed further. A potential benefit of this need may be the initiation of new subsidiary industries.

Scalability: As a continuation to standardization and large-scale production, Mycelium has not yet achieved scalability. Many companies are making advancements in figuring out how to scale up production, but challenges still need to be met for a material like Mycelium to become a functional alternative to traditional production materials. Currently, it is still a material mostly used for hands-on studio-scale production. The small-scale production can start influencing the industry, designer approaches, and consumer habits, but for it to become a mainstream material, researchers and designers need to continue their experiments. Additionally, engineers are developing ways of scaling Mycelium production through different processes like the implementation of Artificial intelligence to optimize growth conditions, experimenting with substrates, and use of biotechnology to genetically modify Mycelium (Waltz, 2016). This is still a costly practice and thus cannot become mainstream yet. But in the future, biotechnology can help modify the entire DNA sequence of mushroom species that are more suitable for scaling up.

Time requirement: Working with Mycelium is a process of taking care of this organism, and requires time to do so. For this reason, it is not a rapid-production material. Rather, it forces the designer to follow the material's own pace and observe the growth over time. In the workshops, this took around two to three weeks. For certain use areas, such as for creating quick mock-ups, this can be perceived as a challenge. However, considering that beyond everything Mycelium is a sustainable alternative to the negative impact of overproduction and overconsumption, the slowness of working with Mycelium can also be seen as a strength. We believe this strength/challenge can be reframed to indeed facilitate behavioral change.

Multi-species engagement: Mycelium and the designer are not the only living organisms involved in working with it. It also provides excellent conditions for bacteria growth. Although the specific types of bacteria that might thrive in these conditions can vary, the process of Mycelium growth can facilitate the growth of mold. The designer needs to be in communication with the material to identify if mold is growing and interfere by using disinfectants if necessary. As a side effect of working with living organisms, Mycelium dictates following certain protocols to prevent potential contamination of black mold or other microbes which can easily be accomplished by working in sterile environments.

Another side effect of multi-species engagement is related to human health. Some species have been reported to produce mycotoxins, which can cause respiratory tract issues. But the fruiting bodies typically produce these mycotoxins, and vary in different species depending on the growth conditions and other factors (Brandhof & Wösten, 2022); fruiting bodies are not normally used in the design processes, so considering the real potential of Mycelium material, sterile conditions and preventive measures can reduce the risk of transferring contaminants.

Overall, our observations and survey answers indicate that Mycelium has great potential even with its challenges. All six challenges that we identified could be seen as opportunities to expand our understanding of using natural materials and changing designer behaviors for more sustainable futures.

The aliveness of the material also shows that the designers need to learn how to balance developing intentional products while being okay with letting the power of control go. Following the material flow is one of the main pillars of creative production (see for example Deleuze & Guattari, Ingold, 2013). When there is a distance between designers and materials, design practice can be alienated from its environmental impact (Anusas & Ingold 2013; Aktaş, 2022). For this reason, we find it important to continue implementing hands-on material explorations in industrial design education.

Existing research typically focuses on Mycelium's mechanical properties and physical properties, such as being a lightweight solution (Verma, et al. 2023), its visual and structural reaction to external force and compression (Attias, et al. 2020), and its unpredictability as well as lack of water absorption (Sydor, et al. 2021). Most research papers observed similar challenges as ours. However, differently from these studies, by framing this study within an explorative and open-ended workshop, we were able to gather insights about more tacit experiences as well. The signifiers that the students preferred hinted how Mycelium might be perceived by users when it comes to its own aesthetics, sensorial experiences it creates, and story. Thus, we believe while our study further explores the technical aspects of this material, it also adds a more humane characterization to this newly emerging material.

Overall, Mycelium has a great potential for becoming a learning tool, especially for sustainability-related issues of design. In our courses, we could observe that this workshop was rejuvenating and concretized the learning outcomes of class discussions. Bringing these explorations to the classroom is crucial for future professionals to stay in contact with materials and comprehend what kind of environmental impact their practice can make.

4 Conclusions

As an emerging biomaterial, Mycelium poses several challenges, like standardization, scalability, and use areas. However, research shows that it is still a promising option to replace existing materials, such as plastics, textiles, and leather. Considering that microplastics can be found even in breast milk, there is an urgent need to replace some of the commonly used materials with more natural ones. Industrial design is considered one of the occupations responsible for harming the environment Papanek (1971/1992). The qualities and use areas of Mycelium hint at promising explorations to tackle unsustainable choices made by designers.

When Mycelium is used for a product, it might be an opportunity to create a full circular production process. The product is used and then regained to the natural system after its use is completed by composting. Reflecting on the current production process of Mycelium, this could even be going further by reducing the size of landfills and improving waste management. There are untapped opportunities and emerging research investigating the use of fungi for tackling issues regarding waste management and overconsumption.

However, critical thinking is required while working with natural capital and natural materials. We cannot put a price tag on nature's wonders, but we can work together with nature to reverse the harm we have already done. One of the most challenging questions for designers in the 21st century is how to contribute to sustainable development and innovative advancement while prioritizing the well being of the planet and all its ecosystems. Industrial design education is an area that could start and take off if we – design educators – embrace these emerging materials and production processes in our teaching.

Acknowledgments. We would like to thank Ecovative for providing us with their supply of mycelium GIY bags, which enabled us to conduct our exploration, as well as all the participants for sharing their insights.

References

Aktaş, B.M.: Entangling with materials: crafting as a way of repairing our relationship with the environment. In: Expanding Environmental Awareness in Education Through the Arts: Crafting-with the Environment, pp. 93–108. Springer Nature Singapore, Singapore (2022)

Alemu, D., Tafesse, M., Mondal, A.K.: Mycelium-based composite: the future sustainable biomaterial. Int. J. Biomater. (2022). https://doi.org/10.1155/2022/8401528

Antinori, M.E., Contardi, M., Suarato, G.: Advanced mycelium materials as potential self-growing biomedical scaffolds. Sci. Rep. **11**, 1–14 (2021)

Anusas, M., Ingold, T.: Designing environmental relations: from opacity to textility. Des. Issues **29**(4), 58–69 (2013)

Ashby, M.F.: Materials and Sustainable Development. Butterworth-Heinemann (2022)

Attias, N., et al.: Mycelium bio-composites in industrial design and architecture: comparative review and experimental analysis. J. Clean. Prod. **246**, 119037 (2020)

Bitting, S., Derme, T., Lee, J., Van Mele, T., Dillenburger, B., Block, P.: Challenges and opportunities in scaling up architectural applications of mycelium-based materials with digital fabrication. Biomimetics **7**(2), 44 (2022)

Brandhof, J.G., Wösten, H.A.B.: Risk Assessment of Fungal Materials – Fungal Biology and Biotechnology (2022)

Camere, S., Karana, E.: Fabricating materials from living organisms: an emerging design practice. J. Clean. Prod. (2018)

Culpepper, J.: New Study Links Microplastics to Serious Health Harms in Humans. Environmental Working Group (2024). https://stevedavieslearningjournal.wordpress.com/. Accessed 09 August 2024

Davies, S. (2022). https://stevedavieslearningjournal.wordpress.com/. Accesed on 09 August 2024

Deleuze, G., Guattari, F.: A Thousand Plateaus: Capitalism and Schizophrenia. Bloomsbury, London (1980/2019)

Doyle, T.P.: NASA Advances Research to Grow Habitats in Space From Fungi. NASA (2024). https://www.nasa.gov/news-release/nasa-advances-research-to-growhabitats-in-space-from-fungi/

Epstein, R.: The Future of Design: Material Evolution. Work Design Magazine (2021). https://www.workdesign.com/2021/01/thefuture-of-design-material-evolution/

Ingold, T.: Making: Anthropology, Archaeology, Art & Architecture. Routledge, London (2013)

Madusanka, C., et al.: A review of recent advances in fungal mycelium based composites. Discov. Mater. **4**(1), 13 (2024)

Papanek, V.J.: Design for the Real World: Human Ecology and Social Change. Academy Chicago, Chicago (1992)

Poblete, S.S.D., Romani, A., Rognoli, V.: Emerging materials for transition: a taxonomy proposal from a design perspective. Sustain. Futures (2024)

Sağlam, S.S., Özgünler, S.A.: An experimental study on production opportunities of biocomposite by using fungal mycelium. J. Des. Resil. Architect. Plan. **3**(2), 237–260 (2022)
Sidhu, S.S., Hossain, M.F., Kaur, S., Kaur, H.: Mycelium-based composites: a review on production, properties, and applications. J. Renewable Mater. **12**(4), 1003–1035 (2024)
Sydor, M., Bonenberg, A., Doczekalska, B., Cofta, G.: Mycelium-based composites in art, architecture, and interior design: a review. Polymers **14**(1), 145 (2021)
Vasarhelyi, K.: Plastic Eating Mushrooms: A Viable Solution to Plastic Decomposition. University of Colorado Boulder (2021). https://www.colorado.edu/ecenter/2021/11/04/plastic-eating-mushrooms-viable-solution-plastic-decomposition
Verma, N., Jujjavarapu, S.E., Mahapatra, C.: Green sustainable biocomposites: substitute to plastics with innovative fungal mycelium based biomaterial. J. Environ. Chem. Eng. **11**(5), 110396 (2023)
Waltz, E.: Gene-edited CRISPR mushroom escapes US regulation. Nature **532**(7599), 293 (2016)
Ørngreen, R., Levinsen, K.: Workshops as a research methodology. Electron. J. eLearn. **15**(1), 70–81 (2017)

Saglam, S.S., Özgünler, S.A.: An experimental study on production opportunities of biocomposite by using fungal mycelium. J. Des. Resil. Architect. Plan. 3(2), 247–260 (2022)
Sidhu, S.S., Hossain, M.E., Kaur, S., Kaur, H.: Mycelium-based composites: a review on production, properties, and applications. J. Renewable Mater. 12(1), 1003–1035 (2024)
Sydor, M., Bonenberg, A., Doczekalska, B., Cofta, G.: Mycelium-based composites in art, architecture, and interior design: a review. Polymers 14(1), 145 (2022)
Vasarhelyi, K.: Plastic-Eating Mushrooms: A Viable Solution to Plastic Decomposition. University of Colorado Boulder (2021). https://www.colorado.edu/ecenter/2021/04/14/plastic-eating-mushrooms-viable-solution-plastic-decomposition
Verma, N., Jujjavarapu, S.E., Mahapatra, C.: Green sustainable biocomposites: substitute to plastics with innovative fungal mycelium based biomaterial. J. Environ. Chem. Eng. 11(5), 110396 (2023)
Waltz, E.: Gene-edited CRISPR mushroom escapes US regulation. Nature 532(7599), 293 (2016)
Ørngreen, R., Levinsen, K.: Workshops as a research methodology. Electron. J. eLearn. 15(1), 70–81 (2017)

Author Index

A. Mateus et al. (Eds.): WAD 2024, What's Around Design?, pp. 323–324, 2026.
https://doi.org/10.1007/978-3-032-08163-6

MIX
Papier aus verantwortungsvollen Quellen
Paper from responsible sources
FSC® C105338

If you have any concerns about our products, you can contact us on
ProductSafety@springernature.com

In case Publisher is established outside the EU, the EU authorized representative is:
Springer Nature Customer Service Center GmbH
Europaplatz 3, 69115 Heidelberg, Germany

Printed by Libri Plureos GmbH
in Hamburg, Germany